Lecture Notes in Computer Science 7960

Commenced Publication in 1973
Founding and Former Series Editors:
Gerhard Goos, Juris Hartmanis, and Jan van Leeuwen

Jacek Cichoń   Maciej Gębala
Marek Klonowski (Eds.)

Ad-hoc, Mobile, and Wireless Network

12th International Conference, ADHOC-NOW 2013
Wrocław, Poland, July 8-10, 2013
Proceedings

 Springer

Volume Editors

Jacek Cichoń
Maciej Gębala
Marek Klonowski
Wrocław University of Technology
Institute of Mathematics and Computer Science
ul. Janiszewskiego 14a, 50-372 Wrocław, Poland
E-mail: {jacek.cichon, maciej.gebala, marek.klonowski}@pwr.wroc.pl

ISSN 0302-9743   e-ISSN 1611-3349
ISBN 978-3-642-39246-7   e-ISBN 978-3-642-39247-4
DOI 10.1007/978-3-642-39247-4
Springer Heidelberg Dordrecht London New York

Library of Congress Control Number: 2013941245

CR Subject Classification (1998): C.2, H.4, D.2, K.6.5, H.3, C.2.4, H.5

LNCS Sublibrary: SL 5 – Computer Communication Networks
and Telecommunications

Typesetting: Camera-ready by author, data conversion by Scientific Publishing Services, Chennai, India

Printed on acid-free paper

Springer is part of Springer Science+Business Media (www.springer.com)

Preface

The 12th International Conference on Ad-hoc, Mobile and Wireless Networks (ADHOC-NOW) took place in Poland in 2013. Since its inauguration in Toronto in 2002, ADHOC-NOW has won high, international reputation as a conference dedicated to wireless and mobile computing. The meetings have demonstrated that the field of wireless and ad hoc computations and communications poses important research questions, deals with urgent practical problems, and is at the center of attention for many scholars.

Previous editions of the conference were held in Canada, France, Mexico, Spain, Germany, and Serbia. In 2013, ADHOC-NOW was organized for the first time in Wrocław – a city of complex and fascinating history. Today Wrocław is seat of a number of high-tech companies and several universities. The conference was hosted by the Institute of Mathematics and Computer Science of Wrocław University of Technology in its Congress Centre.

The 12th ADHOC-NOW conference received 56 submissions of which 27 were accepted for presentation. Paper proposals were evaluated by four independent reviewers. The accepted articles address such diverse topics as routing, rumor spreading, reliability, topology control, security aspects, and the impact of mobility. Some of the submissions contain precise analytical results while others are devoted to solving specific practical problems of implementation and deployment. This range of problems demonstrates that the conference has become a stimulating forum for the exchange of ideas and for discussions between theoreticians and practitioners, whose interaction brings about, on the one hand, more realistic theoretical models and, on the other, more significant practical results.

We would like to thank the members of the Program Committee, the reviewers, and all the people who contributed to organizing the event and putting together an excellent program.

Jacek Cichoń
Marek Klonowski
Maciej Gębala

Organization

General Chair

Jacek Cichon   Wroclaw University of Technology, Poland

Program Chairs

Marek Klonowski   Wroclaw University of Technology, Poland
Evangelos Kranakis   Carleton University, Canada
Jaime Lloret   Polytechnic University of Valencia, Spain
Ivan Stojmenovic   University of Ottawa, Canada

Submission and Proceedings Chair

Maciej Gebala   Wroclaw University of Technology, Poland

Publicity Chairs

Jakub Lemiesz   Wroclaw University of Technology, Poland
Sandra Sendra   Polytechnic University of Valencia, Spain

Web and Local Arrangements

Rafal Kapelko   Wroclaw University of Technology, Poland
Marcin Zawada   Wroclaw University of Technology, Poland

Program Committee

Nael Abu-Ghazaleh   State University of New York at Binghamton,
                                USA
Michel Barbeau   Carleton University, Canada
Zinaida Benenson   University of Erlangen-Nuremberg, Germany
Matthias R. Brust   University of Central Florida, USA
Marcello Caleffi   University of Naples "Federico II", Italy
Juan-Carlos Cano   Technical University of Valencia, Spain
Jiannong Cao   Hong Kong Polytechnic University, Hong Kong
                                SAR
Arnaud Casteigts   LaBRI and Université de Bordeaux, France
Costas Constantinou   University of Birmingham, UK
Ilcinkas David   LaBRI-CNRS and Université de Bordeaux,
                                France
Jose Maria de Fuentes   Universidad Carlos III de Madrid, Spain
Stefan Dobrev   Slovak Academy of Sciences, Slovakia

Additional Reviewers

Gianlorenzo D'Angelo
Mattia D'Emidio
Shantanu Das
Dariusz Dereniowski
Robert Elsaesser
Wenchao Huang
Marek Jawurek
Zhiping Jiang
Rastislav Kralovic
Luca Moscardelli
Dominik Pajak

Tomasz Radzik
Xiaojiang Ren
Peter Rothenpieler
Stefan Schmid
Grzegorz Stachowiak
Przemyslaw Uznanski
Elio Velazquez
Imrich Vrto
Wenzheng Xu
Xiaolan Yao
Xu Zhang

Table of Contents

Design and Assessment of a Reputation-Based Trust Framework in Wireless Testbeds Utilizing User Experience

Aggelos Kapoukakis, Christos Pappas,
Georgios Androulidakis, and Symeon Papavassiliou

Network Management & Optimal Design Laboratory (NETMODE)
School of Electrical & Computer Engineering
National Technical University of Athens (NTUA), Greece
{akapouk,chris,gandr}@netmode.ntua.gr, papavass@mail.ntua.gr

Abstract. In this paper a novel Trust and User Experience Framework (TUEF) that enables trustworthiness between users and testbeds is proposed. The proposed framework utilizes: (1) monitoring data from testbeds and (2) user experience data, stemming from users' capability to provide feedback regarding their Quality of Experience (QoE) and service received. The applicability of the TUEF is demonstrated in a wireless testbed environment as a proof-of-concept implementation, while the performance evaluation is based on both real and simulated experiments. The results presented in this paper demonstrate that the proposed trust mechanism succeeds in delivering accurate reputation values even in the case where a large number of malicious users may provide false feedback.

Keywords: Trust, User Experience, Reputation, Wireless, Testbeds.

1 Introduction

"Trust is the firm belief in the competence of an entity to act dependably, securely, and reliably within a specified context" [1]. The concept of trust has originally been conceived in the social sciences, where several connotations have derived from. In the lapse of time and the rapid evolution of pervasive computing, trust has become an indispensable component in network infrastructures. For example, in a peer-to-peer network where the users perform transactions and exchange data, the establishment of trust between different users is crucial.

Reputation-based trust management systems represent a considerable trend in decision support for service provision applications [2]. The main idea is to let different entities rate each other, after the completion of an interaction, collect the feedback for each entity, aggregate the provided information in a distributed or centralized manner, and produce a reputation score, which can assist other entities in future trust decisions. Reputation mechanisms are employed in web search engines (i.e. Google's PageRank [3]), in online marketplaces – e-Commerce (i.e. eBay, Amazon) [4], in email systems to provide anti-spam functionality [5], etc.

J. Cichoń, M. Gębala, and M. Klonowski (Eds.): ADHOC-NOW 2013, LNCS 7960, pp. 1–12, 2013.

At the same time, along with the huge growth of the Internet and its endued technologies, an increased public awareness of several crucial shortcomings in terms of security, mobility, heterogeneity and performance has led to Future Internet research efforts [6], [7]. Hence, large-scale testbeds are needed for experimenting, integrating and validating next-generation network technologies. Trust in testbeds typically includes a belief in the identity of another party (e.g. Service X is from Provider Y) or in their actions (e.g. Provider Y will allocate resources according to those requested). In this paper the notion of trust in testbeds is extended to include user's Quality of Experience (QoE). Quality of Experience (QoE) is a service level efficiency metric that indicates the level of satisfaction for the received service from the end user perspective [8], [9]. By taking into consideration QoE parameters allows for a more user-centric, rather than service-centric, approach to be adhered. With the expected proliferation of the use of testbeds for experimentation purposes, such an approach would provide an added value service for testbeds owners and users, and especially for wireless infrastructures where conditions produced by the existence of multi-hop routes and dynamic changes in topology, cannot be properly quantified and evaluated. In our approach we take the view that an experimenter's actual needs, requirements and expectations for future wireless testbeds and services, cannot be defined or mapped in static and/or strict way (values and/or thresholds), but rather depend on several metrics, including subjective ones, such as background noise, personal experiences, trust considerations, which are more properly and effectively expressed via QoE rather than traditional QoS. Within such a framework, the access control models, as well as the design and provisioning of a testbed, has to take into account how users perceive trust and what they anticipate from a trusted service/testbed. As a result, the interactions and feedback exchange between testbed owners and experimenters, along with issues related to trust and opinion establishment, are becoming of high importance.

Therefore, we propose a novel Trust and User Experience Framework (TUEF) that enables trustworthiness between users and testbeds, utilizing: (1) monitoring data from relevant systems and testbeds and (2) user experience data [10], stemming from users' capability to provide feedback regarding their Quality of Experience (QoE) and service received. This framework will further support experimenters in the decision making process regarding the potential use of a testbed compared to other similar ones. The performance of the proposed framework is assessed with the deployment of the implemented mechanism in a wireless testbed consisting of 20 802.11a/b/g/n wireless nodes. The evaluation process is based on feedback provided by both real users' experimentation and simulated experiments. In our case, feedback is provided regarding testbed's node availability and Packet Delivery Ratio (PDR). These two metrics are utilized for demonstration purposes only while our proposed trust and user experience framework can employ other metrics, specific to the provided testbed services. To the best of our knowledge this is a first attempt in the literature to create such a Trust and User Experience Framework in order to facilitate and support experimenters in evaluating available testbeds and selecting the most appropriate ones according to their objectives and purposes.

The structure of the paper is as follows: first in section 2 we present the proposed trust and user experience framework, then in section 3 we describe the implemented reputation-based trust mechanism, while in section 4 the performance evaluation of the proposed framework using both real data and simulation results in a wireless ad-hoc testbed is demonstrated. Finally, section 5 provides some concluding remarks.

2   Trust and User Experience Framework

In this section, the proposed framework for the trust and user experience is analyzed. The scope of the framework is to provide mechanisms and tools towards building trustworthy services based on the combination of reputation and monitoring data. The developed mechanisms and tools reflect the end users (experimenters) perspective with the objective of empowering the users to select reliable resources, based on dynamic performance metrics. These metrics would be a dynamic decision making toolbox and a "smart" user support service that provides a unified and quantitative view of the trustworthiness of a facility.

There are two ways towards trust building: (1) users provide feedback from their quality of experience in order to construct a quantitative view of the trustworthiness of the testbed and (2) the testbed provides feedback and insight regarding experimenter motives and behavior, thus creating a service that could point out user trustworthiness. This work mainly focuses on the first kind of trust building. That is building reputation-based trust on a testbed utilizing: (1) monitoring data (e.g. node availability, link quality, delay, bandwidth, packet loss, CPU, memory, disk space) and (2) users feedback regarding their Quality of Experience (QoE) and service received.

2.1   Experiment Lifecycle

The experiment lifecycle is partitioned in three consecutive phases (Figure 1). In the first phase, a testbed user deploys an experiment by reserving the resources that satisfy his requirements. The resources can vary depending on the facility (wired networks, wireless Wi-Fi or WiMAX networks, servers and mobile devices, as well as wireless sensor networks). For example, a wireless testbed may feature wireless (or mobile) nodes, wireless frequencies and bandwidth. On the contrary, a wired testbed may feature memory and network bandwidth, number of dedicated processors etc.

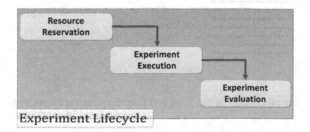

Fig. 1. The experiment lifecycle

In the second phase, the user interacts with the testbed by executing his experiment utilizing his reserved resources at the scheduled timeframe. Afterwards, in the third phase, the experimenter is given the capability to rate the service received at the particular experiment according to his satisfaction both objectively and subjectively. The feedback provided by the experimenters is used to evaluate the testbed services, as it will be described in the following section.

2.2   Testbed Evaluation

A basic scheme for the Trust and User Experience Framework (TUEF) is depicted in Figure 2. Considering a testbed and N users $(U_1, U_2, ..., U_N)$ conducting experiments in it, the testbed will advertise to the users k services $(S_1, S_2, ..., S_k)$. After the execution of an experiment (or during the experimentation phase, if the experiment has a large time span) a user can give feedback regarding his Quality of Experience (QoE) and his perception of the received services from the testbed. Users' feedback for a service is weighted with a value C that is called *Credibility* and refers to the users' honesty on providing feedback.

After each experiment, the credibility value is updated according to the monitoring data retrieved from the testbed via a monitoring tool (i.e. Nagios [17]). If the feedback for a service matches the corresponding monitoring data, then the credibility of the user increases. Otherwise, if for example the user gives negative feedback for a service but according to the monitoring data the service was provided as promised, then the credibility of the user is decreased. Feedback from users is aggregated to form the trust scores for each service $(T_{S1}, T_{S2}, ..., T_{Sk})$. Monitoring data will also be taken into account to form the trust values of the services. The aggregation and computation of the trust scores is performed by a reputation-based trust mechanism, which will be described comprehensively in section 3.

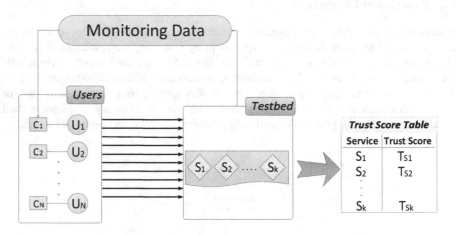

Fig. 2. The general concept of the Trust and User Experience Framework (TUEF)

As mentioned previously, after an experiment the user can provide feedback regarding his Quality of Experience (QoE) and his perception of the received services. In order to enable this feature, QoE metrics must be defined. Our proposition suggests that QoE metrics need to be categorized in technical and non-technical aspects.

The ratings of the technical aspects (i.e. node availability, link quality, bandwidth, CPU, memory, etc.) are given in terms of a *Mean Opinion Score (MOS)*. *MOS* is a typical user-related metric for measuring QoE and can range from 1 (worst) to 5 (best) as portrayed in Table 1. On the other hand, ratings of non-technical aspects are calculated as a simple average along each aspect. This is a similar way to how eBay's traditional feedback system operates and treats users' feedback ratings. The class of non-technical aspects refers to questions whose answers cannot be rated with a numerical value, such as, "did the user face any problems?", "would the user pay for the provided service?" and so on. To conclude this section, it should be mentioned that the feedback used as input for the reputation-based trust mechanism, derives only from the technical aspects. Therefore, we will emphasize only on trust scores for the technical services.

Table 1. Mean Opinion Score for the technical aspects

MOS	Quality
5	Excellent
4	Good
3	Adequate
2	Poor
1	Bad

3   Reputation-Based Trust Mechanism Implementation

A variety of approaches has been proposed for solving the trust management problem using reputation-based methodologies such as EigenTrust [11], ROCQ [12], H-Trust [13], Mate [14] and Strudel [15]. The common target of these trust algorithms is to aggregate feedback provided by users to form a global trust metric which will be used to measure the trustworthiness of an entity. ROCQ's architecture is the best fit for our purpose, as it (i) provides high degree of customization and (ii) allows integration and exploitation of the ground truth that is introduced by Trust and User Experience Framework through monitoring data. In this section we describe the implementation of our proposed Reputation-based Trust Mechanism, which is based on ROCQ [12]. ROCQ mechanism is a reputation-based trust management system, proposed by A.Garg and R.Battiti, which computes the trustworthiness of peers on the basis of transaction feedback. The ROCQ model combines four parameters: Reputation (R) or a peer's global trust rating, Opinion (O) formed by a peer's first-hand interactions, Credibility (C) of a reporting peer and Quality (Q) or the confidence a reporting peer designates on the judgment it provides.

It is clear that ROCQ, as originally proposed, cannot be integrated in testbeds, where no notion of peer-to-peer networking exists. Hence, we implemented a novel reputation-based trust mechanism for testbeds, which has its roots in ROCQ. In the following subsections, the architecture of the system is analyzed.

3.1   Involved Entities

- A testbed that is used to run experiments, consisting of k monitored services.
- N users using the testbed, conducting multiple experiments. M_i is the number of experiments that user i has conducted.
- Users provide feedback for a list of questions that are related to each experiment that has been conducted.
- The core engine that gathers all feedback provided by the users, stores the data, aggregates and computes reputation values for each service along with their representation (Figure 3).

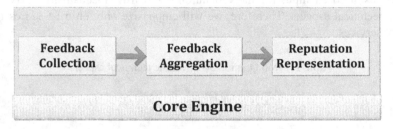

Fig. 3. Core system of the proposed trust mechanism

3.2   Description

The user, after executing an experiment, is requested to provide feedback for the advertised testbed services, according to his own perception and expectations. This feedback -from now on called as opinion- is stored in the core engine for further processing. At the same time, the user is requested to rate the level of his certainty regarding the feedback he provided to the system. This value represents the quality factor.

Afterwards, the core engine estimates the reputation value of each service by aggregating the opinions, weighted by the quality and credibility factors. The credibility factor indicates the level of trustworthiness of a user, regarding the honesty of his opinions. The last step of the process is the representation of the reputation values for each testbed service.

- *Opinion*

Opinion expresses the amount of satisfaction of the user i for his j^{th} experiment and is represented as follows:

$$O_{ij} \{ 1 \leq i \leq N, 1 \leq j \leq M_i \} \tag{1}$$

where N is the number of users in the system and M_i is the number of the experiments conducted by user i. Opinion O_{ij} takes a value in $[0,1]$.

- *Quality*

Quality represents the confidence of the user providing feedback. If a user is not sure about his opinion and provides feedback, his credibility and the final reputation should be influenced proportionally. There are many ways to determine quality metrics. The approach we consider here is to be set manually by the user. Values lie in the [0,1] area.

- *Credibility*

Credibility is used to express whether a user provides true or false judgments for testbed services. Each user has a credibility value assigned, which is adjusted according to the honesty of his ratings. Credibility is used to weigh the opinions that the user provides. If credibility is low – namely the user is not trustful – his opinion plays a minor role in the system evaluation and vice versa. The credibility factor lies in the [0,1] area and is initiated to 0.5 for each user. The credibility of user i after the $k + 1$ experiment is adjusted according to the honesty of his opinion O_{ij} as follows:

$$C_i^{k+1} = \begin{cases} C_i^k + \dfrac{(1 - C_i^k) \cdot Q_{ij}}{2}, & if \ |T - O_{ij}| \le e \\ C_i^k - \dfrac{C_i^k \cdot Q_{ij}}{2}, & if \ |T - O_{ij}| > e \end{cases} \tag{2}$$

Q_{ij} is the quality of the opinion O_{ij} , while T is the real and true value of the service, for which the opinion is being asked, and is provided to the reputation-based algorithm by a monitoring tool. Finally, e is a threshold which is defined as the mean absolute difference of the opinion O_{ij} and the real value of the service T. If a user gives a low (high) rating and the real value is high (low), then credibility is reduced, as it is possibly an attempt to trick the evaluation system. On the contrary, if a user gives a rating and the absolute difference from the real value is not higher than the threshold, then the credibility is increased. At the highest reported quality value of 1, a rating that lies within the absolute difference from the real value, increases the credibility of the reporting user by half the amount required for credibility to reach the maximum value 1. On the other hand, a rating outside this region results in decreasing the credibility value by half.

- *Reputation*

The reputation R of a service is the end result of aggregating user opinions, weighted by credibility and quality values.

$$R = \frac{\sum_i^N \sum_j^{M_i} O_{ij} \cdot C_i \cdot Q_{ij}}{\sum_i^N \sum_j^{M_i} C_i \cdot Q_{ij}} \tag{3}$$

By defining M as the number of all experiments conducted in the system [$M = \sum_i^N M_i$], the computational complexity of R is $O(M)$ (i.e. linear to M), which is considered very efficient.

4   Performance Evaluation

In this section, we evaluate the functionality and performance of Trust and User Experience Framework (TUEF) with the use of real and simulated experiments. The experiments were carried out in NETMODE wireless testbed [16] at National Technical University of Athens (NTUA) which consists of 20 802.11a/b/g/n wireless nodes. The wireless testbed offers to the experimenters the necessary experimentation tools to validate their protocols and applications in real world environments. In our case, for demonstration mainly purposes and without loss of generality, experimenters had the capability to provide feedback for the node availability and Packet Delivery Ratio (PDR) services of the testbed at the end of their experiments. Node availability characterizes the state of a node (up or down). PDR is defined as the ratio of data packets received by the destination node to those generated by the source node and provides an average estimation of the link quality over a period of time. In our wireless testbed, PDR between all nodes is measured in advance and is updated periodically, while node availability is monitored by the Nagios tool [17]. It is worth noting here that these two metrics have been considered as a proof of concept implementation, nevertheless our proposed framework can be utilized with other metrics which are specific to the provided testbed services.

In addition, a rating for the Overall Experience of using the testbed has been included in our evaluation. The algorithm described in section 3 is applied also for the Overall Experience, but due to the subjective nature of the attribute, two amendments are necessary. First, Equation (2) needs proper modification because of the absence of a ground truth, when updating a user's credibility. The real value coming from monitoring data (T) has been substituted with the reputation (R) calculated from the current opinions for the Overall Experience. Second, a separate credibility, from the one for the technical aspects, is stored to avoid intermixing credibility values for aspects of a different nature, as well as for security reasons.

4.1   Real User Experimentation

In order to demonstrate the operation of our trust mechanism, we implemented the algorithm with input instances from real experiments and the respective feedback provided by our wireless testbed users (100 experiments including opinions and their confidence levels about technical aspects and the Overall Experience). In Figure 4, a visual representation of the reputation for each of the aforementioned attributes of the testbed is depicted.

As expected, the reputation value of node availability is close to 1 with negligible variance, due to the fact that all nodes are available under normal circumstances. PDR is difficult to be objectively evaluated from the user perspective, and one would assume that fluctuations would arise; however users' opinions seem to be consistent. For the Overall Experience, each user bases his opinion on different factors with different weights and this subjectivity justifies the fluctuations depicted in the graph.

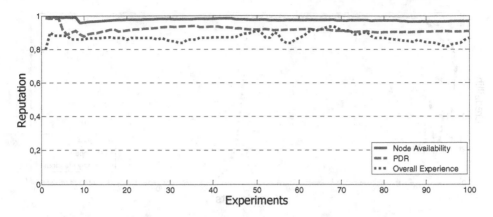

Fig. 4. Reputation of wireless testbed attributes for real world experiments

4.2   Simulated Experimentation

Making sound judgments for our implementation, based solely on real data is imprac-
tical and incomplete. Therefore, we conducted several simulations, with a high number of virtual users and experiments. In order to test our algorithm under abnormal
conditions, we modeled the behavior of four malicious user categories. Low, average
and highly malicious users, provide 20%, 50% and 100% malicious opinions respectively. Sleepers build up their credibility by providing honest opinions for the first
50% of their experiments and then start acting maliciously. By measuring user behavior and testbed performance from the real experiments, we modeled the behavior of
honest users and testbed services. Furthermore, we devised a metric for providing an
evaluation score for trust mechanisms, which does not take into consideration the
specific reputation values, but provides an insight about the effectiveness of malicious
opinions.

$$M = 1 - |R_g - R_T|$$

$$R_g : Total\ reputation\ of\ good\ users$$
$$R_T : Total\ reputation$$

According to this definition, results yielding a metric value close to 1 demonstrate
that the reputation of good users dominates total reputation. Apparently, this value
depends on the amount of malicious opinions and therefore we will conduct simulations with an increasing percentage of malicious users. The credibility graphs of an
honest, a highly malicious and a sleeper user are shown in Figure 5. For this case, a
malicious user population of 10% was considered, while each user conducted 100
experiments.

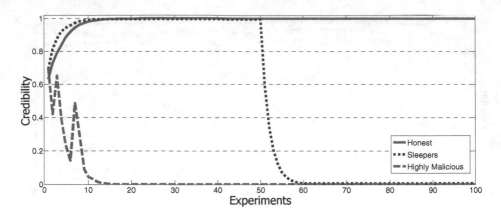

Fig. 5. Credibility graph for various user types

Evidently, the credibility graphs for both the honest and the malicious user converge at values 1 and 0 respectively, after a small number of experiments. The fluctuations that are observed occur because the experiments of the different users are intermixed and conducted in random order and the system needs several iterations to converge. Furthermore, the change in behavior of the sleepers is clearly depicted, as the credibility graph starts to decrease rapidly after the 50th experiment.

The next step towards TUEF evaluation was to conduct a worst case scenario simulation, for an increasing percentage of malicious users. We ran simulations up to a 95% of malicious users, where all of them were highly malicious. In order to have a point of reference, we compared the results of TUEF with those of a simple algorithm that calculates reputation as a simple mean score (Figure 6).

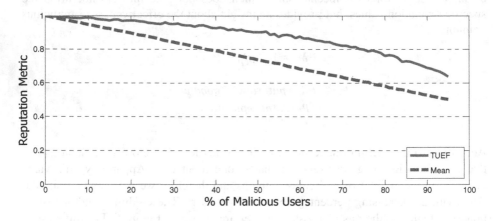

Fig. 6. Comparison between TUEF and Mean Score

Apparently, the TUEF graph has a smoother descend rate than the mean function, because opinions of malicious users play a minor role compared to those of the mean function, where all opinions are equal. However, after a point, malicious users compromise the system and the TUEF graph approaches the mean graph. This is also evident from

Equation (2), where the threshold is the mean absolute difference of the opinions O_{ij} and the monitoring data T. As the number of malicious users proliferates, this threshold increases, meaning that more malicious opinions are treated as honest.

The final simulation is a more moderate approach, where TUEF is tested against an increasing number of malicious users, where each category of malicious users constitutes 25% of the overall malicious population. The reputation metric values for the two aforementioned services (Node availability, PDR) and the Overall Experience are portrayed in Figure 7. We observe that the Overall Experience attribute continuously fluctuates, in contrast to the TUEF services. Considering the fluctuations described in Figure 4, this is something expected and occurs due to the subjective nature of the attribute. Moreover, TUEF has a consistent behavior and manages to suppress highly deviating opinions, even for high percentage of malicious users.

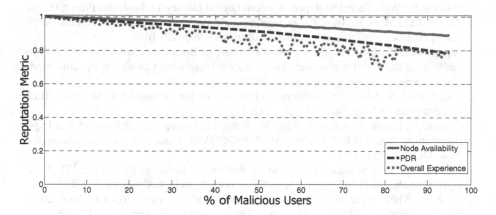

Fig. 7. TUEF Services and Overall Experience in a mixed malicious population

5   Concluding Remarks

In this paper, a novel reputation-based trust mechanism which provides a quantitative view of the trustworthiness of a testbed utilizing monitoring data and experimenters' feedback regarding their Quality of Experience (QoE) was presented. The performance evaluation results demonstrated that the proposed trust mechanism succeeds in delivering accurate reputation values even in the case where a large number of malicious users provides false feedback.

Our future work includes extension of the proposed trust mechanism to include feedback and insight regarding user motives and behavior. Specifically, users' behavior could be monitored based on their previous posed experiments that might have been characterized as illegal or have resulted into testbed malfunctions. For example, a behavior criterion in a wireless testbed could be the commitment of a user to use his reserved wireless channels. Finally, our aim is towards a common federation framework for heterogeneous environments, where reputation based trust management can provide a horizontal paradigm allowing dynamic reputation building and trustworthiness among testbeds and/or users.

Acknowledgements. This work was carried out with the support of the Fed4FIRE project ("Federation for FIRE"), an integrated project funded by the European Commission through the 7th ICT-Framework Programme (318389).

References

1. Grandison, T., Sloman, M.: A Survey of Trust in Internet Applications. IEEE Communications Surveys & Tutorials 3(4), 2–16 (2000)
2. Golbeck, J., Hendler, J.: Accuracy of Metrics for Inferring Trust and Reputation in Semantic Web-Based Social Networks. In: Motta, E., Shadbolt, N.R., Stutt, A., Gibbins, N. (eds.) EKAW 2004. LNCS (LNAI), vol. 3257, pp. 116–131. Springer, Heidelberg (2004)
3. Page, L., Brin, S., Motwani, R., Winograd, T.: The PageRank Citation Ranking: Bringing Order to the Web. Technical Report, Stanford Digital Library Technologies Project (1998)
4. Zacharia, G., Maes, P.: Trust management through reputation mechanisms. Applied Artificial Intelligence. An International Journal 14(9), 881–907 (2010)
5. Alperovitch, D., Judge, P., Krasser, S.: Taxonomy of Email Reputation Systems. In: Proceedings of the 27th International Conference on Distributed Computing Systems Workshops, Toronto, Ontario, Canada, p. 27 (2007)
6. Pan, J., Paul, S., Jain, R.: A Survey of the Research on Future Internet Architectures. IEEE Communications Magazine 49(7), 26–36 (2011)
7. Gavras, A., Karila, A., Fdida, S., May, M., Potts, M.: Future Internet Research and Experimentation: The FIRE Initiative. ACM SIGCOMM Computer Communication Review 37(3), 89–92 (2007)
8. International Telecommunication Union. Definition of Quality of Experience. ITU-T Delayed Contribution D.197, Source: Nortel Networks, Canada, P. Coverdale (2004)
9. Qualinet White Paper on Definitions of Quality of Experience (QoE) (April 2012), http://www.qualinet.eu/images/stories/Newsletter_2012.pdf
10. Law, E., Roto, V., Hassenzahl, M., Vermeeren, A., Kort, J.: Understanding, Scoping and Defining User eXperience: A Survey Approach. In: Proc. of the 27th International Conference on Human Factors in Computing Systems, Boston, MA, USA, pp. 719–728 (2009)
11. Kamvar, S.D., Schlosser, M.T., Molina, H.G.: The EigenTrust Algorithm for Reputation Management in P2P Networks. In: Proceedings of 12th International Conference on World Wide Web, Budapest, pp. 640–651 (2003)
12. Garg, A., Battiti, R.: The Reputation, Opinion, Credibility and Quality (ROCQ) scheme. Technical Report DIT-04-104, University of Trento (2004)
13. Zhao, H., Li, X.: H-Trust: A Robust and Lightweight Group Reputation System for Peer-to-Peer Desktop Grid. In: Proceedings of 28th International Conference on Distributed Computing Systems Workshops, Beijing, pp. 235–240 (2008)
14. Quercia, D., Hailes, S.: MATE: Mobility and Adaptation with Trust and Expected-utility. International Journal of Internet Technology and Secured Transactions (2008)
15. Quercia, D., Lad, M., Hailes, S., Capra, L., Bhatti, S.: STRUDEL: Supporting Trust in the Dynamic Establishment of peering coalitions. In: Proceedings of the 2006 ACM Symposium on Applied Computing, Dijon, France, pp. 1870–1874 (2006)
16. NETMODE Wireless Testbed, http://www.netmode.ntua.gr/testbed
17. Nagios, http://www.nagios.org

Social-Based Routing
with Congestion Avoidance
in Opportunistic Networks

Alexandru Asandei, Ciprian Dobre, and Matei Popovici

University Politehnica of Bucharest, Romania
Faculty of Automatic Control and Computers
alexandru.asandei@cti.pub.ro,
{ciprian.dobre,matei.popovici}@cs.pub.ro

Abstract. In particular types of Delay-Tolerant Networks (DTN) such as Opportunistic Mobile Networks, node connectivity is transient. For this reason, traditional routing mechanisms are no longer suitable. New approaches use social relations between mobile users as a criterion for the routing process. We argue that in such an approach, nodes with high social popularity may become congested. We show that social-based routing algorithms such as Bubble Rap are prone to congestion, and introduce two algorithms *Outer* and *LowerEps*. We present experimental results showing that the latter outperform Bubble Rap and solve the congestion problem.

1 Introduction

The emergence and wide-spread of new-generation mobile devices together with the increased integration of wireless technologies such as Bluetooth and WiFi create the premises for new means of communication and interaction, challenge the traditional network architectures and are spawning an interest in alternative, ad-hoc networks such as *opportunistic mobile networks*.

An opportunistic mobile network (ON) [10] is established in environments where human-carried mobile devices act as network nodes and are able to exchange data while in proximity. Whenever a destination is not directly accessible, a source would opportunistically forward data to its neighbours. The latter act as carriers and relay the data until the destination is reached or the messages expire.

To cope with intermittent connectivity and ON partitioning, the natural approach is to extend the store-and-forward routing to store-carry-forward (SCF) routing [8]. In SCF routing, a next hop may not be immediately available for message forwarding. In this case forwarding will be delayed until a suitable node is encountered. Thus, ON nodes must be (i) capable of buffering data for a considerable duration, and (ii) selecting suitable carriers for a message, from the list of all sighted nodes. A bad forwarding decision may cause the packets to be delayed indefinitely [3].

J. Cichoń, M. Gębala, and M. Klonowski (Eds.): ADHOC-NOW 2013, LNCS 7960, pp. 13–25, 2013.

Routing algorithms for ONs are either mobility-aware or social-aware. The former (which includes protocols such as PRoPHET [9]) takes routing decisions based on the number and duration of node encounters; the latter (which includes Bubble Rap [7]) relies on the knowledge that members (or nodes) of an ON are people carrying mobile devices. Social-aware ONs involve routing decisions based on how people are organized into communities, according to places of living and work, common interests, leisure activities, etc. A network of human relations as well as the overall structure of a community is captured by a social graph. Social networks are popular platforms for interaction, communication and collaboration between friends. The social relations between people can be generally inferred from the user interactions in such networks [12]. This is why in recent years, researchers have started to show an interest in social-based routing algorithms for ONs. However, approaches such as [7] can quickly lead to network congestions, as more popular ON members become flooded by message forwarding requests.

In this context, given that more powerful smarthpones and tablets continue to appear, would such devices be able to successfully carry messages destined for others, as envisioned by the opportunistic approach? Opportunistic networks are designed to support from catastrophic scenarios, where smartphones would take over the entire capacity of the damaged communication facilities in the disaster areas and beyond, all the way to completely distributed social networks, where mobile devices become caches of information in a publish/subscribe approach [10]. A forecast of mobile traffic published by CISCO [1] states that "global mobile data traffic will increase 26-fold between 2010 and 2015. Mobile data traffic will grow at a compound annual growth rate (CAGR) of 92 percent from 2010 to 2015, reaching 6.3 exabytes per month by 2015". Experts agree that at this growth a significant investment in the infrastructure will be needed. Opportunistic networks offer an alternative approach, when some of this wireless traffic can be offloaded directly into the mobile devices within the surrounding area. The current wireless communication infrastructure reached its threshold, as bottlenecks have already been observed - e.g., in [13] the overload of the AT&T infrastructure due to wider deployment of smartphones. Will this happen again when dispersing some of this overwhelming traffic directly to other smartphones? Unfortunately, predictions are not good, as they show that we still have to design communication protocols for opnets that can compensate for potential communication bottlenecks.

The contribution of this paper is twofold. First, we introduce the congestion problem, and show that the social-based Bubble Rap [7] algorithm can frequently produce message buffer overflows in nodes with high popularity, thus leading to lost messages. Second, we introduce two adaptations of the Bubble Rap algorithm, *Outer* which shows significant improvements in scenarios with a high number of ON participants and *Lower-Eps* which outperforms the classic Bubble Rap algorithm in situations where the number of ON nodes is low.

The rest of the paper is structured as follows: In Section 2, we review some of the most prominent social-aware routing algorithms, and especially Bubble Rap, which is considered to exhibit highest performance. Also, we show that

```
 1  begin BubbleProcedure()
 2      if (LabelOf(currentNode) == LabelOf(destination)) then
 3          if (LabelOf(EncounteredNode_i) == LabelOf(destination))
 4              and (LocalRankOf(EncounteredNode_i) > LocalRankOf(currentNode))
 5              and (message.InnerTokens > 0)
 6          then
 7              message.InnerTokens--
 8              EncounteredNode_i.addMessageToBuffer(message)
 9      else
10          if ((LabelOf(EncounteredNode_i) == LabelOf(destination))
11              or (GlobalRankOf(EncounteredNode_i) > GlobalRankOf(currentNode)))
12              and (message.OuterTokens > 0)
13          then
14              message.OuterTokens--
15              EncounteredNode_i.addMessageToBuffer(message)
16  end
```

Listing 1.1. The adjusted Bubble Rap procedure

approaches such as Bubble Rap are prone to congestion. In Section 3 we introduce two algorithms, *Outer* and *LowerEps*, and show that they behave better than Bubble Rap, both in terms of performance, as well as susceptibility to congestions. In Section 4 we present our conclusions.

2 Congestion in ONs with Social-Based Routing

2.1 Social-Based Routing Algorithms

There is a considerable number of works regarding social routing in ONs as well as results showing the prevalence of these methods over traditional routing mechanisms [4].

In [15] a *Socio-Aware Overlay* is used for publish/subscribe communication. ON nodes act as: (i) publishers of certain events (abstracted as messages) and/or (ii) subscribers to certain events they are interested in. Also, there is a chosen event broker responsible for maintaining a high message delivery rate. Instead of relying on criteria such as geographical location for clustering ON nodes, the authors of [15] use communities to build an abstract topology of the ON. Community detection is achieved dynamically, as each node exchanges and updates community-related information (community affiliation, frequently encountered nodes, etc.) upon each encounter with another node. The HiBOp (History-based routing protocol for opportunistic networks) algorithm described in [4] also exploits social information for message routing. Unlike [15], where communities were inferred solely on the cumulative contact duration between nodes, HiBOp holds *context information* as a *set of attributes* which include personal information, device characteristics, residence, hobbies & fun, etc. for each node.

The Bubble Rap algorithm [7], builds on the ideas from [15], but unlike [4] assumes that no information regarding the characteristics of nodes is known in advance. In a manner similar to [15], Bubble Rap organises nodes into communities, and assigns a *local ranking* for each node n and each local community n

is part of. Also, each node receives a *global ranking*. Both the local and global rankings are measures of importance (centrality) a node has either in some community, or in the entire ON. Routing is achieved as follows: messages will be forwarded to nodes with higher global ranking until a node from the destination's local community is encountered. Next, the same forwarding process occurs inside the local community, based on the local ranking instead of the global one. As is the case with [4], Bubble Rap shows significant improvements with respect to traditional (non-social) routing methods [7].

In most ON routing solutions, congestion is often ignored. For instance, in epidemic routing, which rely on broadcasting messages in parts of, or in the entire ON, authors of [11] argue that such protocols are able to ensure message delivery only when the *buffer capacity is sufficient*. In [5], Grossglauser and Tse propose a 2-hop forwarding approach which consider that *nodes have infinite buffer capacity*, which of course is unrealistic. Previous authors assume that nodes with higher popularity (either measured using community rankings or context similarity) are the key elements in routing messages, and are supporting most of the forwarding workload in the ON. Since Bubble Rap does not provide any mechanisms to control the number of copies for each forwarded message, it is hard to distinguish between situations where the entire network or just a subset of popular nodes is being flooded. In order to rule out the complete ON flooding scenario, we add some minor modifications to the original Bubble Rap algorithm which allows us to place an upper limit on the number of generated copies for each message.

```
1  begin Classic()
2    foreach EncounteredNode_i do
3      BubbleProcedure()
4  end
5  Obs: for each generated message, InnerTokens = OuterTokens = MAX_TOKENS/2
```

Listing 1.2. The Classic algorithm

We refer to the adjusted Bubble Rap algorithm as *Classic*. The logic implemented at the encounter of an arbitrary node from the ON is shown in Listing 1.1. The pseudocode for *Classic* is shown in Listing 1.2, and it relies on the *BubbleProcedure* from Listing 1.1. Whenever a message is generated, two values `InnerTokens` and `OuterTokens` expressing the message copy limit in the local and global communities, are associated with it. As seen in Listing 1.1, at lines 5 and 12, whenever the copy limit is reached, the message is no longer forwarded, and deleted from the node's buffer.

2.2 Experimental Setting

We have tested *Classic* on a variety of traces with different features, in order to see whether congestion/saturation can occur. One of the most relevant features of each trace is the *interaction density*, i.e. the number of interactions between distinct users, per period of time. The traces used in the experiments can be roughly divided into three categories, based on interaction density: *low*, *medium* and *high*. *Low* density traces are typical for environments where any two wireless nodes (i.e., cars in traffic, large urban environments) meet with a low probability because of the large covered area. *Medium* density traces are typical for small communities (i.e. suburbs, small towns, university campuses), where node encounters have a higher probability than in case of *low* density scenarios. Finally, *high* density traces correspond to environments where node encounters are frequent (i.e. conferences, employees within a single company). Due to limited space, in this paper we have selected one representative trace for each category. All are publicly available traces, belonging to the CRAWDAD database:

- *Haggle-iMotte-infocom2005* (short *Infocom*): was collected at the IEEE Infocom Conference in Grand Hyatt Miami, USA. It has 41 participants and a duration of 3 days. We consider it to be of high interaction density.
- *Haggle-iMote-content* (short *Content*): resulted from an experiment performed in the city of Cambridge, UK, has 54 participants and lasted 8 weeks. It reflects mobility in a academic environment and also contains a number of 18 devices with a fixed location. We consider it to be of medium interaction density.
- *St. Andrews-sassy* (short *St.Andrews*): was collected in the city of St. Andrews, UK. It has 27 participants and lasted 11 weeks. It reflects mobility in an urban environment. We consider it to be of low interaction density.

The experiments aim at simulating an ON where nodes have the features extracted from a chosen trace. We follow the same assumptions from [14]: we consider an asynchronous communication pattern in which each node can generate a message for some other node(s) in the ON. Out of the total of N nodes in the network, $N/2$ nodes will be randomly chosen as generators of messages. Therefore, the number of messages each generator will send would be distributed according to the normal distribution. In this scenario, any of the $N-1$ nodes different from the generator has an equal chance of being a destination. However, we further enforce the observations from [14] and use a *community bias b* on the distribution of sent messages: all nodes that generate a number greater than b of messages, will use as destination only nodes from the **local** community. For all other nodes, any destination is generated at random. Such a scenario corresponds to the distribution of status messages in a social network as observed in [6].

Messages have an average size of 10KB. We assume each node has a buffer size of approximately 20MB, which means it can store up to 2048 messages. The choice of the buffer size is motivated by the current memory limit of most applications on Android devices, which ranges from 16 to 24 MB, on newer generation phones.

2.3   The Congestion Problem

We say a node becomes saturated whenever a *buffer overflow* event occurs for that particular node. The overflow is experienced when the buffer capacity is

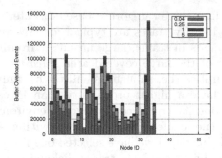

Fig. 1. Out-of-memory events, *Content*, 2 message copies

exceeded - for example, when a node receives messages at a higher rate than it is able to forward, thus causing message buffers to be filled. Whenever a new message is received and the buffer is full, it is discarded. We noticed a surprisingly large number of buffer overflow events during the lifespan of each trace, including those of low density, as can be seen in Figure 1. The results correspond to different message generation rates (nodes generate messages from every hour to as rare as every 5 days).

Next, we were interested to see how pervasive these events are. We therefore monitored the *saturation rate* for each node, i.e. the number of saturation events a node experiences, over a fixed period of time.

Figures 2(a) and 3(a) provide an overview of the saturation rate for the *Content* and *Infocom* traces. As can be noticed in Figure 3(b), even for traces of low density such as *St. Andrews*, high values for the saturation rate are recorded. The charts are plotted for the scenario when half of the total number of nodes generate new messages every 6 hours, which corresponds to a 0.25 message generation rate. The sample period for chart plotting is of 0.33, i.e. every 8 hours. Notice there is a phase-shift between between the moment when messages are generated, and the moment when congestion readings are performed. This allows messages to propagate in the network, and to avoid situations in which only the moments of message generation are recorded.

Finally, we looked at possible connections between saturation rates and the size of the ON. By using the same monitoring technique and running the simulations for subsets ranging from 10 to 50 nodes (or maximum available for each trace if lower than 50) we noticed that no such connection exists. Saturation does not depend on the size of the network, and can appear even in ONs of small size.

Also, we were interested in seeing whether saturation could be avoided by an increase of the message buffer size. The simulations we carried out show that saturation still occurs even with buffer sizes increased up to 10 or 20 times over their initial values (depending on the data set used). This is shown in Figure 2(b), for the *Content* trace.

We therefore argue that node saturation is a *general problem* intrinsic to social-based routing algorithms (and in particularly to our adaptation of Bubble), and does not depend on specific features of the ON at hand.

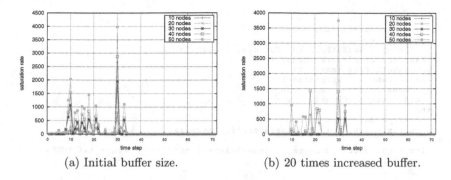

(a) Initial buffer size.   (b) 20 times increased buffer.

Fig. 2. Saturaton rate, *Content*, 2 message copies, 0.25 message generation period

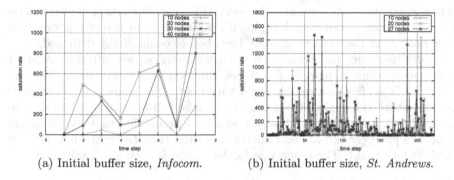

(a) Initial buffer size, *Infocom*.   (b) Initial buffer size, *St. Andrews*.

Fig. 3. Saturaton rate, 2 message copies, 0.25 message generation period

3   Solutions for the Congestion Problem

3.1   The Algorithms *Outer* and *LowerEps*

We examined two approaches for avoiding node saturation. The first one relies on the assumption that once a message has reached a node from it's destination community, it has a high delivery probability. Therefore, after a message copy has been forwarded inside the destination's community, all other copies will be reserved for encounters with nodes from other communities. This restriction is fit for relieving message flooding inside local communities. We denote by *Outer*, the adaptation of the *Classic* algorithm, that takes this into account. It can be seen in Listing 1.3. The restriction is implemented by setting `InnerTokens` to the value of 1, for each message.

Another important feature of *Outer* is related to forwarding messages that are not destined for the local community of the current node. While *Classic* would wait to encounter a node with a higher global ranking before sending the message, *Outer* sends the message to the first node encountered from another community, regardless of its rank. This has the effect of eliminating global rankings as a criterion for forwarding between nodes of different communities.

```
 1 begin Outer()
 2   foreach EncounteredNode_i do
 3     BubbleProcedure()
 4     if (message.wasNotForwardedTo(EncounteredNode_i))
 5       and (LabelOf(currentNode) !=
 6         LabelOf(destination))
 7       and (LabelOf(currentNode) !=
 8         LabelOf(EncounteredNode_i))
 9       and (message.OuterTokens > 0)
10     then
11       message.OuterTokens--;
12       EncounteredNode_i.addMessageToBuffer(message)
13 end
14 Obs: for each generated message, InnerTokens=1 and OuterTokens= MAX_TOKENS-1
```

Listing 1.3. The Outer algorithm

The second approach relies on the observation that, in *Classic*, messages having as destination nodes with low global ranking have to traverse nodes with higher ranking in the local community first, before being finally delivered.

In order to reduce the occurrence of such situations, we introduce another adaptation of *Classic*, namely *LowerEps*, that exploits randomness. The behaviour of *LowerEps* is governed by a threshold ϵ, as follows: for each node encounter there is a probability ϵ that a message will be forwarded to the encountered node, regardless of its rank, and a $1 - \epsilon$ probability that the algorithm will behave precisely as *Classic*. Experiments show that *LowerEps* has an ideal behaviour for $\epsilon = 0.1$. The psedocode for *LowerEps* is shown in Listing 1.4.

```
 1 begin LowerEps()
 2   BubbleProcedure()
 3   if (message.wasNotForwardedTo(EncounteredNode_i))
 4     and (random < 0.1)
 5     and (message.Tokens > 0)
 6   then
 7     message.Tokens--;
 8     EncounteredNode_i.addMessageToBuffer(message)
 9 end
10 Obs: for each generated message, InnerTokens = OuterTokens = MAX_TOKENS/2
```

Listing 1.4. The LowerEps algorithm

3.2   Experimental Results

Our experimental results are aimed at: (i) examining the *performance* of *Outer* and *LowerEps*, with respect to *Classic* and (ii) studying the exposure to *congestion*, for each of the three algorithms. For the experiments we implemented a simulator that parses the traces and then applies an opportunistic routing algorithm at every encounter between two nodes.

For measuring performance, we use the following standard metrics: *hit rate*, i.e. the percentage of successfully delivered messages out of the total number of generated messages, *delivery cost* which represents the ratio between the total number of exchanged messages and the total number of generated messages and *latency* which is the time passed from a message being generated to it being successfully delivered to its destination.

To our knowledge, there is no widely accepted metric for measuring the susceptibility to congestions in ONs with social-based routing. For this reason, we consider *load balancing* to be a relevant criterion for such a task. Intuitively, an ON node i is less prone to congestion if the number of encounters of node i is proportionate to the number of messages sent to node i. This intuition is captured by the *load balancing factor* (LBF) for node i (denoted LBF(i)), which is the difference between the normalized sighting values for node i and those for the normalized number of received messages by node i, i.e.

$$\text{LBF(i)} = \frac{\text{sight}(i)}{\max_{i \in \text{nodes(ON)}} \text{sight}(i)} - \frac{\text{msg}(i)}{\max_{i \in \text{nodes(ON)}} \text{msg}(i)}$$

where nodes(ON) denotes the set of nodes in the network, sight(i) refers to the number of times node i was encountered by other nodes and msg(i) denotes the number of messages received by node i, during the lifespan of the trace.

We use the average value of LBF(i) computed for all nodes i in the ON, to measure the exposure to congestion for each algorithm, on a given trace. We denote this value as *ALBF*.

Performance. As can be observed in Figure 4(a), both *Outer* and *LowerEps* have a better average hit rate. The same is true when increasing the number of copies for each message, as can be seen in Figure 4(b).

The increase of the number of message copies, which is justified by the need of higher hit-rates, increases the delivery costs for *Outer*. However, when the number of generated messages is also increased, the delivery costs become moderate. The same thing is true about the average latency. Therefore, a suitable balance between (i) better hit-rate, and (ii) better latency and delivery costs can be found, depending on the expected traffic of the ON (i.e. number of exchanged messages).

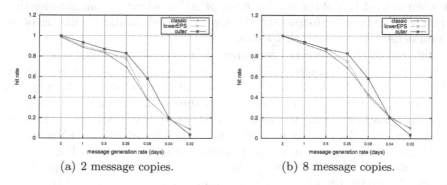

(a) 2 message copies.   (b) 8 message copies.

Fig. 4. Hit rate, *Infocom*

When dealing with a trace of a medium density such as *Content*, *Outer* achieves the best hit rate results.

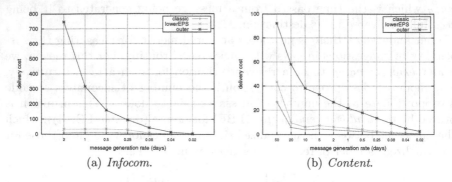

(a) *Infocom.*   (b) *Content.*

Fig. 5. Delivery cost, 2 message copies

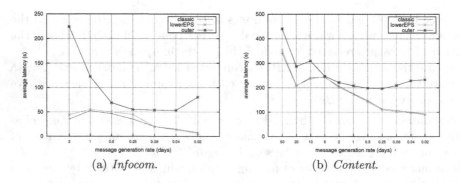

(a) *Infocom.*   (b) *Content.*

Fig. 6. Average latency, 2 message copies

This confirms the performance gain exhibited for high density traces. In terms of delivery cost and latency, the values for *Outer* are still higher but to a lower extent. This tells us that pushing messages towards outside communities works best when at least half of the total number of nodes belong to communities that interact with each other frequently. We have obtained similar results on other medium density traces, such as *UPB2012* [2].

A different picture presents itself when doing the same simulations for the low density trace of *St. Andrews*. The best performance in this case is achieved by *lowerEPS* which has better values for both hit rate and average latency and a delivery cost comparable to *Classic*. This shows that when faced with few interactions in a large geographical space, node centrality is not very reliable as there is not enough data for the computation of a proper value. In such a scenario, an approach which exploits randomness proves to be more useful. Figures 7(a) and 7(b) also show that the routing scheme used by *Outer* is inappropriate for this scenario, having the lowest hit rate. Similar results have been obtained for the low-density trace *UPB2011* [2].

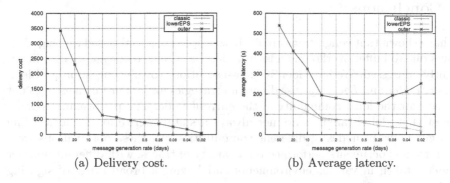

(a) Delivery cost.   (b) Average latency.

Fig. 7. *St. Andrews*, 2 message copies

Exposure to Congestion. In opportunistic networks, *load balancing* depends on the *sighting distribution*. By *sighting distribution* we denote the number of times each node has been sighted (by any other node) during the life span of a trace and is equivalent to the number of interactions. The more times a node interacts with other nodes, the more likely it is to receive messages. Thus, perfect load balancing would result from the use of a totally random routing scheme. Ideally, the distribution of received messages for each node of the ON would be almost identical to the sighting distribution. Under this assumption, we plot the *LBF* value for *Classic*, *Outer* and *LowerEps*, against that of a totally random routing scheme.

Figures 8(a), 8(b) show the result of applying this technique, for the three traces. Due to the nature of opportunistic networks, a critical mass of messages (thus a lower message generation period) in the network is required in order to achieve a close to perfect load balancing even with a random routing scheme.

Figures 8(a) and 8(b) account for the performance of *Outer*. They show that *Outer* achieves the best load balancing, being closest to the random routing algorithm. It is notable that *Classic* has the worst behaviour, which shows that it is unfit to provide good load balancing.

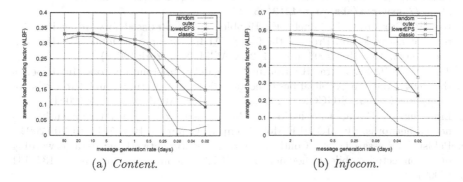

(a) *Content.*   (b) *Infocom.*

Fig. 8. ALBF, 2 message copies

4   Conclusions

We have shown that congestions are likely to appear in social-based routing algorithms due to buffer overflows, an issue which has been, to our knowledge, insufficiently addressed. We have identified two solutions to the congestion problem. *Classic*, *Outer* and *LowerEps* use different strategies for forwarding messages not destined for the local community of the current node. The performance results suggest that global rankings are not always a suitable criterion for forwarding messages between nodes of different communities. All these observations were drawn from experiments performed on a variety of traces, with different number of nodes, taken in various environments and having a broad range of sighting distributions.

Acknowledgment. The research presented in this paper is supported by project "ERRIC - Empowering Romanian Research on Intelligent Information Technologies/FP7-REGPOT-2010-1", ID: 264207.

References

1. Cisco visual networking index: Global mobile data traffic forecast update (2011-2016), http://www.cisco.com/
2. Asandei, A.: Socio-aware routing in opportunistic networks. Bachelor Thesis
3. Bjurefors, F., Gunningberg, P., et al.: Congestion avoidance in a data-centric opportunistic network. In: Proc. of the ACM SIGCOMM Workshop on Information-Centric Networking, ICN 2011, New York, USA, pp. 32–37 (2011)
4. Boldrini, C., Conti, M., Passarella, A.: Exploiting users and social relations to forward data in opportunistic networks: The hibop solution. Pervasive and Mobile Computing 4(5), 633–657 (2008)
5. Grossglauser, M., Tse, D.N.C.: Mobility increases the capacity of ad hoc wireless networks. IEEE/ACM Trans. Netw. 10(4), 477–486 (2002)
6. Huberman, B.A., Romero, D.M., Wu, F.: Social networks that matter: Twitter under the microscope. CoRR, vol. abs/0812.1045 (December 5, 2008), http://ssrn.com/abstract=1313405
7. Hui, P., Crowcroft, J., Yoneki, E.: Bubble rap: social-based forwarding in delay tolerant networks. In: Proc. of the 9th ACM Int. Symp. on Mobile Ad Hoc Net. and Comp., MobiHoc 2008, pp. 241–250. ACM, New York (2008)
8. Jain, S., Fall, K., Patra, R.: Routing in a delay tolerant network. In: Proc. of the 2004 Conf. on Applications, Technologies, Architectures, and Protocols for Computer Comm., SIGCOMM 2004, pp. 145–158. ACM, New York (2004)
9. Lindgren, A., Doria, A., Schelén, O.: Probabilistic routing in intermittently connected networks. SIGMOBILE Mob. Comput. Commun. Rev. 7(3), 19–20 (2003)
10. Pelusi, L., Passarella, A., Conti, M.: Opportunistic networking: data forwarding in disconnected mobile ad hoc networks. IEEE Comm. Magazine 44(11), 134–141 (2006)
11. Vahdat, A., Becker, D.: Epidemic routing for partially connected ad hoc networks. Tech. Rep. CS-2000006, Duke University, Durham, NC (December 2000)

12. Wilson, C., Boe, B., et al.: User interactions in social networks and their implications. In: Proc. of the 4th ACM Eu. Conf. on Computer Systems, EuroSys 2009, pp. 205–218. ACM, New York (2009)
13. Wortham, J.: Customers Angered as iPhones Overload AT&T, The New York Times (September 2, 2009), http://www.nytimes.com/2009/09/03/technology (accessed November 2, 2012)
14. Xu, K., Hui, P., et al.: Impact of altruism on opportunistic communications. In: Proc. of the First Int. Conf. on Ubiquitous and Future Networks, ICUFN 2009, pp. 153–158. IEEE Press, Piscataway (2009)
15. Yoneki, E., Hui, P., Chan, S., Crowcroft, J.: A socio-aware overlay for publish/subscribe communication in delay tolerant networks. In: Proc. of the 10th ACM Symp. on Modeling, Analysis, and Simulation of Wireless and Mobile Systems, MSWiM 2007, pp. 225–234. ACM, New York (2007)

Maximum Lifetime Broadcast
in Mobile Sensor Networks

Bruno Nunes, Frederico Barboza, and Flávio Assis

LaSiD - Distributed Systems Laboratory
DCC - Department of Computer Science
UFBA - Federal University of Bahia
Salvador, Bahia, Brazil
{brpnunes,fred.barboza}@gmail.com, fassis@ufba.br

Abstract. In this paper we investigate the Maximum Lifetime Broadcast (MLB) problem, i.e. the problem of extending network lifetime when a series of broadcast operations is executed, in sensor networks with mobile nodes. For this problem, we present two localized algorithms (nodes use only 2-hop information), namely LPrim and LMCP, which are based on trees with minimum highest cost edges and shortest paths trees, respectively. The algorithms were evaluated through simulations under three distinct mobility models and compared against common broadcasting techniques (broadcasting based on a Connected Dominating Set and Flooding). According to our experiments, LPrim outperformed alternative solutions in terms of the number of successfully executed broadcasts (network lifetime) and in terms of the energy consumed by broadcast.

Keywords: broadcasting, network lifetime, mobile sensor networks.

1 Introduction

The development of energy efficient protocols for wireless sensor networks has been a prolific area of research, as these networks are composed of large sets of low-cost resource-constrained sensor nodes, commonly equipped with non-replenishable batteries. These networks are usually deployed on remote sites, which makes battery substitution unattainable in many cases. Battery lifetime thus becomes one of the most critical resources, as it ultimately determines the total utility time of the sensor node.

In a mobile sensor network (MSN) nodes may change their geographical positions over time. Mobility of nodes may be caused by a number of reasons, such as the action of external agents, like animals or wind, or by the sensor node itself, when equipped with actuators that allow movement. As mobility results in frequent changes in the network topology or might even result in network partition, the design of communication protocols that are energy-efficient and that achieve high message delivery rates in a mobile setting becomes a challenge.

In this paper, we address the *Maximum Lifetime Broadcast* (MLB) problem [10] in the context of mobile networks. This problem consists in maximizing

J. Cichoń, M. Gębala, and M. Klonowski (Eds.): ADHOC-NOW 2013, LNCS 7960, pp. 26–37, 2013.

the lifetime of a wireless sensor network when a series of broadcast operations is executed. Broadcasting is one of the fundamental communication primitives in distributed systems. A message broadcast by a node, called *source node*, is disseminated to all other nodes of the network. Global information update, route discovery, and the dissemination of cryptographic keys are examples of network tasks that are typically executed periodically and that require broadcasting.

The MLB problem has been so far studied in the context of *static* wireless networks. In this paper, we evaluated three algorithms for energy-efficient broadcasting, called respectively LPrim, LMCP and LCDS. LPrim is an algorithm that we designed as a distributed version of the centralized algorithm for maximizing network lifetime described in [10]. LMCP is a distributed algorithm we designed, based on minimum cost paths trees. We refer to LCDS as an algorithm for broadcasting based on the algorithm for connected dominating sets presented in [16]. These algorithms address energy efficiency by controlling the transmission power of each sensor node (*topology control*). The lower transmission power a sensor node uses to transmit, the lower energy it spends while sending messages. These algorithms represent a comprehensive set of specific alternatives to define the transmission power of nodes: calculating trees whose maximum edge weight is the minimum (LPrim); minimum cost paths tree (LMCP); and connected dominating sets (LCDS). Beyond comparing these algorithms with each other, we compared them also against *flooding*, that corresponds to a broadcast strategy with no energy-aware specific control. We evaluated the algorithms considering three representative mobility models, as described in Section 5. The evaluation of the algorithms was performed through simulations.

This paper is organized as follows. Section 2 discusses related work. Section 3 presents the adopted system model and the problem definition. Section 4 describes the evaluated algorithms. Section 5 presents the results of the performance evaluation of the algorithms. Section 6 concludes the paper.

2   Related Work

Network lifetime refers to the time interval within which the network is able to perform the tasks it was designed for. As there exists a number of different applications, each one with its specific requirements, there is also a wide range of network lifetime definitions. A comprehensive survey of these definitions can be found in [6]. The far most commonly used definition in the literature is to consider network lifetime as the time until the first node fails due to energy exhaustion.

A taxonomy for the Maximum Lifetime Broadcast problem was presented in [7], based on the set of source nodes that can initiate a message broadcast and on the number of trees that can be used to extend network lifetime: (1) *Single Source* MLB, when there is only one source node and multiple broadcast trees can be used to extend network lifetime; (2) *Single Source/Topology* MLB, when the number of broadcast trees is restricted to one; and (3) *Single Topology* MLB, when there is a non-empty set of source nodes, and a single undirected tree is used as a broadcast tree by all of them. The existence of efficient solutions or approximation algorithms for these problems is dependent on a set of aspects, such as the type of graph (directed or undirected), symmetry of edge costs, the initial distribution of energy of nodes, among others.

In this paper, we concentrate on the *Single Source* MLB problem. This problem has only been considered for the case of stationary (non-mobile) networks. For these networks, *Single Source* MLB is NP-hard in directed graphs (model assumed in this paper) [2,15], even when: costs are symmetric [2]; all nodes transmit with the same power but the initial energies of all nodes except one are equal [15] or the initial energies of the nodes are equal and the transmission power levels of all nodes except one are equal. The authors in [15] prove that this problem remains NP-hard even if the number of broadcast trees is restricted to a fixed number. This problem can be, however, approximated logarithmically in directed graphs [2,15] or by a constant with high probability in random geometric graphs [1]. Centralized heuristics for this problem in directed graphs are presented in [2,15] and for undirected graphs in [10]. In all these algorithms reception cost is ignored, with the exception of [15], which discusses the problem when each transmitted message is received by a single node. Centralized algorithms, however, are not scalable nor applicable to a mobile setting. A distributed algorithm that provides a guaranteed performance of at least 1/12 of the optimum with high probability on random geometric graphs is presented in [1]. The algorithm does not consider reception costs and depends on the spatial distribution of nodes. In [13], we described a localized distributed algorithm for *Single Source* MLB. In this paper we evaluated a variation of this algorithm called LPrim, as described in Section 4. Further variations of MLB, such as for considering directional antennas, and combinations of unicast, multicast and broadcast, have also been studied (e.g. [4,10,15,7,14,8,12]). Additional work on broadcasting for wireless network exists, but which are focused on different problems, such as, for example, maximizing message delivery rate in the presence of faults.

As far as we know, this paper is the first to consider the problem of extending network lifetime for a series of broadcast operations by controlling transmission power in a mobile setting.

3   Problem Definition

3.1   System Model

A mobile sensor network (MSN) is composed of a set of nodes that might change their positions over time. Each node is equipped with a battery and a variable range radio transceiver, whose transmission power can be adjusted to any value between 0 and a maximum value, P_{max}. The system is modeled as a weighted *time-varying graph* [3] $G = (V, E, \mathcal{T}, w, \rho, IEnergy, Pos)$, where:

- V is a set of $n = |V|$ nodes deployed on a two-dimensional Euclidean plane.
- The relation $E : V \times V$ represents the set of directed edges between nodes.

- As a mobile network is a dynamic system, relations between entities occur over a time span $\mathcal{T} \subseteq \mathbb{T}$, where $\mathbb{T}$ is a sequence of time instants (without loss of generality, we consider it to be a nondecreasing sequence of natural numbers). $\mathcal{T}$ represents the *lifetime* of the network.
- $w : E \times \mathcal{T} \to \mathbb{R}_{\geq 0}$ is the edge weight function, as explained later in this section. The edge costs vary with time.
- $\rho : E \times \mathcal{T} \to \{0, 1\}$ is the *edge presence function*. $\rho(e, t) = 1$ iff the edge e is available at instant $t \in \mathcal{T}$. Otherwise, $\rho(e, t) = 0$. This function models the availability of communication channels between nodes.
- $IEnergy : V \to \mathbb{R}_{\geq 0}$ represents the initial battery supply of each node.
- $Pos : V \times \mathcal{T} \to \mathbb{R}_{\geq 0}^2$ represents the position of a node at a given instant.

Each node is capable of moving in any direction on the two-dimensional space with any speed ranging from 0 to a maximum V_{max}. We assume that nodes do not know their exact movement pattern in advance, and thus future positions cannot be foreseen.

We assume a simple connectivity model, in which all nodes in the transmission range of a node p receives messages transmitted by p. I.e. if the transmission range of node p at an instant t is r, then all nodes $q \in V$ such that $dist(p, q, t) \leq r$ will receive the message, where $dist(p, q, t)$ represents the Euclidian distance between p and q at instant t.

We additionally adopt a usual energy model. The strength of a signal transmitted with power p_t decreases at a rate that is proportional to $1/(p_t)^\alpha$ as the distance from the sender increases, where α is the path loss exponent. Thus, the energy spent by node u to transmit an l-bit message using the minimum power needed to reach a node at a distance d from u ($d \geq 0$) is given by function $tc(u, d, l)$:

$$tc(u, d, l) = l \cdot (cf(u) + \gamma(u) \cdot d^\alpha) \tag{1}$$

where: $cf(u)$ is the (fixed) energy spent by the transmitter electronics at node u; $\gamma(u)$ is a parameter characteristic of the transceiver and the channel; and α is the path loss exponent ($2 \leq \alpha \leq 6$).

We assume that all nodes spend the same amount of energy to receive an l-bit message, as defined by the rc function below, where rcv is a constant:

$$rc(l) = l \cdot rcv \tag{2}$$

The weight of an edge $e = (u, v)$ at instant t is defined by the function $w(e, t)$:

$$w(e, t) = tc(u, dist(u, v, t), 1) + rw(u, v, t) \tag{3}$$

where:

$$rw(u, v, t) = rcv \cdot |\{z : (z \neq u) \wedge (dist(u, z, t) \leq dist(u, v, t))\}| \tag{4}$$

I.e. the weight associated with edge $e = (u, v)$ (Eq. 3) is the transmission energy (for an 1-bit message) spent by u when it transmits using the minimum needed power to reach v (Eq. 1) plus the sum of the reception energy (for an 1-bit message) spent by all nodes that are at a distance from u that is smaller or equal to the distance from u to v at instant t (Eq. 4), i.e. the nodes that would hear transmissions from u to v.

We assume that the system executes in synchronous steps. Each node is equipped with a message queue. At each step, each node removes all the messages from its message queue, processes them, and, if needed, sends messages to other nodes (puts these messages in those nodes' message queues). Nodes thus communicate through reliable communication channels, and each message takes a single step to be transmitted from a node to another.

3.2   Problem Definition

We address the problem of maximizing network lifetime when a series of broadcast operations is executed. As mobility might cause network partition and specific movement patterns might prevent nodes from receiving a broadcast message, just extending the number of steps executed by the network *per se* does not reflect usefulness of an algorithm. We are not interested on algorithms that spend low energy but do not successfully execute a reasonable number of broadcasts.

Thus, we consider maximizing network lifetime as the problem of *maximizing the number of successfully executed broadcast operations*, started from a *source node* $s \in V$, until the first node runs out of energy. A broadcast of a message m is successfully executed when all nodes receive m. Starting from the same network configuration, algorithms that extend the number of successfully executed broadcasts are more effective in usefully spending the energy reservoir of the network. Additionally, this notion of network lifetime is adequate for comparing the algorithms in an asynchronous setting, where there is no explicit notion of time (although we assume a synchronous model in this paper, we are interested in developing algorithms that are suitable for use in asynchronous environments).

4   Evaluated Algorithms

In this paper we evaluated three algorithms: LPrim, LMCP and LCDS. Beyond comparing them with each other, we compared them also against message dissemination without power control (Flooding). LPrim, LMCP and LCDS represent three general approaches to the problem. LPrim and LMCP are based on local computation of trees that minimize the highest cost edge and minimum cost paths, respectively. LCDS is based on the distributed computation of connected dominating sets. Flooding does not apply any energy control. The names of the algorithms come from the fact that LPrim and LMCP are based on local (**L**) computations of, respectively, a variation of **Prim**'s algorithm and a **M**inimum **C**ost **P**aths tree, and LCDS is a localized (**L**) algorithm for calculating a **C**onnected **D**ominating **S**et.

LPrim, LMCP and LCDS are described in the following subsections. LPrim and LMCP are executed whenever the source node starts broadcasting a new message. LCDS is a local algorithm that is reexecuted by each node whenever it receives information about its neighbourhood (*hello* messages).

4.1 LPrim

In [13] we described an algorithm for maximum lifetime broadcast called DLMCA. LPrim is a combination of the centralized algorithm for MLB presented in [10] with the strategy of DLMCA for calculating a broadcast subgraph in the distributed environment. DLMCA is based on local computation of *minimum cost arborescences*, using Edmonds's algorithm [11]. LPrim instead locally executes a variation of Prim's minimum spanning tree (MST) algorithm [5] as described in [10]. DLMCA and LPrim find locally a spanning tree which minimizes the highest edge weight, but LPrim runs locally faster than DLMCA. As described in [10], such trees provide optimal solutions to MLB in the static case (when a single broadcast tree is used).

Prim's algorithm was designed for undirected graphs. The algorithm incrementally constructs a MST from the source node by inserting in the tree a node at a time. Nodes that have already been inserted in the tree are called *covered* nodes. At each step, the algorithm inserts in the tree the edge with the least cost that is adjacent to a covered and an uncovered node. This uncovered node becomes thus covered. As we are considering directed graphs, the variation of this algorithm described in [10], used in LPrim, considers at each step only *directed edges* from covered to uncovered nodes.

LPrim is executed periodically and incrementally. Starting at the source node, each node p_i executes as follows. On a graph that represents p_i's *two-hop neighbourhood*, i.e. p_i's neighbours and the neighbours of p_i's neighbours, p_i uses the variation of Prim's algorithm to calculate a directed tree rooted at p_i that spans all nodes in this graph. This tree is used by p_i to define: (a) its transmission power; (b) the set of nodes that become *covered* by it, i.e. that are reached by p_i when it transmits with the defined transmission power; and (c) the next nodes to continue the algorithm, called *relays*.

Each node p_i determines the relays and covered nodes using the tree as follows. Node p_i chooses a node p_j among its children in the tree for which the weight of edge (p_i, p_j) is the highest. Let us call this node $highest_i$. Node p_i will adjust its transmission power to the minimum power necessary to reach $highest_i$. All nodes p_k in p_i's one-hop neighbourhood for which the weight of edge (p_i, p_k) is less than or equal to the weight of edge $(p_i, highest_i)$ are the *nodes covered by* p_i. The other nodes in the tree are the *uncovered nodes*, according to p_i's view. The relays will be those nodes that are tails of *bridges*, i.e. edges (q, s) in the tree which have a covered node as tail (q) and an uncovered node as head (s)[1].

After having determined the relays, p_i broadcasts a RELAY message, using the minimum power needed to reach $highest_i$. This message contains a list of the nodes that are known by p_i to have been covered and those nodes that were chosen as relays by p_i. When a node, say p_j, receives a RELAY message from

---

[1] For an edge (u, v), we call u and v, resp., the *tail* and the *head* of the edge.

p_i, it acts as follows: (a) if it was chosen by p_i to be a relay and it has not been chosen a relay previously (by some other node), it continues the algorithm (executes the steps above), eliminating from its neighbourhood those nodes that have already been covered; (b) if it was not chosen by p_i to be a relay and it has not been previously chosen a relay by any other node, it sets its transmission power to zero; (c) if it was chosen to be relay by p_i and it has already been chosen to be a relay previously, it ignores the message.

4.2 LMCP

LMCP is a variation of LPrim, where each node calculates locally a minimum cost paths tree, i.e. a tree containing minimum cost paths from it to each other node in its two-hop neighbourhood, instead of the tree described in the previous subsection.

4.3 LCDS

This is an algorithm for broadcasting based on the distributed algorithm for connected dominating set (CDS) described in [16]. The CDS algorithm works as follows. Each node first locally elects itself a member of the CDS if it has two neighbours which are not directly connected, i.e. which cannot communicate with each other in a single hop. Each node in the CDS is said to *cover* its neighbours. After that, each node executes a set of optimization rules, based on the relationships that might exist between the set of nodes that it and its neighbours cover. We evaluated the algorithm with rules 1a and 2a described in [16], which use the id and the degree of nodes to consistently remove nodes from the CDS, when nodes cover overlapping sets of neighbours.

When starting a broadcast, the source node transmits the message with maximum power. At least one of the nodes in the CDS will hear it (even if the source node is not in the CDS). After that, messages are disseminated by retransmissions by each node in the CDS using maximum power. In the particular case of a static network, all nodes will receive the message (by the definition of CDS).

5 Evaluation and Analysis

5.1 Performance Metrics

The algorithms were evaluated according to the following metrics:

(a) Number of Successful Broadcasts: this metric represents the number of *successfully executed broadcasts*, performed until the first node fails due to battery exhaustion. A message is successfully broadcast when it is received by all nodes.

(b) Average Consumed Energy: this metric is the average of the total energy spent by successful broadcasts. The total energy consumed by a successful broadcast is the sum of the energy spent by each single node due to transmissions and receptions.

5.2 Description of Experiments

The evaluation of the algorithms was performed through simulations, using Sinalgo, a Java-based framework for testing and validating network algorithms [9]. The mobility scenarios were generated with BonnMotion v2.0, a tool that generates mobility traces according to a specific mobility model. The same traces were used to evaluate all the algorithms. In order to properly initialize the network, the first 3600 simulation steps were disconsidered. Experiments were performed with the number of nodes varying from 40 to 100 (different node densities), moving on a two-dimensional 500m X 500m area. For each network density, 20 different scenarios were generated for three distinct mobility models:

(a) Random Waypoint Model: In this model, each node randomly chooses a speed and a direction, which are independent from past moves. In our simulation, the minimum and maximum speeds were set to 1 m/s and 5 m/s, respectively.

(b) Gauss-Markov Mobility Model: This model assumes that the mobility of each node is temporarily correlated. It is based on a Gauss-Markov stochastic process which represents the speed of a node on time instant t as dependent on its speed on time instant $t - 1$. The degree of dependency on past behaviour is determined by a parameter, referred to here as $\delta, 0 \leq \delta \leq 1$. When $\delta = 0$, nodes are memoryless, whereas when $\delta = 1$ the speed of a node on instant t is equal to its speed on time $t - 1$. In our experiments, $\delta = 0.5$, and the maximum speed was 5 m/s.

(c) Manhattan Grid Model: In this model, nodes move on a grid, mimicking the movement of vehicles on streets. In our experiments, the simulation area was structured as a 10 x 10 grid of blocks, on which nodes moved with a mean speed of 5 m/s, with a standard deviation of 0.2. For each node, the speed was updated every time the node ran for 5 meters. We assumed a 50% probability for a node to maintain its direction and 25% to turn either right or left.

For the sake of simplicity, each scenario is composed of a homogenous set of sensor nodes (i.e. all the nodes have the same transmission and reception costs, and the same initial battery energy), even though this is not a premise for the correctness of the evaluated algorithms.

We used the following values for the variables in Equations (1)-(4): $cf(u) = 48$ nJ/bit, $\gamma(u) = 16.1$ pJ/bit/m^2, $\alpha = 2$ and $rcv = 236.4$ nJ/bit. These values are based on characteristics of the CC2420 transceiver, used in many typical sensor nodes. The maximum transmission range was 100m (outdoor range). We assumed that each node has an ideal battery, i.e. energy is linearly consumed. The initial battery charge was 5 J.

LPrim and LMCP require each node to know the neighbours in its two-hop vicinity and distances between them. This information is periodically exchanged in beacon packets by each node. In the messages we have used 7-bit identifiers (sufficient for 100 nodes) and 7 bits for each distance value (which provides a precision of approximately 1 meter - recall that the maximum transmission range assumed is 100m). Each "hello" message is composed of a 7-bit source node identifier and an id/distance pair to each of the source node's neighbours. For LCDS, the size of the control messages is reduced, as each node needs to know the neighbour relationship between nodes in its two-hop vicinity, but not the distances between them. Flooding does not use any control messages. We assumed that nodes are capable of estimating the distance to its one-hop neighbours by using some signal-strength-based technique. Thus, there is no need for nodes to exchange their geographic coordinates. In the experiments, nodes transmit one control message per time unit. This was a conservative choice and of course the frequency with which control messages are sent can be tunned according to the dynamicity pattern of each network, in order to balance control overhead and accuracy of the view of the network by each node. An improved beaconing strategy would favor LPrim, LMCP and LCDS, which are the algorithms that need propagation of neighbourhood information.

We assumed that the data to be broadcast has 1024 bytes. Thus, an LPrim or LMCP RELAY message containing broadcast data (piggybacked) contains 1024 bytes plus 7 bits for each of the identifiers on the lists of relays and covered nodes. LCDS and Flooding packets size lengths only 1024 bytes.

5.3   Experiment Results and Discussion

In this section, we present our evaluation of the algorithms under the three chosen mobility models, for the metrics defined in Section 5.1.

Number of Successful Broadcasts. Two distinct factors are responsible for limiting the total number of successful broadcasts. First, as mobile networks are susceptible to partitions, some nodes may be prevented from receiving messages, what is especially true in sparse settings. On the other hand, at the same time that dense networks hardly face disconnections, their nodes suffer from rapid battery draining due to the high communication traffic, which in turn results in a shorter operational time of nodes. As the number of nodes grows, failed broadcasts due to disconnections become rarer, whereas unsuccessful broadcasts caused by the depletion of nodes' batteries are more likely to happen. From this tradeoff, we derive an expected behaviour of network lifetime in mobile networks. For lower densities, the total number of successful broadcasts grows as density grows. From some specific density on, the impact of the high communication overhead on the lifetime of nodes becomes more significant and acts as a limiter to network lifetime, which begins to decrease (with the increase in network density).

For the Random Waypoint model (Fig. 1), LPrim performs significantly better than other approaches. In scenarios with 100 nodes, LPrim successfully executes 63.2% more broadcasts than LCDS, 82.5% more than LMCP and 153.8% more than Flooding.

A similar behaviour is observed for the Gauss-Markov mobility model (Fig. 3). For $density = 100$, LPrim achieved 148 successfull broadcasts, 41.9 % better than LMCP, 48.6% better than LCDS and 86.4% better than Flooding. Observing the behaviour of the algorithms in Fig.3, this difference is expected to still increase for networks with more than 100 nodes. As in Random Waypoint,

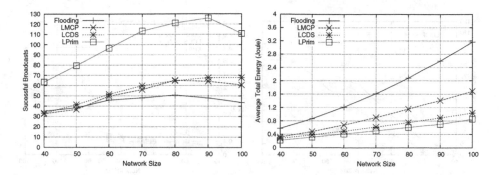

Fig. 1. Random Waypoint - Absolute number of successfully performed broadcasts achieved by each algorithm

Fig. 2. Random Waypoint - Mean total energy consumed by the network during a sucessful broadcast operation

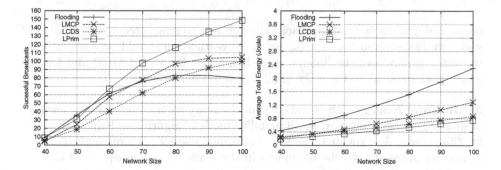

Fig. 3. Gauss Markov - Absolute number of successfully performed broadcasts achieved by each algorithm

Fig. 4. Gauss Markov - Mean total energy consumed by the network during a sucessful broadcast operation

LMCP and LCDS had a close performance. LCDS tends to perform asymptotically better than LMCP for these mobility models.

For the Manhattan Grid model, the algorithms exhibited a different behaviour than for the Random Waypoint and Gauss-Makovian models, as can be seen in Figure 5. For sparse networks (less than 90 nodes) Flooding performs better than the other algorithms. This can be explained by the fact that the particular pattern of movement in the Manhattan Grid model contributes for network disconnections, especially in sparse settings. As density grows and partitions becomes more unlikely to happen, LPrim and LMCP outperformed Flooding due to their lower communication cost. For 100 nodes, LPrim and LMCP performed approximately 13% better than Flooding and 115% better than LCDS. LCDS poor performance can be justified by the reduced set of retransmitter nodes (which reaches 30%, while for LMCP and LPrim this ratio can reach up to 70% of nodes). For networks having this specific pattern of mobility, the

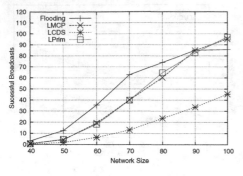
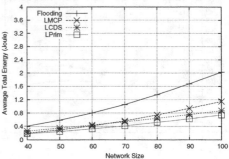

Fig. 5. Manhattan - Absolute number of successfully performed broadcasts
achieved by each algorithm

Fig. 6. Manhattan - Mean total energy consumed by the network during
a sucessful broadcast operation

connected dominating set approach is insufficient to guarantee a high number of
successfully delivered broadcasts. Further investigation should be done in order
to better understand the differences in behaviour between LMCP and LPrim,
which performed almost equally in all simulated densities.

Mean Total Energy. The total consumed energy metric is commonly used for
evaluating energy-efficiency. In all mobility models, algorithms behave similarly
for this metric. As expected, the total consumed energy per broadcast increases
as density grows due to the increment in the communication overhead and the
impact of overhearing. LPrim presented the lowest energy expenditure per successful broadcast among the simulated algorithms. For the Random Waypoint
Model (Fig. 2), in networks with 100 nodes, LPrim spent 22.6% less energy than
LCDS, 97.6% less than LMCP and 271% less than Flooding. For the Gauss-
Markov model (Fig. 4), the advantage of LPrim is around 14.6% in comparison
to LCDS, 72% in comparison to LMCP and 205% in comparison to Flooding.
For the Manhattan Grid model, the gain obtained by LPrim is lower but still
significant (Fig. 6). For 100 nodes, LPrim spends 16.2% less energy than LCDS,
54% less energy than LMCP and 172% less energy than Flooding.

6   Conclusion

In this paper we investigated the maximum network lifetime problem (MLB) in
mobile sensor networks. Commonly used approaches for broadcasting (Flooding,
use of Connected Dominating Sets, Shortest Paths Trees and trees with the
minimum highest weight edge) were evaluated under three different mobility
models (Random Waypoint, Gauss-Markov and Manhattan Grid). We presented
the LPrim algorithm, which locally uses a variation of Prim's MST algorithm
for directed graphs (as introduced in [10]), and LMCP, which is based on local
computations of shortest path trees. Simulations showed that, in all mobility

models, LPrim outperformed other solutions both in terms of the number of successfully delivered broadcasts and in terms of the average energy consumed per broadcast.

References

1. Calamoneri, T., Clementi, A., Fusco, E., Silvestri, R.: Maximizing the number of broadcast operations in random geometric ad hoc wireless networks. IEEE Trans. Parallel Distrib. Syst. 22, 208–216 (2011)
2. Calinescu, G., Kapoor, S., Olshevsky, A., Zelikovsky, A.: Network lifetime and power assignment in ad hoc wireless networks. In: Di Battista, G., Zwick, U. (eds.) ESA 2003. LNCS, vol. 2832, pp. 114–126. Springer, Heidelberg (2003)
3. Casteigts, A., Flocchini, P., Quattrociocchi, W., Santoro, N.: Time-varying graphs and dynamic networks. In: Frey, H., Li, X., Ruehrup, S. (eds.) ADHOC-NOW 2011. LNCS, vol. 6811, pp. 346–359. Springer, Heidelberg (2011)
4. Chang, J.-H., Tassiulas, L.: Energy conserving routing in wireless ad-hoc networks. In: IEEE INFOCOM, pp. 22–31 (2000)
5. Cormen, T.H., Leiserson, C.E., Rivest, R.L., Stein, C.: Introduction to Algorithms, 3rd edn. The MIT Press (2009)
6. Dietrich, I., Dressler, F.: On the lifetime of wireless sensor networks. ACM Transactions on Sensor Networks 5(1) (January 2009)
7. Elkin, M., Lando, Y., Nutov, Z., Segal, M., Shpungin, H.: Novel algorithms for the network lifetime problem in wireless settings. In: Coudert, D., Simplot-Ryl, D., Stojmenovic, I. (eds.) ADHOC-NOW 2008. LNCS, vol. 5198, pp. 425–438. Springer, Heidelberg (2008)
8. Elkin, M., Lando, Y., Nutov, Z., Segal, M., Shpungin, H.: Novel algorithms for the network lifetime problem in wireless settings. Wireless Networks 17, 397–410 (2011)
9. Distributed Computing Group ETH. Sinalgo - simulator for network algorithms, http://disco.ethz.ch/projects/sinalgo/
10. Kang, I., Poovendran, R.: Maximizing network lifetime of broadcasting over wireless stationary ad hoc networks. Mobile Networks and Applications 10, 879–896 (2005)
11. Korte, B., Vygen, J.: Combinatorial Optimization: Theory and Algorithms, 4th edn. Springer (2008)
12. Lee, S.H., Radzik, T.: Improved approximation bounds for maximum lifetime problems in wireless ad-hoc network. In: Li, X.-Y., Papavassiliou, S., Ruehrup, S. (eds.) ADHOC-NOW 2012. LNCS, vol. 7363, pp. 14–27. Springer, Heidelberg (2012)
13. Nunes, B., Barboza, F., Assis, F.: A localized algorithm for the maximum lifetime broadcast problem with asymmetric edge costs. In: Proc. of the 12th IEEE Int. Symp. on Network Computing and Applications, NCA (August 2012)
14. Nutov, Z., Segal, M.: Improved approximation algorithms for maximum lifetime problems in wireless networks. In: Dolev, S. (ed.) ALGOSENSORS 2009. LNCS, vol. 5804, pp. 41–51. Springer, Heidelberg (2009)
15. Orda, A., Yassour, B.-A.: Maximum-lifetime routing algorithms for networks with omnidirectional and directional antennas. In: Proc. of the 6th ACM Int. Symp. on Mobile Ad Hoc Networking and Computing, MobiHoc 2005, pp. 426–437 (2005)
16. Wu, J., Dai, F., Gao, M., Stojmenovic, I.: On calculating power-aware connected dominating sets for efficient routing in ad hoc wireless networks. IEEE/KICS Journal of Communications and Networks 4, 59–70 (2002)

On the Reliability of Wireless Sensor Networks Communications*

Alexandre Mouradian and Isabelle Augé-Blum

Université de Lyon, INRIA, INSA Lyon, CITI, F-69621, France
firstname.lastname@insa-lyon.fr

Abstract. More and more Wireless Sensor Networks (WSNs) applications and protocols are proposed. Notably, critical applications, which must meet time and reliability requirements. Works on the real-time capability of WSNs have been proposed [1]. In this paper we propose to study the achievable reliability of WSNs, tacking into account the probabilistic nature of the radio link. We define the reliability of a WSN to be the probability that an end-to-end communication is successful (i.e. the packet is received by the sink). We propose a theoretical framework inspired by a reference model [5]. We use the framework to derive the reliability of two types of routing schemes: unicast-based and broadcast-based. We show that in the case of broadcast-based, the sink is a reliability bottleneck of the network. We also discuss the impact of the MAC scheme on the reliability.

1   Introduction

Wireless Sensor Networks (WSNs) are multihop large scale networks composed of up to thousands of sensor nodes. They are usually deployed to monitor environment parameters of an area, or to monitor some equipments (such as power meters or gas meters). Sensor nodes run on batteries so they should consume as little energy as possible in order to increase the network lifetime. Because WSNs can contain lots of nodes, the financial cost of a node should be as low as possible, this leads to design nodes with poor capabilities (computation, radio, memory, etc...). For these reasons, research on WSNs mainly focused on self-organization and energy consumption efficiency.

Nevertheless, new applications appear. Notably critical applications on which human life and environment may depend. For example forest fire detection application must send an alarm to the sink when a forest fire is detected. Such applications require the respect of time constraints and a high reliability. In our example, the fire alarm must reach the sink before a known time bound and the probability that it reaches the sink must be high. This leads to ask which constraints can be handled by WSNs. Part of the answer is given by previous work [1], which derives a bound on real-time capacity of WSNs. However, this

---

* This work has been partially founded by French Agence Nationale de la Recherche under contract VERSO 2009-017.

J. Cichoń, M. Gębala, and M. Klonowski (Eds.): ADHOC-NOW 2013, LNCS 7960, pp. 38–49, 2013.

work does not take into account the probabilistic nature of the radio link. A protocol that can meet end-to-end deadlines loses its value if its delivery ratio is very low. The goal of this paper is to give an insight on what reliability is achievable in WSNs. We provide a theoretical framework used to model the reliability of WSNs. This framework is inspired form previous works [5] [10]. In this paper, we define the reliability to be the probability that a packet is received by the sink after H hops. The main contributions of this paper are the theoretical framework and its application, with different sets of hypotheses, to derive the achievable reliability of WSNs. Notably with a link reliability that depends on the emitter-receiver distance and a broadcast-based routing scheme.

In section 2 we describe related works. In section 3 we present our theoretical framework. In section 4 and 5 we apply the framework, respectively with a basic and realistic propagation model. In section 6, we discuss the impact of MAC on reliability results. In section 7, we conclude and give future works.

2 Related Work

In the literature, some reliability models for WSNs have been proposed with different definitions of reliability. The authors of [2] define the reliability of a WSN to be the probability to have a minimum rate of information delivered to the sink. They propose an algorithm in order to compute it. Nevertheless their approach is not focused on failures coming from the probabilistic radio link but more on node failures. In [9], the authors present a theoretical framework to compute the reliability of transport protocols in WSNs. This framework uses an elegant block diagram approach and includes several possible faults. Nevertheless, this is a high level model which not able to capture the complexity of realistic communications in WSNs. In [3] authors study the reliability of multi-path routing in ad-hoc networks. They take into account the probabilistic radio link, but the link quality is not dependent on the emitter-receiver distance (it is a parameter of the model) and they derive reliability equation for peer-to-peer traffic, not convergecast as in WSNs. Our proposition is complementary to these approaches. It focuses on the reliability of the end-to-end communications by modeling physical, MAC and routing layers.

Several reliable communication protocols for WSNs have been proposed [13][11] [12][7]. Our framework allows to understand better why these communication protocols are reliable and also to develop new, and even more reliable, schemes.

3 Theoretical Framework

3.1 Unreliable Links

Usually, the signal transmitted through the wireless channel is not only attenuated by the distance between the emitter and receiver, but it also experiences random attenuation coming from changes in the environment (moving objects, etc). Due to this random attenuation, the reception of a packet is probabilistic, i.e. there is a probability that a transmission between two nodes fails. In the remainder of this paper we will consider two link models: a basic model in which a node has a predefined set of neighbor and a probability that a packet sent is correctly received by a neighbor, and a more realistic model in which the node can potentially communicate with every node of the network with a probability that depends on the emitter-receiver distance.

In the basic model, each node is provided with a set of neighbor nodes which it can communicate with. Nevertheless the communications between a node and its neighbors are unreliable. Indeed, a neighbor of a transmitting node can receive the packet with a probability P_{bcr}. In this basic model we assume that the probability is the same for all neighbor nodes.

The realistic model is based on the log-normal propagation model. In [14], the authors advocate that it provides a realistic propagation model for WSNs. In this case, the reception probability depends on the emitter-receiver distance (at constant transmission power). The probability to receive correctly a packet is a function: $P_{cr}(d)$ with d the emitter-receiver distance. More information on this model can be found in [4]. It is possible to consider packet retransmissions in the probability formula as in [6].

We can notice that, in both models, we consider the probability for a node to receive a packet is independent from the probability for another node.

3.2 Network Topology

Our network model is based on [5] and [10]. N nodes are placed randomly and uniformly on a disk of area 1. The sink is placed at the center of the disk. Messages are generated by nodes in the network and must be routed to the sink in a multihop fashion. We assume that the nodes are distributed uniformly on the disk, thus on average there is one node in a $\frac{1}{N}$ surface. We can also determine the radius of a disk that contains d nodes on average:

$$R = \sqrt{\frac{d}{\pi N}} \qquad (1)$$

We also assume that a gradient has been constructed in the network. As in [13], a node is given a number that corresponds to the number of hops a packet originated from this node has to do to reach the sink. In the case of the basic model, the number of hops a packet has to do depends on the size of the neighbors set of a node which is linked to the nodes range by equation 1. In the case of unreliable links, the hop-count is difficult to evaluate because any node can potentially communicate with the sink or with any other nodes (with a given probability). We thus assume that the gradient is constructed using a threshold. Above a given probability of reception threshold two nodes are considered to be neighbors. Since we assumed the nodes are uniformly distributed, the nodes with the same hop-count form concentric rings of width R centered on the sink, as depicted on the left-hand side of Fig. 1 (In the case of realistic model, the ring width depends on the probability threshold).

3.3 Protocols

In this work, we focus more on the influence of routing on reliability because we are interested in end-to-end reliability. Nevertheless the impact of MAC operation is discussed in section 6. We divide MAC schemes into two categories regarding reliability, in the first, the MAC mechanism ensure that the packet is received (no collisions or resolved collisions). In the second, the packet may not be correctly received (unresolved collisions).

We use the number of hops as a routing metric. The shortest path is the path composed of the less possible number of hops according to the gradient. When the gradient is constructed, several metrics can be used (probability of reception, energy,...). Thus many routing metrics can be mimiced by the number of hops metric, by carefully constructing the gradient. We present two ways of routing which lead to different reliability as it is shown in the remainder of this paper.

In unicast-based routing, a shortest path from the source to the sink is selected (using the gradient). A node sends the packet to the neighbor which is next in the path. In the remainder of the paper we derive the reliability formulas of the presented schemes.

With a broadcast-based scheme, the packet is broadcasted to the neighbors of a node (as in [13] for example). A set of potential forwarders compete to relay the packet. In our case it is the nodes with a smaller hop-count (known thanks to the gradient presented in section 3.2). The selection of the forwarder can be made based on several criteria (signal strength, energy level, random, etc). In this paper we do not detail this process and we assume that a unique node is arbitrarily chosen among all potential forwarders which have correctly received the packet.

4 Case 1: The Basic Radio Link Model

In this section we derive the end-to-end communication probability from the presented theoretical framework with the basic propagation model. Here we assume that the first MAC scheme is used, i.e. no packet is lost due to MAC operation.

4.1 Unicast Scheme

We consider that the routing scheme selects a path from the sender to the sink. A node belonging to the path forward the packet to the neighbor that is next in the path. The probability that one hop is successful is P_{bcr}. The probability that the packet is correctly received by the sink after H hops is given by equation 2.

$$P_{e2e\_1u} = \prod_{i=0}^{H} P_{bcr} = (P_{bcr})^H \tag{2}$$

The number of hops depends on the range of a node:

$$H = \left\lceil \frac{1}{R\sqrt{\pi}} \right\rceil \tag{3}$$

$\frac{1}{\sqrt{\pi}}$ is the radius of the disk of area 1. The probability P_{e2e_1u} goes to zero when R goes to zero (because H goes to $+\infty$ and $0 \leq P_{bc} \leq 1$) and it is equal to P_{bc} when $R \geq \frac{1}{\sqrt{\pi}}$. It means that, in the case of the basic model, when there are less hops, the communications are more reliable. This is true because P_{bcr} does not decrease in function of the emitter-receiver distance. The basic model fails to capture the fact that long communications are less reliable. Nevertheless it is still useful when the difference between the furthest neighbor and the closest neighbor of a node is small.

Using a large value for R is not only a problem because long range communications are unreliable (notice that increasing the power of transmission increases the reliability) but also because it reduces the possible spatial reuse (simultaneous communications in the network) and thus reduce the achievable throughput as mentioned in [5].

4.2   Broadcast Scheme

In the case of broadcast-based routing, there is not only one forwarder but there is a set of potential forwarders. This set is defined by the routing protocol used. In our case, one of the shortest paths must be selected, the set of potential forwarders is thus the nodes with a smaller hop count than the sender.

With the gradient information we can define the set of forwarders by determining the percentage of the neighbors of a node which have smaller hop-count. If no gradient information is provided this percentage has to be determined according to the routing scheme used.

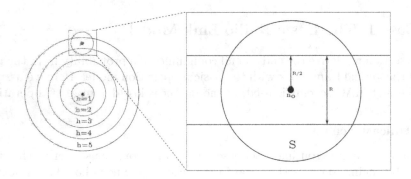

Fig. 1. Surface S

As shown on Fig. 1, we assume that if a ring h has a sufficiently large radius, it can be considered as two parallel straight lines at the scale of a node. This is not true for rings close to the sink, thus a correction factor might have to be considered. For exact calculation, see [8]. With this assumption, a node is on average placed at equal distance from the ring edges. It means it has, on average, an overlapping surface S with the lower ring. The surface S is the surface of a segment:

$$S = \left(\frac{\pi}{3} - \frac{\sqrt{3}}{4}\right) R^2 \qquad (4)$$

The ratio between the surface of the disk defined by the range of the node and the surface S is given by equation 5.

$$r = \frac{S}{\pi R^2} = \left(\frac{1}{3} - \frac{\sqrt{3}}{4\pi}\right) \qquad (5)$$

We know there are d nodes in the range of the node and, as nodes are distributed uniformly, the number of nodes in an area is proportional to the surface of this area. We thus define m the number of potential forwarders to be: $m = rd$.

The probability that one hop is successful, in the case of broadcast-based routing, is the probability that at least one potential forwarder receives the packet. The probability that the end-to-end communication is successful is the probability that every hop is successful. It is given by equation 6 for a H hops path.

$$P_{e2e\_1b} = P_{bcr} \prod_{h=1}^{H-1} (1 - \prod_{j=1}^{m}(1 - P_{bcr}))$$
$$= P_{bcr}(1 - (1 - P_{bcr})^m)^{H-1} \qquad (6)$$

For the last hop the only potential forwarder is the sink so h ranges from 1 to $H - 1$. The last hop is thus less reliable because if the sink fails to receive the packet, it is completely lost as in the case of unicast-based routing. The sink can thus be seen as a reliability bottleneck. We can notice that the number of potential forwarders m should be smaller for rings closer to the sink, because these rings are smaller. In order to take this effect into account the surface S should have to be computed using the technique from [8] where the exact surface formula is given. Then m would be a function of h.

When m goes to $+\infty$, $(1 - P_{bcr})^m$ goes to 0 thus P_{e2e_1b} goes to P_{bcr}. It means that the more potential forwarder there is, the more reliable the end-to-end communication is. But, the sink is a reliability bottleneck so P_{e2e_1b} does not go to 1 when m goes to $+\infty$. It means that by putting several sinks instead of one at the center of the disk area we can increase the reliability. Similarly to the unicast case, the reliability increases also when there are less hops. Nevertheless same issues as unicast case, about long range communications and spatial reuse appear.

Broadcast-based routing is more reliable than the unicast-based scheme. We do not give a formal proof here, but intuitively, we observe that in the case of broadcast-based scheme, to have a successful communication we need that, at least, one node receives the packet. So if all communications but one fails, the communication is successful. In the case of unicast, if the unique communication fails the packet is lost. On the other hand, broadcast-based routing involves more nodes. The potential forwarders have to be awake and wait for packets. A mechanism for forwarder election has to be implemented in order to select the forwarder. These mechanisms consume energy. This highlight a trade-off between energy consumption and reliability.

5   Case 2: The Realistic Radio Link Model

In this section, we use a more realistic propagation model in order to derive the end-to-end communication probability. In this case, the probability that a packet is correctly received depends on the emitter-receiver distance.

5.1   Unicast Scheme

With the unicast-based routing scheme, a path from a node to the sink is predefined. Nevertheless in the case of the realistic model, every node in the network can potentially communicate with every other node (in some cases this probability is very low). So two cases are possible, either the forwarder of a packet sent by a node in the path must be the next hop in the path, or it can be any node of the path closer to the sink, as in [7].

In the former case, a packet originated from a node in ring H has to travel a distance of $H \times R$ by doing H hops. The probability to reach the sink is thus obtained by substituting $P_{cr}(R)$ for P_{bcr} in equation 2. Here the construction of the gradient is important. Indeed, if the probability threshold (mentioned in section 3) is low the probability to receive correctly the packet for one hop is low but a hop has a longer distance. On the contrary if probability threshold is high the probability for one hop is high but a hop has a shorter distance. This latter alternative is better for reliability (for example if a node is three hops away from the sink with $P_{cr}(R) = 0.95$ then $P_{e2e} = 0.86$, with another gradient threshold it is six hops away from the sink and $P_{cr}(R) = 0.99$ then $P_{e2e} = 0.94$) but more hops lead more transmissions and thus to higher end-to-end delays.

In the latter case, every node on the path closer to the sink than the sender can relay the packet. We assume that a node forwards a packet it has just received only if no node closer to the sink has received it (only nodes belonging to the path can receive the packet). We can notice that it is not a strictly unicast solution because there are more than one receiver. A protocol which implements this behavior is provided in [7]. For a packet originated in ring H, there are $H-1$ potential forwarders (plus the sink). For each potential forwarder, the probability that it is elected is equal to the probability that it receives the packet and that no node closer to the sink receives it. The end-to-end communication probability is the sum of the probabilities that each node of the path is a forwarder and the probability to reach the sink from the forwarder. The probability to reach the sink from the forwarder is defined similarly to the end-to-end communication, but with H being the ring of the forwarder. The end-to-end communication probability is thus a recursive function defined by equations 7 and 8.

$$P_{e2e\_2u}(0) = 1 \tag{7}$$

$$P_{e2e\_2u}(H) = \sum_{h=0}^{H-1} \left\{ \prod_{j=0}^{h-1} [1 - P_{cr}((H-j)R)] \times P_{cr}((H-h)R) \times P_{e2e\_2u}(h) \right\} \tag{8}$$

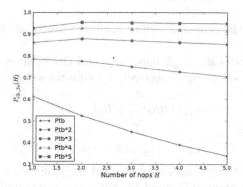

Fig. 2. P_{e2e_2u} in function of H for different transmission powers

Fig. 2 is a plot of equation 8 with H ranging from 1 to 5 and different transmission powers (the transmission power here depends on a basis value P_{tb}). We observe that, for the lower transmission powers, the probability of a successful end-to-end communication decreases when H increases. For higher transmission powers, probability can be higher for $H = 2$ than for $H = 1$. This is due to the fact that the probability to communicate from the ring $H = 2$ to the sink is high when the transmission power increases. The reliability in this case is higher than in the strictly unicast case (intuitively a packet has more options to reach the sink). Nevertheless, the former solution requires that all nodes of the path closer to the sink than the sender are awake at each hop. This increases energy consumption.

5.2 Broadcast Scheme

With this scheme, a node can relay a packet if it is closer to the sink than the sender and no node closer to the sink receives the packet. But unlike the unicast second case, in this case, there are several potential forwarders in each ring h. We thus have to determine the number of potential forwarders at a given ring, the number of nodes closer to the sink and the distance of those nodes from

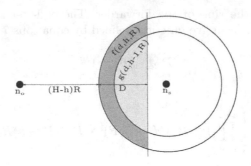

Fig. 3. Link between surfaces and D

the sender (because the probability of reception depends on the emitter-receiver distance).

The number of nodes in ring h noted m_h and the number of nodes closer to the sink noted m_{h-} are given respectively by equations 9 and 10

$$m_h = \lfloor (\pi(Rh)^2 - \pi[R(h-1)]^2)N \rfloor \tag{9}$$

$$m_{h-} = \lfloor (\pi[R(h-1)]^2)N \rfloor \tag{10}$$

In order to compute the probabilities, we need to evaluate the distance between the sender and the potential forwarders. Since nodes are uniformly distributed on the disk, the number of nodes in an area is proportional to the size of the area. The range of a node defines a disk area. Thus the first neighbor of a node is at least at a distance that allows the two nodes to be in the same circle. In our case we define areas as depicted on Fig. 3. The segments areas f and g are defined in function of the distance D, so the emitter-receiver distance is linked to the surface and the surface is linked to the number of nodes. We can thus determine the distance between the sender and the nth node in the ring h (notice that in this case again the circle corresponding to the range of node n_o is approximated by a straight line).

The function f is the surface of segment of ring h, and is defined as follows:

– if $0 < D < R$:

$$f(D,h,R) = \cos^{-1}\left(1 - \frac{D}{hR}\right)(hR)^2 - (Rh-D)\sqrt{2hRD - D^2} \tag{11}$$

– if $R \le D \le (2h-1)R$:

$$f(D,h,R) = \cos^{-1}\left(1 - \frac{D}{hR}\right)(hR)^2 - (Rh-D)\sqrt{2hRD - D^2}$$
$$- \cos^{-1}\left(1 - \frac{D-R}{(h-1)R}\right)((h-1)R)^2 \tag{12}$$
$$+ (R(h-1) - D + R)\sqrt{2(h-1)R(D-R) - (D-R)^2}$$

– if $D \geq (2h - 1)R$:

$$f(D, h, R) = \cos^{-1}\left(1 - \frac{D}{hR}\right)(hR)^2 - (Rh - D)\sqrt{2hRD - D^2}$$

$$- \cos^{-1}\left(1 - \frac{D - R}{(h - 1)R}\right)((h - 1)R)^2 \qquad (13)$$

The inverse function with respect to D gives the distance in function of the surface (with $S = n/N$ for the nth neighbor):

$$f^{-1}(S, h, R) = D \qquad (14)$$

Similarly, the surface for nodes closer to the sink is given by equation 15:

$$g(D, h, R) = \cos^{-1}\left(1 - \frac{D}{hR}\right)(hR)^2 - (Rh - D)\sqrt{2hRD - D^2} \qquad (15)$$

and $g^{-1}(S, h, R)$ is defined similarly to f^{-1}.

As in the second case of the unicast scheme, the probability of a successful end-to-end communication is a recursive function defined by equations 16 and 17.

$$P_{e2e\_2b}(0) = 1 \qquad (16)$$

$$P_{e2e\_2b}(H) = \sum_{h=0}^{H-1}\left\{\left[1 - \prod_{j=1}^{m_h}(1 - P_{cr}(f^{-1}(\frac{j}{N}, h, R) + R \times (H - h)))\right]\right.$$

$$\times \left[\prod_{k=1}^{m_{h-}}(1 - P_{cr}(g^{-1}(\frac{k}{N}, h - 1, R) + R \times (H - (h - 1))))\right] (17)$$

$$\left. \times P_{e2e\_2b}(h)\right\}$$

The probability that an end-to-end communication, originated at ring H, is successful is the sum of the probabilities for each ring that at least one node of the ring h receive the packet, and no node closer to the sink receives the packet, and the probability to reach the sink from ring h.

We can notice that when $h = 0$, f^{-1} and g^{-1} are equal to 0, m_h is equal to 1 (the sink node) and m_{h-} is equal to 0 (similarly, for $h = 1$, m_{h-} is equal to 1 and g^{-1} is equal to 0).

Fig. 4(a) is the plot of equation 17 with N ranging from 50 to 2000 and H from 1 to 5 with a power transmission of $P_{tb} \times 2$. The probability of a successful end-to-end communication increases with N up to a maximal value that depends on H. The maximal value is less than one. By increasing the number of nodes N, the number of potential forwarder in any ring increases but there is still only one sink so the probability cannot converge to one. As in the case of the basic model, the sink can be seen as the reliability bottleneck of the network.

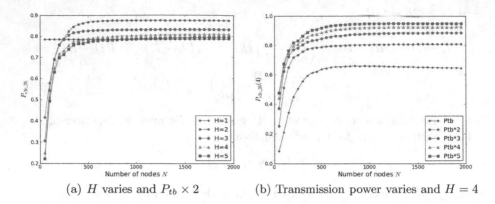

(a) H varies and $P_{tb} \times 2$ (b) Transmission power varies and $H = 4$

Fig. 4. P_{e2e_2b} in function of N

Fig. 4(b) is the plot of equation 17 with N ranging from 50 to 2000 and the transmission power from $1 \times P_{tb}$ to $5 \times P_{tb}$ with $H = 4$. In this case, the probability also reaches a maximal value. This maximal value increases with the transmission power because communications are more reliable. We can remark that, by adding sinks (enlarging the bottleneck), we improve the reliability without increasing the transmission power, thus without spending more energy (indeed new sinks consume energy but it does not reduce the network lifetime).

6   Impact of MAC

In the previous sections we consider that no packet is lost due to MAC operation. The probability to lose a packet only comes from the radio propagation. In this section, we consider that the probability to receive a packet is the probability that the packet is not lost because of the MAC (unresolved collision) and that it is not lost due to the propagation.

In the case of the unicast-based routing and the basic model the only change is that the probability P_{bcr} in equation 2 is replaced by $P_{bcr\_nodet} = P_{bcr} \times P_{no\_coll}$ with P_{no_coll} the probability that there is no packet collision at the receiver.

In the case of broadcast-based routing and the basic model, if there is a collision, it may prevent several potential forwarders from receiving the packet. In the worst case, all the potential forwarders cannot decode the packet because of the collision. In this case, equation 6 becomes:

$$P_{e2e\_1b} = (P_{bcr} \times P_{no\_coll}) \times ((1 - (1 - P_{bcr})^m) \times P_{no\_coll})^{H-1} \qquad (18)$$

The broadcast-based scheme reliability is thus highly reduced by MAC unresolved collisions. Similarly, these results can be easily extended to the realistic propagation model case with the same conclusions.

7 Conclusion and Future Works

In this paper we develop a theoretical framework which aims at studying the reliability of WSNs. We apply it to two types of propagation models with different routing and MAC schemes. Our results show that the broadcast solutions are more reliable than unicast based ones. Nevertheless, this comes at a price in terms of energy consumption and throughput. We also conclude that the sink is a reliability bottleneck in the case of the broadcast-based scheme. The presented framework allows a better understanding of the reliability of end-to-end communications in WSNs and thus guide protocol designers' decisions in order to meet applications reliability requirements.

More information on WSNs reliability can be derived from our model, for example we plan to further investigate the effect of the gradient construction probability threshold on the reliability. We also plan to enhance our model by taking into account correlations between reception probabilities of different nodes of the same area. The influence of MAC and routing acknowledgment packet shall also be considered.

References

1. Abdelzaher, T., Prabh, S., Kiran, R.: On Real-Time Capacity Limits of Multihop Wireless Sensor Networks. In: RTSS 2004, Lisbon, Portugal, pp. 359–370 (2004)
2. AboElFotoh, H.M., ElMallah, E.S., Hassanein, H.S.: On the reliability of wireless sensor networks. In: IEEE ICC 2006, pp. 3455–3460 (2006)
3. Caleffi, M., Ferraiuolo, G., Paura, L.: A reliability-based framework for multi-path routing analysis in mobile ad-hoc networks. International Journal of Communication Networks and Distributed Systems 1(4), 507–523 (2008)
4. Goldsmith, A.: Wireless Communications. Cambridge University Press (2005)
5. Gupta, P., Kumar, P.: The capacity of wireless networks. IEEE Transactions on Information Theory 46(2), 388–404 (2000)
6. Jaffrès-Runser, K.: Worst case delay analysis for a wireless point-to-point transmission. In: RTN 2012 (colocated with ECRTS 2012), Pisa, Italy (2012)
7. Lampin, Q., Barthel, D., Augé-Blum, I., Valois, F.: Qos oriented opportunistic routing protocol for wireless sensor networks. In: Wireless Days, pp. 1–6 (2012)
8. Mouradian, A., Augé-Blum, I.: 1-D coordinate based on local information for MAC and routing issues in WSNs. In: Li, X.-Y., Papavassiliou, S., Ruehrup, S. (eds.) ADHOC-NOW 2012. LNCS, vol. 7363, pp. 42–55. Springer, Heidelberg (2012)
9. Shaikh, F.K., Khelil, A., Suri, N.: On modeling the reliability of data transport in wireless sensor networks. In: IEEE PDP 2007, pp. 395–402 (2007)
10. Watteyne, T.: Energy-efficient slef-organization for wireless sensor networks. Phd thesis manuscript (2008)
11. Yang, F., Augé-Blum, I.: Constructing virtual coordinate for routing in wireless sensor networks under unreliable links. In: IWCMC, pp. 815–819 (2009)
12. Yang, F., Augé-Blum, I.: Delivery ratio-maximized wakeup scheduling for ultra-low duty-cycled WSNs under real-time constraints. Com. Netw. 55(3), 497–513 (2011)
13. Ye, F., Zhong, G., Lu, S., Zhang, L.: Gradient broadcast: a robust data delivery protocol for large scale sensor networks. Wirel. Netw. 11, 285–298 (2005)
14. Zuniga, M., Krishnamachari, B.: Analyzing the transitional region in low power wireless links. In: IEEE SECO 2004, pp. 517–526 (2004)

Fault Repair Schemes for Static Wircless Sensor Networks Driven by an Analytical Energy Dissipation Model

Skander Azzaz and Leila Azouz Saidane

RAMSIS Team - CRISTAL Lab., ENSI, Manouba University, Tunis, Tunisia
{skander.azzaz,leila.saidane}@ensi.rnu.tn

Abstract. In this paper, we introduce three proactive maintenance strategies for static Wireless sensor Networks (WSNs) using a limited number of mobile maintainer robots: the Centralized Proactive Maintenance Strategy (CPMS), the Fixed Distributed Proactive Maintenance Strategy (FDPMS) and the Adaptive Distributed Proactive Maintenance Strategy (ADPMS). The proposed maintenance strategies are based on a simple analytical energy dissipation model to estimate the occurrence times of the expected sensor failures in the network. Once identified, the anticipated failures are replaced by the available robots before they happen. Simulation results have shown that CPMS gives the minimal network dysfunction time representing the interruption time of the service provided by the network. But, due to its significant signaling cost, we have remarked that CPMS can be deployed only in small scale WSNs. In large scale ones, we recommend using the ADPMS maintenance strategy. However, in particular cases, when sensor failures are uniformly distributed on the network map, FDPMS has given the best performances.

1 Introduction

Providing a continuous service is the main requirement for many Wireless Sensor Network (WSN) applications [1]. To achieve this goal, the WSN must deploy a set of mechanisms to protect the network coverage from the eventual sensor failures [2]. In literature, many approaches have been proposed to conserve the initial WSN Quality of Service (QoS) parameters and restore the coverage and the connectivity upon a sensor failure such as: the use of mobile wireless sensors [3], and the exploiting of the nodes redundancy in the network [4]. However, and in order to reduce the maintenance strategy deployment cost, [5] proposes to use a small number of mobile robots to deal with the detected sensor failures in a static WSN. Indeed, in [5], Y. Mei and all have presented a set of algorithms to detect, report and handle the occurred sensor failures in the WSN. Thus, three approaches are introduced to coordinate the robot motions: the Centralized Manager Algorithm (CMA), the Fixed Distributed Manager Algorithm (FDMA) and the Dynamic Distributed Manager Algorithm (DDMA). To detect failures, classical strategies use a guardian-guardee relationship established between the network sensors. Indeed, each sensor (guardee node) must broadcast

J. Cichoń, M. Gębala, and M. Klonowski (Eds.): ADHOC-NOW 2013, LNCS 7960, pp. 50–62, 2013.

periodically a signaling message. If a one-hop neighbor (a guardian node) has not received any message from a guardee for a certain amount of time, it deduces that the guardee has failed and notifies a manager robot. In CMA, one robot is selected to operate as a manager. Upon the reception of the failure report, the manager robot schedules the closest maintainer robot to handle the failed sensor. With FDMA, different manager robot identities are manually configured on the network sensors. The report failure of a given sensor is transmitted to its corresponding manager robot. Upon the reception of the failure report, the manager moves to the failure to replace it with a functional one. In FDMA, a network robot is both a manager and a maintainer. To share the sensors repair load and reduce the travel robots distance, FDMA subdivides the WSN sensing area in subareas with equal surface. The same robot manager identity is attributed to the sensors of an obtained subarea. However, with DDMA, the manager robot is selected dynamically by the guardian node as the closest robot. By reacting after the sensor failures detection, the reactive WSN maintenance approachesrisk causes the service interruption during the failure recovery times. The induced network dysfunction time ratio cannot be tolerable by many real-time WSN application types such as surveillance and military applications [6]. In this paper, we introduce newer proactive maintenance strategies dedicated to provide a QoS support in WSNs using a limited number of mobile maintainer robots. The proposed maintenance strategies are based on an analytical model that represents the network sensor energy consumption. With this energy model, the WSN maintainer robots can anticipate the energy depletion failures and replace them before they occurred. Since the sensors faults sources are variants, the introduced proactive maintenance strategies use also the guardian-guardee relationship to detect the unpredictable sensor destructionsin the network. To coordinate the maintainer robots movement, the Centralized Proactive Maintenance Strategy (CPMS) uses a centralized scheduling approach. With this strategy, all scheduling robot operations are done by a selected robot designated by the manager robot. However, as their names indicate, the Fixed Distributed Proactive Maintenance Strategy (FDPMS) and the Adaptive Distributed Proactive Maintenance Strategy (ADPMS) opt for a distributed scheduling approach.

The present paper is organized as follows: the section 2 introduces our analytical energy dissipation model presented in [7]. In the following section, we focus on the centralized WSN maintenance approach CPMS. Then, in section 4 (respectively 5), we present the distributed maintenance approach strategy FDPMS (respectively ADPMS). Proposed strategies are evaluated and compared in section 6. Finally, we conclude the paper in section 7.

2   The Analytical Energy Dissipation Model

The energy dissipated by a sensor can be subdivided in several parts: the communication energy consumption, the sensing energy consumption and the unit process energy consumption (neglected in this study). In [7], we have used a Markov Chain to represent the communication energy consumption of a sensor node. Indeed, a sensor MAC layer can be modeled with a 4-states Discrete-time Markov Chain. We choose a time step (unit) such that the duration of any action (e.g. transmission/reception of a frame) is a multiple of this time step and we suppose that all state transitions occur at the beginning of the time step. With the IEEE 802.11 MAC protocol, a sensor MAC layer transits between three states: *transmit* (state 1), *receive* (state 2) and *idle* (state 3). An additional state *sleep* (state 4) is introduced with IEEE 802.15.4 or SMAC protocol. Each MAC state (j) is characterized by its power consumption denoted by ε_j $(1 \leqslant j \leqslant 4)$. We have, then:

$$\varepsilon_{Tot}^{Rad^{(n)}} = \sum_{i \in [1..n]} \left(\sum_{j \in \{1,2,3,4\}} \sum_{k \in \{1,2,3,4\}} \Pi_k^{(i-1)} \Pi_j^{(i)} (\Delta t\ \varepsilon_j) \right) \tag{1}$$

Where: $\varepsilon_{Tot}^{Rad^{(n)}}$ is the communication energy cost spent by a sensor during n time steps, Δt is the length in seconds of a single time slot and $\Pi_j^{(n)}$ $(1 \leqslant j \leqslant 4)$ is the probability that the MAC layer is in state j at time step n (with $\Pi_3^{(0)} = 1$ and $\Pi_j^{(0)}\ \forall j \in \{1, 2, 4\}$). The vector of probabilities $\Pi^{(n)} = [\Pi_1^{(n)}, \Pi_2^{(n)}, \Pi_3^{(n)}, \Pi_4^{(n)}]$ is determined recursively as follows [7]:

$$\Pi^{(n)} = [P_{ij}]_{1 \leq i,j \leq 4}\ \Pi^{(n-1)} \tag{2}$$

Where P_{ij} represents the probability that a node in state i will enter in state j at the next transition.

In addition to $\varepsilon_{Tot}^{Rad^{(n)}}$, the sensor node spent an amount of energy for event sensing. $\varepsilon_{Tot}^{Sens^{(n)}}$ denotes the consumed sensing energy in n time slots. We have:

$$\varepsilon_{Tot}^{Sens^{(n)}} = n\Delta t \frac{\lambda}{\mu} \varepsilon_{sensing} \tag{3}$$

Where $\varepsilon_{sensing}$ is the energy consumed by a node in sensing mode (the energy it costs to take one measurement), λ is the average rate of events detected by a sensor node and $\frac{1}{\mu}$ is the sensing mean time of an event.

Finally, the total consumed energy of a given sensor until the instant $(t = n.\Delta t)$ denoted by $(\varepsilon_{Tot}(n))$ can be computed with the equation 4:

$$\varepsilon_{Tot}(n) = \varepsilon_{Tot}^{Rad^{(n)}} + \varepsilon_{Tot}^{Sens^{(n)}} \tag{4}$$

The estimated sensor lifetime (ttl) corresponding to:

$$\begin{cases} \varepsilon_{Tot}(ttl) < \varepsilon_0 \\ \varepsilon_{Tot}(ttl + 1) > \varepsilon_0 \end{cases} \tag{5}$$

Where ε_0 is the sensor residual energy (at $t = 0$).

3   The CPMS Strategy

We begin by presenting our centralized proactive maintenance strategy where robot coordination is scheduled by a central element to repair the expected and detected sensors faults. In CPMS, we distinguished three WSN components types: a selected manager robot, a set of maintainer robots and evidently the sensors nodes. One robot is manually configured to function as a manager. The robot manager broadcasts periodically its identity in a Robot Manager Advertisement Message (RMAM). After the initialization stage, each network sensor must measure the different transitions probabilities between its MAC layer states [(1) transmit, (2) receive, (3) idle and (4) sleep] (P_{ij} , $0 \leq i, j \leq 4$) to be communicated to the robot manager with a Sensor Probability transition Matrix Message (SPMM). Each sensor determines its probability P_{ij} based on its activity history. In fact, the sensor divides firstly a fixed interval Ti of its past time in a set of unitary time slots TS. The size of the TS is taken so small to avoid that a sensor MAC layer transits from more than two states in a single TS. For each TS of the interval time Ti, the sensor determines the corresponding status MAC layer. P_{ij} can be computed as the number of time slots with state i followed by a time slot with the state j ($\left|TS_{i \rightarrow j}^{T}\right|$ in equation 6) divided by the total number of time slots with state i ($\left|TS_i^T\right|$ in equation 6)

$$P_{ij} = \frac{\left|TS_{i \rightarrow j}^{T}\right|}{\left|TS_i^T\right|} \tag{6}$$

In addition to the probability transition matrix, SPMM contains the following parameters: the sensor residual energy (ε_0), the length in seconds of a time slots (Δt), the average rate of events detected by sensor node (λ) and the sensor position coordinates. These parameters enable to compute the estimated network sensors life time according to the equation 5.

With the SPMM messages received from the network sensors, the manager robot can determine a fixed number (N) of the first coming expected failures using the proposed analytical energy dissipation model. The set of the N identified failures is designed by the anticipated failures window denoted by: $< f^i, i = 1..N >$. Each anticipated failure f^i is characterized by: (i) its position on the two dimensions network map (included in the SPMM message) represented by the pair ($f^i.x, f^i.y$) and (ii) its calculated life time (that represents also its reparation required time) obtained with the analytical model and denoted by ($f^i.t$). To reduce the network dysfunction time, the manager robot must schedule the available maintainer robots to replace sensors before the total depletion of their energies. In other words, the manager robot must find a robot scheduling solution denoted by ($< M^i, i = 1..N >$, where M^i represents the maintainer robot identity to deal with the anticipated failure f^i) that minimize as possible the sum of the off-service time of the expected failed sensors $< f^i, i = 1..N >$. This problem is known in literature by Multi-Robot Task Allocation (MRTA) Problem. In this paper, we propose the use of the Genetic Algorithms (GA) to obtain the adequate robot scheduling solution with a reduced complexity. The adopted approach (GA) has demonstrated its efficiency to resolve similar problems [8]. In General, GAs are designed to simulate the survival of the better individuals in a population over successive generations. In our case, each individual represents a robot scheduling solution (a combination of

N robots to be scheduled to handle the N anticipated failures). The individuals can be made, between generations, using a crossover and/or a mutation genetic operators. In a first step, CPMS generates an initial population of (Pop_{size}) individuals. The initial population representing a set of candidates scheduling solutions is composed with: (i) An incremental scheduling solution obtained with a heuristic method: *the maintainer robot of the failure f^i is selected as the nearest robot that minimizes its off-service time* and (ii) Random scheduling solutions. Any individual is encoded in a chromosome. We choose to represent a chromosome as a string of robots identities designed to repair an anticipated failures window. In this case, the ith chromosome encodes the maintainer robot identity (M^i) of the anticipated failure f^i. Like the general case of a GA, each chromosome is evaluated with a fitness function. For CPMS, we have introduced two metrics to evalute a given chromosome: (i) the network dysfunction time that the scheduling solution (described by the chromosome) can provide and (ii) the robot traveling distance metric. After evaluating the population's individuals, CPMS selects a set of P_{size} parental chromosomes using a selection scheme such as random or fitness-biased selection. The selection operator must ensure that better members with higher fitness (lower network dysfunction time and lower robot traveling distance) in the population have a greater probability of being selected for reproduction. Several researchers have studied the impact of different selection strategy on the GA performances, and [9] demonstrates the efficiency of tournament selection compared to others techniques such as Proportional Roulette Wheel Selection, Rank-based Roulette Wheel Selection. In tournament selection, n individuals are selected randomly from the population, and the winner individual is selected as the individual with the highest fitness. CPMS opts for the binary tournament selection ($n = 2$) to give more chance to all individuals to be selected and preserves diversity in the selected population. To compare the two parental chromosomes, the expected network dysfunction time metric is considered first. The chromosome with the lowest estimated network dysfunction time metric is retained. If the two chromosomes share the same expected network dysfunction time metric, we choose the chromosome that provides the lowest robot traveling distance. In the following step, CPMS applies, on each consecutive selected chromosomes pair, a crossover operator to produce two new children with a probability p_c (typically in the range $[0.6, 0.9]$ [9]). Otherwise (with a probability $(1 - p_c)$, the parental chromosomes are copied in the children's chromosomes. The crossover genetic operator built the two new offspring chromosomes with genes taken from the two parental chromosomes. [9] shows that in majority of cases, uniform crossover operator (used by CPMS) gives the better performances. This crossover operator generates randomly a mask (sequence of bits) with the same length as the individual's chromosome (N). The parity of the ith bits in the mask indicates for a given child form which parent it will inherit his i^{th} allele. Once the children's chromosomes are produces, CPMS uses a mutation genetic operator applied on each chromosome of the two new chromosomes, to guarantee the population's diversity. In this step, each offspring allele is flipped by CPMS, with a probability p_m, to a random maintainer robot identity. The mutation probability is typically chosen between $1/(population\ size)$ and $1/(chromosome\ length)$ [9]. At the last step (insertion), CPMS selects (P_{size}) individuals from the ($2P_{size}$) (parents + children) to be injected in the new population and replace the (P_{size}) parental individuals. To protect the best-fit individuals, CPMS sorts firstly the ($2P_{size}$) individuals by their network estimated dysfunction time. Thereafter, the (P_{size}) best individuals will join the new population. CPMS reiterates the procedure until a fixed number of generations ($NGmax$) or when the best-fit individual don't change through a fixed number of successive generation (the fitness convergence).

4   The FDPMS Strategy

FDPMS splits statically the WSN sensing area in a set of subareas. The number of subareas is equal to the number of robots denoted by NR. Each robot is assigned to a single subarea to deal with the detected and anticipated failures in its corresponding subarea. FDPMS uses a dynamic procedure to select the appropriate robot manager for a given sensor. Indeed, each robot ($R_i, i \in [1, NR]$) broadcasts periodically (each τ seconds) a RMAM Message containing its identity (M^i) and the center coordinates of its supervised subarea (c_i). With the RMAMs, any sensor (s) can calculate the distance (d_i^s) that it separates to each subarea center. The robot manager (R_m^s) of the sensor (s) is then selected as the robot that gives the minimal distance (d_i^s). After selecting the convenient maintainer robot, the sensor node must measure periodically the transition probabilities between its MAC layer states (Similar to CPMS). Once the probability matrix is determined, it's sent (with the sensor characteristics such as residual energy, sensor position coordinates) to the selected maintainer robot. With the analytical model introduced in section 2 parameterized by the received transition probabilities, each maintainer robots can estimate the energy depletion failure time (sensor life time) of any supervised sensor according to the equation 5.

The maintainer robots must determine first the N first expected failures in its subarea $< f^i, i = 1..N >$. To provide a null network dysfunction time, the optimal robot scheduling solution involves replacing each expected sensor failure f^i at its requested replacement deadline ($f^i.t$). However, we don't have any guaranties that the difference between the replacement deadlines of two successive expected failures ($f^{i+1}.t - f^i.t$) is sufficient for the maintainer robot to move from the f^i position to that of f^{i+1}. For this reason, we have introduced, for each expected failure f^i, a fictive replacement deadline denoted by ($f^i.Ft$) with ($f^i.Ft < f^i.t$). The new computed deadlines ($f^i.Ft, 1 \leqslant i \leqslant N$) must respect the following constraint:

$$f^i.Ft - f^{i-1}.Ft \geqslant \frac{||f^i, f^{i-1}||}{S} \tag{7}$$

Where: f^{i-1} is the failure handled before f^i, $||f^i, f^{i-1}||$ is the distance between the locations of the failures f^i and f^{i-1} and S represents the robot speed.

Each failure f^i is repaired at ($f^i.Ft$) instead of ($f^i.t$). With the new reparation deadlines, the maintainer robot can then ensure a null network dysfunction time.

At the beginning, $(f^i.Ft, 1 \leqslant i \leqslant N)$ are initialize respectively with $(f^i.t, 1 \leqslant i \leqslant N)$. Usually, the deadline $f^{i+1}.Ft$ is computed before $f^i.Ft$ (initially $i = N - 1$). Indeed, if the time requested by the maintainer robot to move from the location of failure f^i to the location of f^{i+1} is less than the difference between the fictive replacement times of the two failures f^i and f^{i+1} $(f^i.Ft - f^{i+1}.Ft)$, then the replacement of f^i is advanced to: $(f^{i+1}.Ft - \frac{||f^i, f^{i+1}||}{S})$.

CPMS tries, in a following step, to sort the first i failures of the current anticipated failures window according to their fictive replacement deadlines. If the time period $(f^i.Ft - f^{i+1}.Ft)$ becomes sufficient for the robot to move from the location of failure f^i to the location of f^{i+1}, CPMS uses the same procedure to fix the fictive replacement time of the $(i - 1)^{th}$ failure, in the next iteration.

5   The ADPMS Strategy

The major drawback of FDPMS is eventually the fixed subdivision of the total sensing area. As a direct consequence, FDPMS can overload particular maintainer robots affected to subareas characterized by a high failures density (since in general cases, failures aren't uniformly distributed over the network map). To equilibrate the failures density of the obtained subareas, the network map subdivision technique of any distributed maintenance strategy must take into consideration the node density distribution (over the network map) and the sensors traffic topology. In particular, ADPMS opts for a partition mechanism called the Center Growth Algorithm initially designed to control the air traffic [10]. Indeed, in a National Airspace System (NAS), many controllers must monitor radar screens to track aircraft report messages (messages that describe weather conditions in a particular position). To balance the load among the controllers, the Center Growth Algorithm is used to divide the NAS plan in NR (number of controllers) subareas with equal aircraft report message load. To achieve this goal, the Center Growth Algorithm decomposes first the NAS plan in a fine grid of hexagonal cells. Then, a traffic load mass is computed for each cell equal to the number of aircraft report messages observed in the current cell. Thereafter, NR subarea centers are chosen randomly on the map. Each center belongs to a single hexagonal cell representing the initial NAS plan subdivision. The subarea with lowest traffic load (equal to the sum of its cell costs) is allowed to grow one cell layer (all neighboring cells). This procedure is repeated until the total coverage of the NAS plan. We have adapted the Center Growth Algorithm to partition the WSN map in NR subareas with equal failures density. ADPMS decomposes then the network map in a hexagonal grid. A cost is computed for each cell $(Cell_i)$ equal to the number of anticipated failures expected (in the current cell) in a coming interval of T seconds (obtained with the analytical energy dissipation model). It's clear that the cell cost depends on the cell node density and the cell nodes traffic topology of the evaluated cell. In a following step, ADPMS fixes NR subarea centers $(C_i, i \in [1, NR])$. C_i can be chosen as the centers of the subareas obtained with a fixed subdivision of the WSN map (as FDPMS proceeds). We denote by $Cell_{c_i}$ the cell containing the center C_i and by $SubA_i$

the ith subarea. Initially, $SubA_i$ is composed only by the cell $Cell_{c_i}$. Through successive iterations, $SubA_i$ are growing up to cover the total WSN sensing area. In each iteration, ADPMS evaluates firstly the cost of each subarea equal to the sum of its elementary cells costs. Then, the subarea with the lowest cost is enlarged with all neighboring cells. In ADPMS, the maintainer robots broadcast (periodically) signaling messages (RMAMs) containing their identities. With the RMAMs messages, any sensor can identify the available maintainer robots list. Once the sensor analytical model parameters are available, they are sent to a selected Collector Robot (CR) with the highest identity. The CR must execute, in consequence, the Center Growth Algorithm to determine the limits of each subareas and communicate the results to the others maintainer robots.

6   Experiments

In this section, we evaluate the performances of our proposed proactive maintenance strategies (CPMS, FDPMS and ADPMS) by simulation. Obtained results are compared to those of the classical reactive strategies: CMA and FDMA.

6.1   Experimental Setup

The proposed maintenance strategies has been implemented in the Network Simulator 2 (NS-2) [11]. We have selected the following simulation parameters: a 2D square sensing area map with the dimensions (1600×1600) m^2 covered by a total of (32×32) sensor nodes, the position coordinates of a given sensor is $(15\ i, 15\ j)$, $0 \leq$ i <32 and $0 \leq$ j <32. Each sensor has a communication radius equal to 25 meters and a coverage radius equal to 15 meters. The network sensors generate the same load of traffic with a constant bit rate and a constant packet size (128 bytes) for the sink node (node 0 at the position (0,0)). In the presented scenarios, we have varied the node traffic bit rate from 1 to 8 Kb/s. Adhoc On-demand Distance Vector (AODV) which is used in the ZigBee Stack [12] is selected as the routing protocol. And, the total simulation time is equal to one day. IEEE 802.15.4 is used as a sensor MAC Layer in the WSN. The power consumption in each sensor MAC layer state [sleep, transmit, receive and idle] is equal to [0.1404, 0.1404, 0.0018, 0.000018] Watt taken from a ZigBee node implementing IEEE 802.15.4 medium access [12]. We have also varied the number of the maintainer robots (from 1 to 8 robots). Each robot is characterized by: a speed equal to 1m/s, and a communication radius equal to 60 meters, based on the specification of Pioneer 3DX robots [5].

6.2   Performance Results

We consider the following performance metrics in evaluating the WSN maintenance strategies: The provided network dysfunction time, the efficient of the sensors repair load sharing, the robot traveling distance per failure and the induced messaging overhead.

The Network Dysfunction Time: To
provide a QoS support, the WSN maintenance strategies must reduce, as much as
possible, the induced network dysfunction
time using a minimal number of robots.
Based on this metric, we give, in this section, a comparison between the proactive
maintenance strategies (CPMS, FDPMS
and ADPMS) and the reactive ones (CMA
and FDMA). In figure 1, we present the
variation of the network dysfunction time
provided by each maintenance strategy
versus a variable number of maintainer
robots, a node bit rate equal to 8 Kbit/s
and a null rate of unpredictable failures.
With a high failure rate density (greater

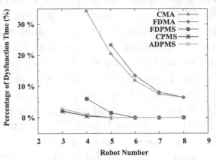

Fig. 1. The network dysfunction time
ratio versus the maintainer robots
number for CMA, FDMA, CPMS,
FDPMS and ADPMS

than the robots capacity), the robots fail to repair the occurred failures and
the network can eventually disappear. Figure 1 shows that the centralized (respectively distributed) reactive strategy CMA (respectively FDMA) request a
minimum of 4 (respectively 5) robots to maintain the persistence of the WSN.
In addition, we can remark that the network dysfunction time provided by the
reactive maintenance strategies remains always greater than a given limit. Indeed, even if we increase the number of maintainer robots; the off-service time
of a faulted sensor is always greater than the failure detection time threshold
plus the maintainer robot motion time. However, by introducing the proactive
maintenance strategies, we obtain better performances with a reduced number
of robots (only a minimal of 3 (respectively 4) robots for CPMS (respectively
FDPMS). With a sufficient number of maintainer robots, the reactive strategies
(CPMS, FDPMS and ADPMS) are able to provide a null network dysfunction
time. For CPMS (respectively FDPMS), presented results are obtained with a
failure window length equal to 150 (respectively 50) failures.

Robot Load Sharing: We present in figures 2, 3 and 4 the repairing load
percentage of each maintainer robot equal to the number of handled failures
per robot devided by the total number of failures. We have considered: three
maintainer robots and a varied traffic node bit rate (from 1 to 6 Kbit/s). In
the presented figures, we designate by robot I: the maintainer robot assigned to
the sink vicinity area and robot II (respectively III): the maintainer robot of the
middle (respectively the last) subarea. In our simulation scenarios, a many-to-
one traffic model is adopted (like the general case in WSNs). For this reason, we
obtain a non-uniform distribution of failures over the network map and consequently a high failure rate neighboring the sink node. Since the sensor failures
are dense around the sink node and in order to reduce the total traveling distance achieved by the maintainer robots, CPMS overloads the robot I in case of
a low failures rate (node bit rate equal to 1Kb/s in figure 2). In such a case, the
repairing load of the robot I is still lower than its total capacity.

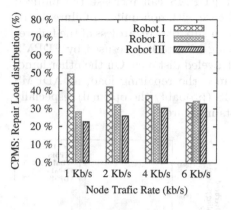

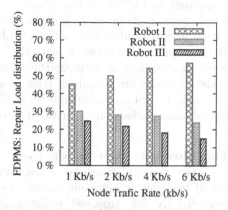

Fig. 2. CPMS: Robot load distribution versus the node bit rate

Fig. 3. FDPMS: Robot load distribution versus the node bit rate

However, if the node traffic rate increases (high failure rate), CPMS will balance the repairing load and all maintainer robots will function with their maximum capacity to ensure a minimal network dysfunction time. With FDMA, and since it uses a fixed WSN area partitioning, figure 3 shows that if the failures rate increases, the robot I will bear alone the majority of the repairing load. In this case, we risk an overtaking of the robot I capacity with an under-utilization of the robot II and III. Contrary, figure 4 proofs that ADPMS provides a balanced load sharing among the maintainer robots independently of the node bit rate.

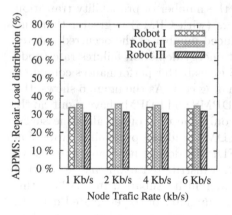

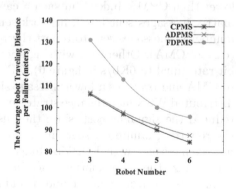

Fig. 4. ADPMS: Robot load distribution versus the node bit rate

Fig. 5. The mean traveling distance per failure versus the number of robots for CPMS, FDPMS and ADPMS

The Average Robot Traveling Distance per Failure: In figure 5, we report the average distance traveled by robots to repair an anticipated failure for the three preventive WSN maintenance strategies: CPMS, FDPMS and ADPMS.

Compared to CSMA, figure 5 shows that FDPMS can increase the induced robot traveling distance. Indeed, in general case with non-uniform failure distribution and by limiting the motion of robots in subareas regardless of the failures distribution on the network map, the sharing load technique used by FDPMS becomes inefficient to optimize the robots traveled distance. On the other hand, if the load sharing technique tries to balance the repairing load, as ADPMS proceeds, we obtain similar results as CPMS (probably the optimal), especially with an overloaded robots (case of 3 robots in figure 5).

The Messaging Overhead: The messaging overhead is measured as the number of transmitted signaling messages introduced by the proposed maintenance strategies. We report in figure 6 the signaling cost for each maintenance strategy with a node bit rate equal to 1 and 6 kb/s. CMA introduces two types of signaling messages: the reporting failure messages and the robots communication messages induced by the manager-maintainers communication. CPMS adds a third message type: the probability transition matrix messages sent by sensors to their manager

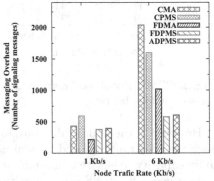

Fig. 6. The Messaging overhead versus the node bit rate for CMA, CPMS, FDMA, FDPMS and ADPMS

robot. In a case of low density sensor failures (node bit rate equal to 1 Kb/s in figure 6), simulation results show that CMA provides a signaling messages ratio lower than CPMS. Indeed, in such a case, the number of probability transition matrix messages sent by all network sensors (with CPMS) is greater than the number of messages sent to report the reduced number of the occurred failures (case of CMA). Otherwise, with significant energy depletion failures rate (node bit rate equal to 6Kb/s in figure 6), CPMS gives better performances compared to CMA and provides a lower message signaling cost. As the figure 6 shows, the distributed WSN maintenance strategies ADPMS and FDPMS have significantly reduced the signaling cost, since they use only one type of signaling messages: the reporting failure messages in case of FDMA and the probability transition matrix messages with FDPMS. Indeed, FDPMS doesnt need any message to synchronize the robot motion since in such a distributed maintenance solution a robot is both: a manager and a maintainer in its subarea. In ADPMS this communication is limited: (i) to exchange a summary of the estimated network life time (between the collector robot and the maintainer robots) and (ii) to establish the auctioneer negotiation upon the reception of a failures report. In addition, FDPMS still gives a low signaling message cost compared to FDMA in a case of a high sensors failures rate for the same reasons as the centralized maintenance strategies: CMA versus CPMS.

7   Conclusion

In this paper, we have proposed three proactive maintenance strategies for WSNs to satisfy the QoS requirement of deployed applications. We have used an analytical energy dissipation model to estimate the life time of a sensor and a robot coordination technique to handle the determined expected failures. Presented results have shown that the centralized approach (CPMS) gives the better result in term of the network dysfunction time. Howover, this technique is inadequate for the large scale WSNs, since it introduces a heavy signaling cost to synchronize the mobile robot displacement. To remedy this problem, we have proposed two distributed preventive maintenance strategies denoted by: FDPMS and ADPMS. On one hand, FDPMS subdivides the total sensing area in equal surface subareas. Each maintainer robot deals with the failures anticipated/detected in a single subarea. Simulation shows that FDPMS optimizes the signaling cost. But, with its partitioning technique, we risk an unbalanced load sharing among the available robots. Obtained results show that FDPMS requests a minimal number of maintainer robots greater than CPMS. On the other hand, ADPMS introduces an adaptive partitioning technique based on the Center Growth Algorithm to balance the repairing load of the predictable failures among the robots. Compared to FDPMS, simulation results have shown that ADPMS reduces not only the minimal number of used robots, but also the total robot traveling distance. In conclusion, given its performance, we recommend CPMS as a maintenance strategy for the small scale WSNs and ADPMS for the large-scale ones. In particular cases, when the failures are uniformly distributed over the network map FDPMS becomes the optimal maintenance solution.

References

1. Velazquez, E., Santoro, N.: Neighbour Selection and Sensor Knowledge: Proactive Approach for the Frugal Feeding Problem in Wireless Sensor Networks. In: Simplot-Ryl, D., Dias de Amorim, M., Giordano, S., Helmy, A. (eds.) ADHOCNETS 2011. LNICST, vol. 89, pp. 162–176. Springer, Heidelberg (2012)
2. Szczytowski, P., Khelil, A., Ali, A., Suri, N.: TOM: Topology oriented maintenance in sparse Wireless Sensor Networks. In: SECON 2011 (2011)
3. Li, M., Cheng, W., Liu, K., He, Y., Li, X., Liao, X.: Sweep Coverage with Mobile Sensors. IEEE Transactions on Mobile Computing (2011)
4. Sakib, K., Tari, Z., Bertok, P.: An analytical framework for identifying redundant sensor nodes from a dense sensor network. In: ICCIT 2010 (2010)
5. Mei, Y., Xian, C., Charlie, Y.: Repairing Sensor Network Using Mobile Robots. In: Workshop on Wireless Ad hoc and Sensor Networks (2006)
6. Tafa, Z., Dimisic, G., Milutinovic, V.: A survey of military applications of wireless sensor networks. In: MECO 2012 (2012)
7. Azzaz, S., Saidane, L., Minet, P.: Repairing sensors strategies in fault-tolerant wireless sensor networks. In: PE-WASUN 2011 (2011)

8. Yun, S., Parasuraman, S., Ganapathy, V.: Dynamic path planning algorithm in mobile robot navigation. In: ISIEA 2011(2011)
9. Hu, X.B., Di Paolo, E.: An efficient Genetic Algorithm with uniform crossover for the multi-objective Airport Gate Assignment Problem. In: CEC 2007 (2007)
10. Klein, A.: An Efficient Method for Aispace Analysis and Partitioning Based on Equalized Traffic Mass. Atm Seminar 2005 (2005)
11. The Network Simulator ns-2, http://www.isi.edu/nsnam/ns/
12. Liu, H., Leung, Y., Chu, X.: Ad Hoc and Sensor Wireless Networks: Architectures, Algorithms and Protocols. Bentham Science Publishers (2009) eISBN: 978-1-60805-018-5

I-VDE: A Novel Approach to Estimate Vehicular Density by Using Vehicular Networks

Javier Barrachina[1], Piedad Garrido[1], Manuel Fogue[1],
Francisco J. Martinez[1], Juan-Carlos Cano[2],
Carlos T. Calafate[2], and Pietro Manzoni[2]

[1] University of Zaragoza, Spain
{barrachina,piedad,mfogue,f.martinez}@unizar.es
[2] Universitat Politècnica de València, Spain
{jucano,calafate,pmanzoni}@disca.upv.es

Abstract. Road traffic is experiencing a drastic increase in recent years, thereby increasing the every day traffic congestion problems, especially in cities. Vehicle density is one of the main metrics used for assessing the road traffic conditions. Currently, most of the existing vehicle density estimation approaches, such as inductive loop detectors or traffic surveillance cameras, require infrastructure-based traffic information systems to be installed at various locations. In this paper, we present I-VDE, a solution to estimate the density of vehicles that has been specially designed for Vehicular Networks. Our proposal allows Intelligent Transportation Systems to continuously estimate the vehicular density by accounting for the number of beacons received per Road Side Unit, as well as the roadmap topology. Simulation results indicate that our approach accurately estimates the vehicular density, and therefore automatic traffic controlling systems may use it to predict traffic jams and introduce countermeasures.

Keywords: Vehicular Networks, vehicular density estimation, Road Side Unit, VANETs.

1 Introduction

Enhancing transportation safety and efficiency has emerged as a major objective for the automotive industry in the last decade [13]. However, road traffic is experiencing a drastic increase. Hence, vehicular traffic congestion is becoming a major problem, especially in metropolitan environments throughout the world. Traffic congestion: (i) reduces the efficiency of the transportation infrastructure, (ii) increases travel time, fuel consumption, and air pollution, and (iii) leads to increased user frustration and fatigue [15].

Some of the factors that cause traffic congestion are badly managed roads, poorly designed roads, or bad traffic lights sequencing [14]. These factors provoke that vehicles are not uniformly distributed on the roads, making it possible to find extremely high congested areas where vehicles travel very slowly or even get stuck.

J. Cichoń, M. Gębala, and M. Klonowski (Eds.): ADHOC-NOW 2013, LNCS 7960, pp. 63–74, 2013.

In vehicular environments, wireless technologies enable peer-to-peer mobile communication among vehicles (V2V) [10], and communication between vehicles and the infrastructure (V2I) [12]. Vehicles can broadcast warning messages in case of an accident, and also periodically exchange other messages (beacons) that contain information about their position, speed, route, etc. These messages are received by the rest of vehicles and by the *Road Side Units* (RSUs), which are communication nodes installed to create a vehicular infrastructure.

Traditionally, vehicle density has been one of the main metrics used for assessing the road traffic conditions. A high vehicle density usually indicates that the traffic is congested. However, the density of vehicles circulating in a city highly varies during the day depending on the area and the time.

Currently, most of the vehicle density estimation approaches are designed for using infrastructure-based traffic information systems, which require the deployment of vehicle detection devices such as inductive loop detectors or traffic surveillance cameras. However, these approaches are limited since they can only be aware of traffic density in a very specific and reduced area (i.e., the streets and junctions in which these devices are already located), making it difficult to estimate the vehicular density of a neighborhood, or a whole city. In addition, some of these approaches are not able to perform the density estimation process in real time (e.g., using cameras involves hard image treatment and analysis).

We consider that a vehicular communications system able to estimate the traffic density in real time could mitigate or even solve traffic congestion problems. In this work, we present a solution to estimate the traffic density on the roads that relies on the V2I communication capabilities offered by Vehicular Networks. In particular, we intend to estimate the density, taking into account the number of beacons received by the RSUs and the characteristics of the topology of the selected area. Hence, real-time traffic controlling systems can precisely estimate the vehicular density in a specific area, and then redirect vehicles to lower traffic density areas in order to avoid traffic jams. This could be possible by using the in-vehicle communication capabilities and navigation systems.

The rest of this paper is organized as follows: Section 2 reviews previous approaches related to our work, focusing on infrastructure-based solutions to estimate traffic density. Section 3 details our proposal for real-time RSU-based vehicular density estimation, assessing its effectiveness. Additionally, we discuss the obtained results and measure the estimated error. In Section 4 we validate our proposal. Finally, Section 5 concludes this paper.

2 Related Work

In this section we review previous works related to our proposal. In particular, we focus on the infrastructure-based solutions to estimate traffic density.

Despite the importance of determining the vehicular density to reduce traffic congestion, so far there have been few studies that explored the density estimation process.

Tyagi et al. [15] considered the problem of vehicular traffic density estimation, using the information cues available in the cumulative acoustic signal acquired from a roadside-installed single microphone. This cumulative signal comprises several noise signals such as tire noise, engine noise, engine-idling noise, occasional honks, and air turbulence noise of multiple vehicles. The occurrence and mixture weightings of these noise signals are determined by the prevalent traffic density conditions on the road segment. Based on these learned distributions, they used a Bayes' classifier to classify the acoustic signal segments. Using a discriminative classifier, such as a *Support Vector Machine* (SVM), results in further classification accuracy compared to a Bayes' classifier. Tan and Chen [14] proposed a novel approach based on video analysis which combines an unsupervised clustering scheme called AutoClass with *Hidden Markov Models* (HMMs) to determine the traffic density state in a *Region Of Interest* (ROI) of a road. Firstly, low-level features were extracted from the ROI of each frame. Secondly, an unsupervised clustering algorithm called AutoClass was applied to the low-level features to obtain a set of clusters for each predefined traffic density state.

These works established the importance of vehicular density awareness for neighboring areas, but none has deepened in the analysis of the accuracy of the method used to estimate this density, or the effect of the topology in the results obtained. Moreover, this estimation does not take place in real time.

Regarding the use of Vehicular Networks (VNs), Garelli et all. [6] proposed a fully-distributed approach to the online estimation of vehicle traffic density. Their approach makes communicating vehicles to cooperate in order to collect density measurements through a uniform sampling of the road sections of interest. The proposed scheme does not require the presence of any network infrastructure, central controller or devices triggered by the passage of vehicles, and it is suitable for both highway and urban environments. Results derived through simulations show that their solution is very effective, providing accurate, on-line estimates of the traffic density with minimal protocol overhead. More recently, Akhtar et al. [1], proposed a fully distributed and infrastructure-free mechanism for the density estimation in vehicular ad hoc networks. Unlike previous distributed approaches, that rely either on group formation or on vehicle flow and speed information to calculate density, their proposal is inspired by the mechanisms proposed for system size estimation in peer-to-peer networks. Authors adapted and implemented three fully distributed algorithms, namely Sample & Collide, Hop Sampling, and Gossip-based Aggregation. The simulations of these algorithms at different vehicle traffic densities and area sizes for both highways and urban areas reveal that Hop Sampling provides the highest accuracy in least convergence time and introduces least overhead on the network, but at the cost of higher load on the initiator node.

Although these works studied the use of Vehicular Networks to estimate vehicular density in real time, authors did not account for the effect of obstacles in the wireless signal propagation which can make results very inaccurate, especially in urban scenarios. Moreover, they only accounted for the number of beacons received, while omitting the map features where the vehicles are located.

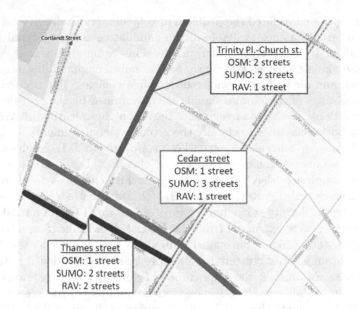

Fig. 1. Different criteria when counting the number of streets

3   Real-Time Vehicular Density Estimation

In this work we propose a method able to accurately estimate the density of vehicles, which is based on the number of beacons received by RSUs and the roadmap topology. We made a total of 900 experiments. These experiments involved the simulation of controlled scenarios (i.e., scenarios where the actual density is known). According to the results obtained, and using a regression analysis, we propose a density estimation function capable of estimating the vehicular density in every urban environment at any instant of time.

In this section we first present a discussion about the most important features of the different city roadmaps. Later, we present the parameters and the methodology used in our simulations. Finally, we detail our proposed density estimation function, and estimate its error.

3.1   Features of the Cities Studied

The roadmaps used during the experiments to achieve the density estimation were selected in order to have different profile scenarios (i.e., with different topology characteristics).

The first step before starting the simulations was to obtain the main features of each roadmap (i.e., the number of streets, the number of junctions, the average distance of segments, and the number of lanes per street). As for the streets, we realized that different alternatives could be selected to obtain the number of streets of a given roadmap. Basically, they are: (i) the number of streets

Table 1. Number of Streets obtained depending on the approach used

City	SUMO	OSM	RAV
New York	700	827	257
Minnesota	1592	105	459
Madrid	1387	1029	628
San Francisco	1710	606	725
Amsterdam	3022	796	1494
Sydney	1668	315	872
Liverpool	3141	1042	1758
Valencia	5154	1050	2829
Rome	2780	1484	1655

obtained in SUMO [7], where each segment between two junctions is considered a street, (ii) the number of streets obtained in *OpenStreetMap* (OSM) [11], where each street has a different "name", and (iii) the number of streets according to our *Real Attenuation and Visibility* (RAV) radio propagation model, where the visibility between vehicles is taken into account when identifying the streets [5].

Figure 1 shows a small portion of New York City to depict the different criteria when counting the number of streets. For example, Thames Street is considered only one street in OSM, whereas the SUMO and RAV models consider that there are two different streets instead. However, if we observe Cedar Street, the RAV visibility model and the OSM approaches consider a single street (as expected), whereas it is represented by three different streets according to SUMO, since it has three different segments. Finally, according to both the OSM and SUMO approaches, Trinity Place and Church Street are represented as two different streets, whereas the RAV model considers that only one street exists.

Table 1 shows the values obtained according to each criterion to count the number of streets for the cities studied. As shown, the differences between these approaches are significant (e.g., New York has 700, 827, or 257 streets when considering SUMO segments, OSM streets, or the RAV visibility approach, respectively, whereas Sydney has 1668, 315, or 872 streets, depending on the selected criterion). Therefore, it is important to decide which one to use in order to obtain accurate results. After some experiments, we realized that the third approach better correlated with the real features of cities, since the other two present some drawbacks: they are not accurate enough, or they present some errors. So, we choose this approach for the analysis that follows.

Table 2 shows the main features of each map of the cities under study (i.e., the number of streets according to the RAV algorithm, the number of junctions, the average distance of segments, and the number of lanes per street). We also added a column labeled as *SJ Ratio*, which represents the result of dividing the number of streets between the number of junctions. As shown, the first city (New York) presents an SJ ratio of 0.5130, which indicates that it has a simple topology, whereas the last cities in the table present a greater value, which indicates a more complex topology. This aggregated factor correlates well with the obtained results.

Table 2. Map Features

Map	Streets	Junctions	avg. segment distance (m.)	lanes/street	SJ Ratio
New York	257	500	45.8853	1.0590	0.5140
Minnesota	459	591	102.0652	1.0144	0.7766
Madrid	628	715	83.0820	1.2696	0.8783
San Francisco	725	818	72.7065	1.1749	0.8863
Amsterdam	1494	1449	44.8973	1.1145	1.0311
Sydney	872	814	72.1813	1.2014	1.0713
Liverpool	1758	1502	49.9620	1.2295	1.1704
Valencia	2829	2233	33.3653	1.0854	1.2669
Rome	1655	1193	45.8853	1.0590	1.3873

3.2   Simulation Environment

Simulations were done using the ns-2 simulator [3], where the PHY and MAC layers have been modified to follow closely the IEEE 802.11p standard, which defines enhancements to the 802.11 required to support ITS applications. We assume that all the nodes of our network have two different interfaces: (i) an IEEE 802.11n interface tuned at the frequency of 2.4 GHz for V2I communications, and (ii) an IEEE 802.11p interface tuned at the frequency of 5 GHz for V2V communications. In terms of the physical layer, the data rate used for packet broadcasting is 6 Mbit/s, as this is the maximum rate for broadcasting in 802.11p. The MAC layer was also extended to include four different priorities for channel access. Therefore, application messages are categorized into four different *Access Categories* (ACs), where AC0 has the lowest and AC3 the highest priority.

To prove how maps affect the performance of vehicular communications, [9], we selected nine street maps, each one representing a square area of 4 km^2. Figure 2 shows the topology of the maps used in the simulations. In order to deploy RSUs in the maps, we use the Uniform Mesh deployment policy [2], that consists on distributing RSUs uniformly on the map. The advantage of this deployment policy is that it achieves a more uniform coverage area since the distance between RSUs is the same, preventing RSUs to be positioned too closely, or too sparsely.

As for the mobility of the vehicles, it has been performed with *CityMob for Roadmaps* (C4R) [4], a mobility generator able to import maps directly from OpenStreetMap [11], and generate ns-2 compatible traces. Table 3 shows the parameters used for the simulations.

We tested our proposal by evaluating the performance of a Warning Message Dissemination mechanism, where each vehicle periodically broadcasts information about itself or about an abnormal situation (traffic jams, icy roads, etc.). To increase the realism of our results, we include the possibility that vehicles share accident notification messages in our simulations. In fact, we consider that vehicles can operate in two different modes: (a) warning, and (b) normal. Vehicles in warning mode inform other vehicles about their status by sending warning messages periodically (every second). Normal mode vehicles enable the diffusion of these warning packets and, every second they also send beacons with information such as their positions, speed, etc. These periodic messages are not propagated by other vehicles. All the results represent an average of over 10 repetitions with different scenarios, and each simulation run lasted for 30 seconds.

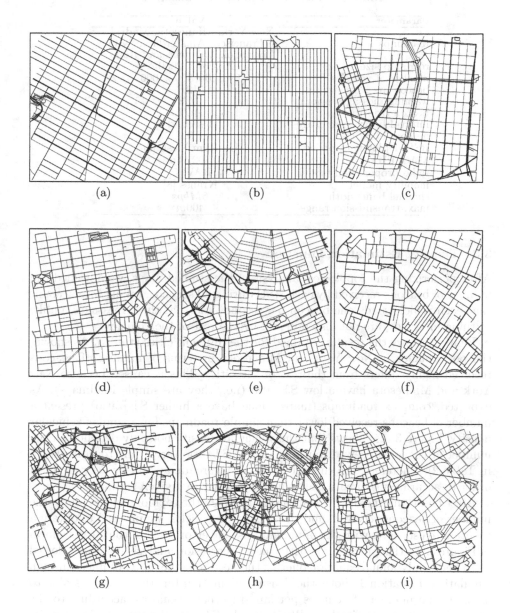

Fig. 2. Scenarios used in our simulations. Fragments of the cities of: (a) New York (USA), (b) Minnesota (USA), (c) Madrid (Spain), (d) San Francisco (USA), (e) Amsterdam (Netherlands), (f) Sydney (Australia), (g) Liverpool (UK), (h) Valencia (Spain), and (i) Rome (Italy)

Table 3. Parameters used for the simulations

Parameter	Value
roadmaps	New York, Minnesota, Madrid, San Francisco, Amsterdam, Sydney, Liverpool, Valencia, and Rome
roadmap size	$2000m \times 2000m$
number of vehicles	$[100, 200, 300...1000]$
beacon message size	$512B$
warning messages priority	$AC3$
beacon priority	$AC1$
interval between messages	1 second
number of RSUs	9
RSU deployment policy	Uniform Mesh [2]
MAC/PHY	802.11p
radio propagation model	RAV [5]
mobility model	Krauss [8]
channel bandwidth	$6Mbps$
max. transmission range	$400m$

3.3   Density Estimation Function

After performing the topological analysis of the studied maps, we obtained the number of beacons received by each RSU during 30 seconds, taking into account that each vehicle sends one beacon per second, and that these messages, unlike warning messages, are not disseminated by the rest of the vehicles.

Figure 3 shows the results obtained for the different cities studied. As shown, the performance in New York and Minnesota in terms of number of beacons received highly differs from the rest of the cities. This is caused because New York and Minnesota have a low SJ ratio (i.e., they are simple roadmaps). As expected, complex roadmaps (maps which have a higher SJ Ratio) present a number of beacons received lower than simple roadmaps for a similar vehicular density. Figure 3 also shows that the vehicular density not only depends on the number of beacons received, but also on the SJ ratio (according to data shown in Table 2). Therefore, the characteristics of the roadmap will be very useful in order to accurately estimate the vehicular density in a given scenario.

After observing the direct relationship between the topology of the maps, the number of beacons received, and the density of vehicles, we proceed to obtain a function to estimate, with the minimum possible error, each of the curves shown in Figure 3. To this purpose, we performed a regression analysis [16] that allowed us to find a polynomial equation offering the best fit to the data obtained through simulation. Equation 1 shows the density estimation function, which is able to estimate the number of vehicles per km^2 in urban scenarios, according to the number of beacons received per RSU, and the SJ ratio (i.e., streets/junctions).

$$f(x,y) = a + b \cdot ln(x) + \frac{c}{y} + d \cdot ln(x)^2 + \frac{f}{y^2} + \frac{g \cdot ln(x)}{y} \qquad (1)$$

In this equation, x is the number of beacons received by each RSU, and y is the SJ ratio obtained from the roadmap. The values of the polynomial coefficients $(a, b, c, d, f, and\ g)$ are listed in Table 4.

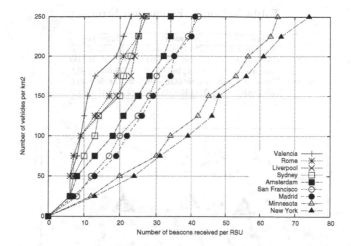

Fig. 3. Number of beacons received when varying the vehicular density and the roadmap

Table 4. Proposed equation coefficients

Coeff.	Value
a	2.4328753582642619E+02
b	8.8667060945557523E+00
c	-4.2340086242746855E+02
d	3.2563178030488615E+01
f	1.8200236614892370E+02
g	-6.4626326366022894E+01

To determine the accuracy of our proposal, it is necessary to measure the estimated error. Table 5 shows the different types of errors calculated when comparing our density estimation function with the values actually obtained. Note that the average relative error is of only 3.63%. We consider that this error can be neglected in the majority of traffic congestion mitigation applications, thus validating our proposed function.

4   Validation of Our Proposal

To assess our proposed density estimation function, we simulated a new particular case. Specifically, we chose Mexico D. F., a city with a small SJ Ratio (0.7722), and we simulated a density of 200 vehicles per km^2. Figure 4 shows the RSU deployment strategy and the vehicles' location at the end of the simulation for the studied example, and Table 6 shows the obtained results. As shown, the average number of beacons received per RSU is 47.56. According to I-VDE (i.e., applying the polynomial function as shown in Equation 2), we estimate a density of 196.91 vehicles. In this example, the estimation of vehicular density obtained an error of 3.09 vehicles, which only represents the 1.55% of the total vehicles.

Table 5. Density Estimation Error

Error	Absolute	Relative
Minimum	-5.736426E+01	-1.218902E+00
Maximum	5.135632E+01	1.784647E+00
Mean	-1.642143E-14	3.634060E-02
Std. Error of Mean	2.596603E+00	3.592458E-02
Median	-1.914503E+00	-2.313015E-02

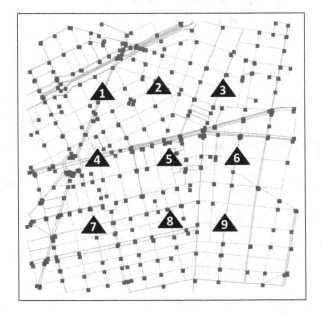

Fig. 4. RSUs deployment and vehicles location at the end of the simulation

$$f(x,y) = a + b \cdot ln(47.56) + \frac{c}{0.7722} + d \cdot ln(47.56)^2 + \frac{f}{0.7722^2} + \quad (2)$$
$$+g \cdot \frac{ln(47.56)}{0.7722} = 196.91$$

Moreover, using our system, we are able to estimate the vehicular density in more specific areas. For example, using the data included in Table 6, our I-VDE can identify areas where the traffic is more congested (i.e., areas where the RSUs receive a higher percentage of beacons). In our experiment, RSUs 4 and 1 received a higher number of beacons compared to RSUs 6 and 9. According to these results, an automatic traffic control system could take advantage from V2I communication capabilities, to adapt the vehicles' routes in order to redirect vehicles traveling in more congested areas to those areas where the RSUs receive a lower number of messages (i.e., less congested), thus avoiding traffic jams.

Table 6. Received Beacons when simulating 200 vehicles/km^2 in Mexico D.F.

RSU	Received beacons	% of received beacons
1	54	12.62
2	46	10.75
3	43	10.05
4	68	15.89
5	48	11.21
6	38	8.88
7	48	11.21
8	46	10.75
9	37	8.64
Total	428	100
Average	47.56	-

5 Conclusions

This paper proposes I-VDE, a method that allows estimating the vehicular density in urban environments at any given time by using the communication capabilities between vehicles and RSUs. Our proposal allows improving traffic congestion mitigation mechanisms to better redistribute the vehicles routes, adapting them to the specific traffic conditions.

Unlike existing works, our vehicular density estimation algorithm takes into account not only the number of beacons received by the RSUs, but also the topology of the map where the vehicles are located. As a result of a large number of simulations, using maps from different cities, we have obtained an equation that is able to accurately predict the vehicular density. Results show that our proposal allows estimating the vehicular density for any given city, thereby allowing governments to improve their traffic control mechanisms.

Acknowledgments. This work was partially supported by the *Ministerio de Ciencia e Innovación*, Spain, under Grant TIN2011-27543-C03-01, as well as by the Fundación Universitaria Antonio Gargallo (FUAG), and the Caja de Ahorros de la Inmaculada (CAI).

References

1. Akhtar, N., Ergen, S., Ozkasap, O.: Analysis of distributed algorithms for density estimation in VANETs. In: IEEE Vehicular Networking Conference (VNC), pp. 157–164 (November 2012)
2. Barrachina, J., Garrido, P., Fogue, M., Martinez, F.J., Cano, J.C., Calafate, C.T., Manzoni, P.: D-RSU: A Density-Based Approach for Road Side Unit Deployment in Urban Scenarios. In: International Workshop on IPv6-based Vehicular Networks (Vehi6), Collocated with the 2012 IEEE Intelligent Vehicles Symposium, pp. 1–6 (June 2012)
3. Fall, K., Varadhan, K.: ns notes and documents. The VINT Project. UC Berkeley, LBL, USC/ISI, and Xerox PARC (February 2000),
http://www.isi.edu/nsnam/ns/ns-documentation.html

4. Fogue, M., Garrido, P., Martinez, F.J., Cano, J.C., Calafate, C.T., Manzoni, P.; A Realistic Simulation Framework for Vehicular Networks. In: 5th International ICST Conference on Simulation Tools and Techniques (SIMUTools 2012), Desenzano, Italy, pp. 37–46 (March 2012)
5. Fogue, M., Garrido, P., Martinez, F.J., Cano, J.C., Calafate, C.T., Manzoni, P.: Evaluating the impact of a novel message dissemination scheme for vehicular networks using real maps. Transportation Research Part C: Emerging Technologies 25, 61–80 (2012)
6. Garelli, L., Casetti, C., Chiasserini, C., Fiore, M.: MobSampling: V2V Communications for Traffic Density Estimation. In: IEEE 73rd Vehicular Technology Conference (VTC Spring), pp. 1–5 (May 2011)
7. Krajzewicz, D., Rossel, C.: Simulation of Urban MObility (SUMO), Centre for Applied Informatics (ZAIK) and the Institute of Transport Research at the German Aerospace Centre (2012), http://sumo.sourceforge.net
8. Krauss, S., Wagner, P., Gawron, C.: Metastable states in a microscopic model of traffic flow. Physical Review E 55(5), 5597–5602 (1997)
9. Martinez, F.J., Toh, C.K., Cano, J.C., Calafate, C.T., Manzoni, P.: Determining the representative factors affecting warning message dissemination in VANETs. Wireless Personal Communications 67(2), 295–314 (2012)
10. Martinez, F.J., Cano, J.C., Calafate, C.T., Manzoni, P., Barrios, J.M.: Assessing the feasibility of a VANET. In: ACM Workshop on Performance Monitoring, Measurement and Evaluation of Heterogeneous Wireless and Wired Networks (PM2HW2N 2009, held with MSWiM), pp. 39–45. ACM, NY (2009)
11. OpenStreetMap: Collaborative project to create a free editable map of the world (2012), http://www.openstreetmap.org
12. Soldo, F., Lo Cigno, R., Gerla, M.: Cooperative synchronous broadcasting in infrastructure-to-vehicles networks. In: Fifth Annual Conference on Wireless on Demand Network Systems and Services (WONS), pp. 125–132 (January 2008)
13. Stanica, R., Chaput, E., Beylot, A.: Local density estimation for contention window adaptation in vehicular networks. In: IEEE 22nd International Symposium on Personal Indoor and Mobile Radio Communications (PIMRC), pp. 730–734 (September 2011)
14. Tan, E., Chen, J.: Vehicular traffic density estimation via statistical methods with automated state learning. In: IEEE Conference on Advanced Video and Signal Based Surveillance (AVSS), pp. 164–169 (September 2007)
15. Tyagi, V., Kalyanaraman, S., Krishnapuram, R.: Vehicular traffic density state estimation based on cumulative road acoustics. IEEE Transactions on Intelligent Transportation Systems 13(3), 1156–1166 (2012)
16. ZunZun: Online Curve Fitting and Surface Fitting Web Site (2013), http://www.zunzun.com

Repairing Wireless Sensor Network Connectivity with Mobility and Hop-Count Constraints

Thuy T. Truong, Kenneth N. Brown, and Cormac J. Sreenan

Mobile & Internet Systems Laboratory and Cork Constraint Computation Centre,
Department of Computer Science, University College Cork, Ireland
{tt11,k.brown,cjs}@cs.ucc.ie

Abstract. Wireless Sensor Networks can become partitioned due to node failure or damage, and must be repaired by deploying new sensors, relays or sink nodes to restore some quality of service. We formulate the task as a multi-objective problem over two graphs. The solution specifies additional nodes to reconnect a connectivity graph subject to network path-length constraints, and a path through a mobility graph to visit those locations. The objectives are to minimise both the cost of the additional nodes and the length of the mobility path. We propose two heuristic algorithms which prioritise the different objectives. We evaluate the two algorithms on randomly generated graphs, and compare their solutions to the optimal solutions for the individual objectives. Finally, we assess the total restoration time for different classes of agent, i.e. small robots and larger vehicles, which allows us to trade-off longer computation times for shorter mobility paths.

Keywords: Sensor Network, Connectivity Repair, Sink Placement.

1 Introduction

Wireless Sensor Networks are becoming increasingly important for monitoring phenomena in remote or hazardous environments, including pollution monitoring, chemical process sensing, disaster response, and battlefield monitoring. As these environments are uncontrolled and may be volatile, the network may suffer damage, from hazards, direct attack or accidental damage from wildlife and weather. They may also degrade through battery depletion or hardware failure. The failure of an individual sensor node may mean the loss of particular data streams generated by that node; more significantly, node failure may partition the network, meaning that many data streams cannot be transmitted to the sink. This creates the network repair problem, in which we must place new radio nodes in the environment to restore connectivity to the sink for all sub-partitions.

In this work, we assume a survey has been completed, and so we know which nodes have failed, which radio links have been blocked, and which routes between positions can no longer be traversed. The tasks that remain are to decide on the positions for the new radio nodes, and to plan and follow a route through the environment to place those nodes. We assume possible locations for new radio

J. Cichoń, M. Gębala, and M. Klonowski (Eds.): ADHOC-NOW 2013, LNCS 7960, pp. 75–86, 2013.
© Springer-Verlag Berlin Heidelberg 2013

nodes are limited to a finite set of positions where a node can be securely placed and which can be accessed. Radio nodes are expensive, and so solutions which require fewer nodes are preferred. In addition, the users of the WSN may require data to be transmitted from the sensors quickly, to allow a timely response, and so there will be limits on the number of radio hops allowed between the sensors and the wider network. To achieve this, we may prefer to deploy some expensive sink nodes which provide their own network connection, in addition to relay nodes. Physically moving around the environment may be expensive in energy use, may take significant time, or may expose the agent placing the nodes to danger, and so solutions which allow cheaper path plans are also preferred. Depending on the application, either one of the two objectives may be more important: placing expensive nodes in, for example, agricultural pollution monitoring favours solutions with fewer nodes, while restoring connectivity during disaster response favours solutions that can be deployed quickly even if they require more nodes. Thus the network repair problem is multi-objective.

We introduce the problem of simultaneous network repair with hop count limits and route planning with limited mobility. We assume a set of desired locations from which sensor data is required by the network, and we assume the agent knows the state of the network and accessibility. The objective is to connect as many as possible of these locations, placing extra sensors, relays and sinks as required, minimising the relay and sink costs and the mobility costs, while obeying the constraint on the number of allowed radio hops. We consider two different heuristic approaches for the multi-objective problem, each prioritising a different objective: minimising mobility costs, and minimising the relay and sink costs. We evaluate the two algorithms on randomly generated problems, and analyse their effectiveness under different assumptions. Finally, we consider the total estimated time to restore the network, for two different classes of agent (a small robot and a larger vehicle), and we show that the choice of priority should be dependent on the performance of the agent.

2   The Network Repair Problem

Given a damaged sensor network and set of terminal locations from which we require sensed data, our goal is to place new nodes to ensure that each terminal is connected to a sink within a given number of radio links, and to find a mobility path through the environment to place the nodes, while minimising both the cost of the radio nodes and the length of the path.

Let V be a set of possible radio locations. $E_c \subseteq V \times V$ is the set of possible radio links, $E_m \subseteq V \times V$ is a set of traversable edges, and $w{:}E_m \to \mathbb{N}$ specifies the length of each edge. A path p in graph $G{=}(V, E)$ is a sequence $[x_1, x_2, x_3, \ldots, x_{k-1}, x_k]$ where each $\{x_i, x_{i+1}\} \in E$. The *hop count* of a path in the *connectivity graph* $G_c{=}(V, E_c)$ is one less than the number of nodes in the path, while the length of a path p in the *mobility graph* $G_m{=}(V, E_m)$ is $\sum_{(x_i, x_{i+1}) \in p} w(\{x_i, x_{i+1}\})$. L_r is the set of locations with existing relay (and sensor) nodes, while L_s is the set of locations with existing sink (and sensor) nodes. T is the set of terminal nodes which must be reconnected within the hop count limit k, and $\alpha \in V$ is the initial starting location of our agent. c_r is the cost of a relay node, while c_s is the cost of a sink node. The problem is to find two new subsets $R \subseteq V$ and $S \subseteq V$ of relay nodes and sink nodes, such that for each terminal $t \in T$ there is a sink $s \in S$ and a path through the connectivity graph $(L_r \cup L_s \cup R \cup S, E_c)$ to s with a hop count $\leq k$, and a tour p in the mobility graph G_m that starts and finishes at α and visits each element of $R \cup S$, which minimises the pair $(\text{length}(p), (|R| * c_r) + (|S| * c_s))$ of the mobility tour length and the node cost.

Note that the two objectives may conflict. As an example, Figure 1 shows (a) a connectivity graph and (b) a mobility graph for a set of terminals $T = \{t_1, t_2, t_3\}$ and a set of candidate locations $\{a,b,d,e,f,g,h,j\}$. Assuming $c_s = 3 * c_r$, $k = 2$ and the current location of the agent is at f, a minimal cost node deployment to reconnect all terminals within a hop count limit of 2 is $S = \{d, t_3\}, R = \{t_1\}$, with cost $7 * c_r$. The shortest path in the mobility graph is $[f, t_3, f, e, b, h, t_1, h, a, d, a, e, f]$ with length 45. For a deployment of $S = \{t_1, t_3, g\}, R = \{\}$, the node cost is $9 * c_r$, but there is a path $[f, g, f, t_3, f, e, b, h, t_1, h, b, e, f]$ with length 34. Which of these solutions should be selected will depend on the relative cost of the sink and radio nodes compared to the cost of traversing the path. High node costs and low mobility costs will prefer the first solution, while high mobility costs will prefer the second solution.

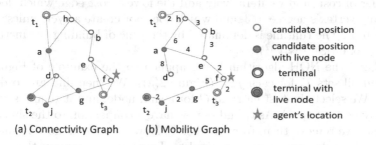

(a) Connectivity Graph   (b) Mobility Graph

○ candidate position

● candidate position with live node

◎ terminal

◉ terminal with live node

★ agent's location

Fig. 1. Example Network Repair Problem

The objective of minimising the mobility path length has a travelling sales-man path problem embedded inside it, and so we choose to investigate heuristic approaches. To address the two objectives, we consider two approaches, which each prioritise one of the objectives.

3   The Node Optimisation Heuristic Algorithm

Our first approach prioritises the node cost, by searching for a low cost set of relay and sink nodes which connect the terminals to sinks within the hop count limit (Algorithm 1). Given a set of nodes, we then search for the cheapest mobility path that visits those nodes (Algorithm 2). We start by finding, for each terminal j the set T_j of all possible sink locations (i.e. locations within k hops of j in the connectivity graph). For each sink location s_i which is associated with two or more terminals, we determine the valuation (x_i, y_i), where x_i is the number of terminals which could be connected by s_i, and y_i is an upper bound on the number of new relay nodes that would be required to connect them. The set $O = \{s_i | x_i * c_s \geq c_s + y_i * c_r\}$ then contains all such sinks that could connect all of their terminals for less than the cost of placing a separate sink for each terminal. We order O by the expected total cost of deploying that sink, $(|T| - x_i) * c_s + y_i * c_r + c_s$, where $y_i * c_r + c_s$ is the cost of placing that sink with extra radio nodes y_i to reconnect the expected x_i terminals, and $(|T| - x_i) * c_s$ is an upper bound on the cost of placing sinks at the remaining terminals in T. We then select the first sink location in O, add it to S, the set of new sinks, compute relay node locations to connect the terminals and add them to R, the new relays. We then recompute the valuations and reorder the set O to reflect the changes in the unconnected terminals, and repeat. Once O is empty, for each remaining unconnected terminal j we place a sink at a random location in T_j. The most expensive operation is the calculation of (x_i, y_i) where we use a best-first search to find the shortest connectivity path from a terminal to each location $s_i \in O$, and we re-use those paths when we select the relay node locations.

For the problem of finding a short tour for the selected nodes (Algorithm 2), we create from the mobility graph G_m a metric closure graph for the new nodes in $S \cup R$. We then apply [1]'s Greedy-TSP heuristic - we sort the edges in increasing order of cost, and we iteratively add the lowest cost edge which does not increase any vertex's degree to 3, and which does not create a cycle unless it completes the tour. The runtime is dominated by the time of building the metric closure graph, i.e. $O(|S \cup R| * |V|^2)$.

Figure 2 shows the NOH algorithm being applied to the example of Figure 1. First we find all sets T_j for each $j \in T$ (Figure 2(a)). We then find and order the set O (b). We select the first entry in O for a sink node, and it requires one additional relay node at t_1 (c). As t_1 and t_2 now have a connection to the sink at d within 2 hops, we remove them from the list T. Now a connects no terminals in T, while b and g only connect one terminal t_3. Therefore, we remove them all from the list O and terminate the *while* loop. We then select t_3 for a new sink (d) and finish the algorithm. Finally, we apply Greedy TSP to find a tour visiting those selected locations, giving $P = [f, t_3, f, e, b, h, t_1, h, a, d, a, e, f]$ which costs 45 units for 2 sinks and 1 relay node.

4   The Path Optimisation Heuristic Algorithm

Our second approach prioritises the mobility cost. First, for each terminal $j \in T$, we find the set C_j of all locations within the hop count limit from j but that require at most one extra node to connect j. Since we must guarantee connectivity within k hops for each terminal, we must place at least one node in each C_j. Placing a sink node in C_j ensures that no other node is required in C_j. Since the aim is to minimise mobility cost, we now search for a set of nodes that cover the C_j and which can be visited with the shortest possible tour.

Algorithm 1. Node Selection

Data: $G_c=(V, E_c)$, $G_m=(V, E_m)$, L_s, L_r, T, k
Result: S (new sinks), R (new relays)
begin

  foreach j *in* T **do**
    $\lfloor$ Find the set $T_j \subseteq V$ within k hops of j in G_c

  $O=\{\}$

  foreach o_i *in more than one* T_j **do**
    Compute the values (x_i, y_i)
    if $c_s * x_i \geq c_r * y_i + c_s$ **then**
      $\lfloor$ add o_i to O in increasing order of $(|T|-x_i)*c_s + y_i*c_r + c_s$

  while O *and* T *are not empty* **do**
    Move first o_i from O into S
    foreach $j \in T$ *such that* $o_i \in T_j$ **do**
      Find connectivity path p for o_i to j
      foreach *location* l *in* p *not in* $L_s \cup L_r$ **do**
        $\lfloor$ add l into R
      Remove j from T
    Re-calculate (x_i, y_i) for all affected entries in O
    foreach $o_j \in O$ *with* $x_j=1$ **do**
      $\lfloor$ Remove o_j from O
    Re-order O

  foreach $j \in T$ **do**
    $\lfloor$ select a location l from T_j and add to S

Algorithm 2. GreedyTour

Data: A set of vertices V', a graph $G_M = (V, E_M)$
Result: a tour in G_M visiting all nodes in V'
begin

  $G'' = (V'', E'', w'') = metric\_closure(V', G_M)$;
  $P = Greedy\_TSP(G'')$;
  return $[V'', P]$;

We build a new graph $G=(V', E')$ where $V'=\bigcup_{j \in T} C_j$, and $E'=\{\{u, v\}|u \in C_i, v \in C_j, i \neq j\}$, i.e. the graph of locations in the cluster sets C_j with an edge between every pair of locations from different cluster sets. We associate a weight to each edge in E' equal to the shortest path length in G_m between the two endpoints. Note that any location which appears in two cluster sets will have a 0-weighted self-edge. The path optimisation problem can now be modeled as a Generalized Travelling Salesman Problem (GTSP) on G where the agent needs to visit exactly one node in each cluster C_i ([2]). We use the memetic algorithm for the GTSP proposed by Gutin and Karapetyan in [3]. The algorithm

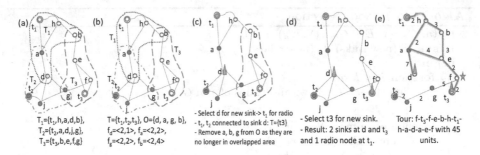

$T_1=\{t_1,h,a,d,b\}$,
$T_2=\{t_2,a,d,j,g\}$,
$T_3=\{t_3,b,e,f,g\}$

$T=\{t_1,t_2,t_3\}$, $O=\{d, a, g, b\}$,
$f_d=<2,1>$, $f_a=<2,2>$,
$f_g=<2,2>$, $f_b=<2,4>$

- Select d for new sink-> t_1 for radio
- t_1, t_2 connected to sink d: T=\{t3\}
- Remove a, b, g from O as they are
  no longer in overlapped area

- Select t3 for new sink.
- Result: 2 sinks at d and t_3
  and 1 radio node at t_1.

Tour: f-t_3-f-e-b-h-t_1-
h-a-d-a-e-f with 45
units.

Fig. 2. A sample execution of NOH

creates a first generation of $2*|T|$ tours by creating random permutations of the clusters and then finds the best vertex in each cluster using the Cluster Optimisation Heuristic (CO). The CO heuristic uses the shortest (s, t)-path for acyclic digraphs to find the best vertex for each cluster when the order of clusters is fixed (see [4]). It then runs a local improvement procedure on each solution. The procedure runs several local search heuristics sequentially including *Swaps* (swap every non-neighboring pair of vertices), *k-Neighbor Swap* (try all the permutations which are not covered by any of i-Neighbor Swap, $i = 2, 3, ..., k - 1$), *2-opt* (try to replace every non-adjacent pair of edges (s_i, s_{i+1}) and (s_j, s_{j+1}) by the edges (s_i, s_j) and (s_{i+1}, s_{j+1})), *Direct 2-opt* (a modification of 2-opt, which only selects some of the longest edges in the solution), and *Insert* (remove a vertex from the solution and insert it in different position). The procedure applies all those local search heuristics in a loop, removing any heuristic that fails to improve the solution. Once the loop terminates, it applies the CO heuristic again, and stops. The next generation is created by reproduction, crossover, and mutation operators applied in parallel to the previous generation. Reproduction simply copies the best solutions from the previous generation. The crossover operator is a 2-point crossover producing a single child, by selecting a fragment from the first parent, and then completes the tour by copying the order of the 2nd parent's nodes starting with the node at the end of the selected fragment, and deleting any repeated nodes. The parents are selected randomly from the top 33% of individuals. The mutation operator modifies each selected parent (selected from the top 75%) by removing a random fragment and inserting it randomly in a new position. Reproduction, crossover and mutation generate new children in the ratio $(1 : 8 : 2)$. The local improvement moves are then applied to each individual. The algorithm repeats until a time limit is reached. When running the algorithm in the experiments, we use the same parameter values given in [3]. Note that the overall running time is dominated by the creation of the initial weighted graph.

The memetic algorithm results in a sequence of locations to be visited, and we obtain a feasible solution if we place a sink at each location. We now improve that solution by replacing as many sinks with relay nodes as possible (Algorithm 4). We start by assuming all locations in the set are occupied by relay nodes. We order the set in decreasing order of the number of terminals within k hops

---

Algorithm 3. Memetic Algorithm

Data: $G = (V', E')$ (graph), a set of clusters $C_i, i \in T$
Result: a tour visiting exactly one node in each C_i
begin

    Initialize, construct first generation of solutions;
    Improve the first generation by local search, eliminate duplicate solutions;
    while *not termination condition* **do**

        Produce next generation by genetic operators (reproduction, crossover, mutation);
        Improve the next generation by local search, eliminate duplicate solutions;

Algorithm 4. Node Selection Algorithm

Data: N (set of locations), $G_c = (L_s \cup L_r \cup N, E_c)$ (graph), T (set of terminals)
Result: S (locations for sinks), R (locations for relays)
begin

    $S \leftarrow \{\}$;
    while T *is not empty* **do**

        Sort N in decreasing order of number of $j \in T$ within k hops in G_c;
        Move first element n_0 from N to S;
        Remove all j from T where j is connected to n_0 within k hops in G_c;
    $R \leftarrow N$

---

of each location. We select the first location, convert it into a sink, and remove from T all terminals connected to that sink in $\leq k$ hops. We repeat until T is empty. Any locations remaining in N are left as relay nodes.

Figure 3 shows the POH algorithm being applied to the example of Figure 1. First, we find a set of clusters for each terminal in G_c (Figure 3(a)). We then calculate the costs of moving from each node in each cluster to other nodes in different clusters (b). We apply the memetic algorithm to this new graph and it produces the tour $[f, t_3, g, t_1, f]$ (c) which is then mapped into G_m as $P=[f, t_3, f, g, f, e, b, h, t_1, h, b, e, f]$ with a mobility cost of 34 (d). Finally, we apply the Node Selection algorithm in G_c for those visited locations $\{t_1, t_3, g\}$ which results in 3 new sinks.

5   Evaluation

Both proposed algorithms (NOH and POH) are heuristic, and take different approaches to the multi-objective problem. Therefore, we evaluate them empirically on randomly generated graphs, to compare the quality of their solutions on both objectives, and also on their runtime. For the graphs, our aim is to represent a physical area rather than abstract random graphs, and so we use a grid to generate the graphs. Connectivity is based on the distance between two

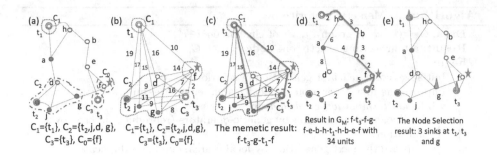

Fig. 3. A sample execution of POH

locations. To represent a landscape or a building interior, we add obstacles into the grid, which may hinder or forbid access. The mobility graph is then based on line-of-sight, with some limited ability to cross the obstacles.

We generate graphs within a rectangular area consisting of $n \times m$ squares of size 10 units. Within this space, we place o mobility obstacles, where each obstacle is a random polygon contained within a randomly selected pair of neighbouring cells. For each square, we generate a random position within it; if that position is inside an obstacle, we discard it, otherwise we designate it as a candidate location. Each obstacle is given a random weight w between 0 and 1, representing the difficulty it creates for the agent to traverse it, and such that any obstacle with a weight greater than 0.2 is assumed not able to be traversed.

We then create the connectivity graph by adding edges indicating that two candidate locations are within transmission range. For each pair of locations, we add with probability 0.85 an edge if they are within 10 units apart; we add with probability 0.2 an edge for each pair of locations which is between 10 and 20 units apart. This is to simulate the radio obstacles where we don't have uniform communication ranges. For the mobility graph, we add an edge between any pair of locations which are less than 25 units apart and which can be connected by a straight line that does not cross an obstacle. The weight of the edge is simply the length of the connecting line. For any pair of locations separated by a distance of less than 25 and which has a straight line that traverses all obstacle with a weight less than or equal to 0.2, we add those edges into the mobility graph. The cost of the edge is the distance plus 10*weight for each obstacle it crosses.

We consider the problem size: (i) a 10×10 grid, and thus a maximum of 100 candidate locations, and 20 possible obstacles[1]. We perform two sets of experiments: varying the number of terminals and varying the hop count limit k. For each data point, we generate 50 instances, and present the average solution cost (mobility cost, number of nodes needed) and runtime. For each instance, we randomly select candidate locations as terminals or live nodes. Finally, we calculate the total time to restore the network for different agent speeds, by combining the runtime with the estimated travel time.

---

[1] We also experimented with the problem size 5×10 grid, and thus a maximum of 50 candidate locations, and 10 possible obstacles, and got similar results.

To assess the quality of the solutions, we compare the results against the optimal solutions for each individual objective, generated by an exhaustive search. That is, OPT-N first finds the set of sinks and relay nodes with lowest cost that connect all the terminals with the hop count. For that set, a tour is then generated using Algorithm 2. OPT-P first generates the optimal mobility tour that could connect all terminals within the hop count (by selecting the optimal tour that visits each cluster set). For the locations on the tour, we then select the locations to be used as sinks using Algorithm 4.

First, we vary the number of terminals and fix the hop count limit as $k = 2$. Figure 4 shows the number of nodes placed by each algorithm. As the number of terminals to be connected increases, the node cost rises as expected. The two algorithms that prioritise mobility incur 25% higher node costs than their node equivalents. The mobility costs rise as we increase the number of terminals. The heuristic POH is within 30% of the exact OPT-P. As above, the algorithms that prioritise node cost create approximately 40% higher mobility costs.

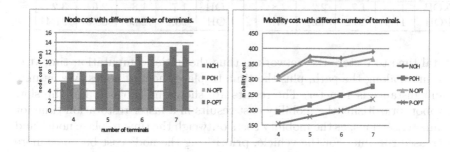

Fig. 4. Varying number of terminals

In the second set of the experiments, we fix the number of terminals at 5, and vary the hop count limit k. As the hop limit increases, the node costs decrease, as the connectivity problem becomes easier (Figure 5). The performance gap between the node-based algorithms and the path-based algorithms increases. The mobility costs also decrease as the hop limit increases. We believe this is because the reduction in the cost is due to placing a small number of sinks in the centre of the map, thus requiring a shorter path to visit those locations.

Table 1 shows the runtimes for the algorithms in both experiments. All increase with the number of terminals and the hop count k. The running time of NOH is significantly faster that of POH due to POH takes time to compute the clusters and path cost between clusters.

Finally, we note that the mobility costs are associated only with the distance travelled. For real scenarios, there is a tradeoff between the cost of the extra nodes and the speed at which connectivity is restored, and so we should consider the combination of runtime and the estimated time to execute the solution. We assume that it takes an agent 30s to position a new node. We then consider two scenarios, the first representing a small robot which moves at 0.1ms^{-1}, the second representing a larger vehicle moving over rough terrain at 4ms^{-1}.

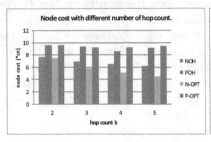
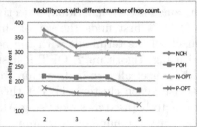

Fig. 5. Varying hop count limit k

Table 1. Runtime

(a) Varying number of terminals

	4	5	6	7
NOH	1.1	1.7	2.2	2.8
POH	17.9	23.2	43.1	57.7

(b) Varying hop count limit

	2	3	4	5
NOH	1.7	1.8	2.6	3.2
POH	23.2	28.8	88.6	91.4

The total time to restore the network is thus the initial computation time, the time to move along the path, plus the time to place the new nodes. We assume each unit distance is 1 meter. The results are shown in Tables 2. For the slow small robot, prioritising the mobility cost results in a faster restoration time for all parameter settings, as the mobility costs outweigh the time to place nodes and the increased runtime. For the vehicle, prioritising the node cost becomes more important, since the reduction in mobility cost by the path-based algorithms has difficulty compensating for the increased runtime and the increased node-placement cost. Thus the WSN restoration problem is subtle, with the choice of approach clearly dependent on the details of the specific problem. Solution methods must take into account the main objectives (minimising infrastructure and minimise time), but also consider the capabilities of the agent that will implement the eventual solution.

6   Related Work

Wireless sensor networks are prone to failures, which can lead to a loss of connectivity when the network becomes partitioned and/or nodes become isolated. Several researchers have addressed this problem by devising appropriate planning methodologies such that a deployed network will be resilient in the face of limited failures, or can be altered to avoid anticipated failures, for example due to node power exhaustion. In contrast, our work seeks to repair a network after failures have occurred. Some other papers have addressed this problem by using specialised nodes that can be moved into position and restore connectivity, e.g. [5], [6], [7] but such solutions are not attractive because mobile nodes are expensive and in large networks many such nodes may be required.

Table 2. Total Restoring Time

(a) $V=0.1ms^{-1}$, vs number of terminals

	4	5	6	7
NOH	3275.1	3978.3	3969.8	4212.8
POH	2187.9	2478.5	2873.7	3225.8

(b) $V=0.1ms^{-1}$, vs hop count limit

	2	3	4	5
NOH	3978.3	3391.8	3549.9	3505.1
POH	2478.5	2419.1	2478.6	2055.2

(c) $V=4ms^{-1}$, vs number of terminals

	4	5	6	7
NOH	252.6	326.3	373.4	404.7
POH	300.3	365.3	456.0	521.3

(d) $V=4ms^{-1}$, vs hop count limit

	2	3	4	5
NOH	326.3	288.3	281.4	270.5
POH	365.3	363.5	399.9	407.9

Many papers address the repair problem by placing relay nodes that reconnect partitions in the network, minimising number of required nodes. For example, using centralised solutions, [8] uses a spider web approach while [9] forms a connectivity chain toward a centre point of the network. In contrast, we optimise both the number of additional nodes and the mobility path length needed for their deployment. We also explicitly take into account the impact of obstacles that can impede both the available paths and the ability of nodes to communicate directly, and we bound the number of network hops by judiciously deploying additional sink nodes in the interest of quality of service constraints.

In regard to the problem of multiple sink deployment, some other authors have addressed this for the purpose of improving network performance. There are two main strategies: to reduce energy consumption so as to maximise the network lifetime, e.g. [10], [11] and to minimise the average data latency or maximum hop distance between a node and its nearest sink, e.g. [12]. These papers do not deploy sink nodes in the act of network restoration, and thus do not consider the mobility cost to deploy sink nodes, and the implications of obstacles on the choice of possible sink locations.

7   Conclusion

In this paper we address the problem of restoring connectivity in a wireless sensor network, using a mobile agent to place relay and sink nodes, avoiding obstacles and respecting bounds on quality of service (defined in terms of network hop-count). We formulate the problem as one of searching over two linked graphs which share the vertex set. The multi-objective problem requires us to minimise both the cost of additional nodes and the path to be taken by the mobile agent. We propose two heuristic algorithms which prioritise the different objectives. We evaluate them on randomly generated networks and compare their solutions to the optimal solutions. Finally, we also evaluate the total restoration time for each solution as a function of the mobile agent's speed, quantifying a trade-off with computation time. In future work, we will consider different algorithms for the two different priorities, and we will investigate the Pareto frontier. We will also consider the more general problem, in which we must discover the damage to the network as we repair it. Finally, we will consider the problem of continually spreading damage, and the use of teams of agents cooperating to repair the network.

Acknowledgment. This work was funded by the HEA PRTLI4 project NEM-BES, and by the SFI centre CTVR (10/CE/I1853).

References

1. Lawler, E.L., Lenstra, J.K., Rinnooy Kan, A.H.G., Shmoys, D.B.: The Traveling Salesman Problem. John Wiley & Sons (1985)
2. Fischetti, M., Salazar-Gonzalez, J.J., Toth, P.: The generalized traveling salesman and orienteering problems. In: Gutin, G., Punnen, A.P. (eds.) The Traveling Salesman Problem and Its Variations. Combinatorial Optimization, vol. 12, pp. 609–662. Springer (2004)
3. Gutin, G., Karapetyan, D.: A memetic algorithm for the generalized traveling salesman problem. Journal of Natural Computing 9(1), 47–60 (2010)
4. Fischetti, M., Gonzlez, J.J.S., Toth, P.: A branch-and-cut algorithm for the symmetric generalized travelling salesman problem. Operations Research 45, 378–394 (1997)
5. Akkaya, K., Senel, F., Thimmapuram, A., Uludag, S.: Distributed recovery from network partitioning in movable sensor/actor networks via controlled mobility. IEEE Transaction on Computing 59, 258–271 (2010)
6. Abbasi, A.A., Younis, M.F., Baroudi, U.A.: A least-movement topology repair algorithm for partitioned wireless sensor-actor networks. International Journal of Sensor Networks 11(4), 250–262 (2012)
7. Abbasi, A.A., Younis, M.F., Baroudi, U.A.: Recovering from a node failure in wireless sensor-actor networks with minimal topology changes. IEEE Transaction on Vehicular Technology 62(1), 256–271 (2013)
8. Senel, F., Younis, M., Akkaya, K.: A robust relay node placement heuristic for structurally damaged wireless sensor networks. In: LCN 2009, pp. 633–640 (2009)
9. Lee, S., Younis, M.: Recovery from multiple simultaneous failures in wireless sensor networks using minimum steiner tree. Journal of Parallel Distributed Computing 70, 525–536 (2010)
10. Kim, H., Kwon, T., Mah, P.S.: Multiple sink positioning and routing to maximize the lifetime of sensor networks. IEICE Transactions on Communications 91-B(11), 3499–3506 (2008)
11. Gu, Y., Ji, Y., Li, J., Chen, H., Zhao, B., Liu, F.: Towards an optimal sink placement in wireless sensor networks. In: ICC 2010, pp. 1–5 (2010)
12. Kim, D., Wang, W., Sohaee, N., Ma, C., Wu, W., Lee, W., Du, D.Z.: Minimum data-latency-bound k-sink placement problem in wireless sensor networks. IEEE/ACM Transaction on Networking 19(5), 1344–1353 (2011)

Distributed Energy Efficient Data Gathering without Aggregation via Spanning Tree Optimization

Lenka Carr-Motyčková and David Dryml

Department of Computer Science
Faculty of Science, Palacky University
17. listopadu 12, CZ-771 46 Olomouc, Czech Republic
{lenka.motyckova,david.dryml}@upol.cz

Abstract. A distributed algorithm that solves energy efficient data gathering Weighted Spanning Tree Distributed Optimization (WSTDO) is proposed in this paper. It is based on an optimization performed locally on the data gathering spanning tree. WSTDO algorithm is compared to two centralized spanning tree optimization algorithms MITT and MLTTA. The performance of WSTDO achieves between one half and one third of the MITT performance and proves to be better than MLTTA. The performance depends on the density of the network. It works better for sparse networks. WSTDO has lower overhead than MITT and MLTTA for sparse networks. Though the proposed algorithm has a worse performance than MITT it has other features that over-weights this fact. It is able to perform optimization parallely in disjoint sub-trees and also during data gathering which allows a short data sampling period. It is also prone to link and node failures that can be solved locally.

Keywords: Data gathering, Wireless sensor networks, Maximum lifetime, Convergecast tree, Spanning tree optimization.

1 Introduction

The Wireless Sensor Networks (WSNs) are one of the fast growing type of networks. This is due to advancements in the development of cheap micro-controllers and transceivers with the low energy consumption that allow deployment of the massive number of sensor nodes for a long period of time. Sensor Network is composed of a single powerful node (base station or sink) and simple cheap nodes with limited resources.

Typically sensor networks perform a periodic sampling of data (temperature, noise, luminosity, ...) from attached sensors followed by its transfer to the sink. Intermediate nodes are often used to transfer data from nodes that are out of a sink communication distance. The scenario is usually called convergecast. The transfer can be combined with a perfect data aggregation in the intermediate nodes. Nevertheless the data aggregation is not suitable for all deployment scenarios. This is the reason why the convergecast algorithms with imperfect or

J. Cichoń, M. Gębala, and M. Klonowski (Eds.): ADHOC-NOW 2013, LNCS 7960, pp. 87–98, 2013.

without aggregation must be developed. They constitute a more complicated problem. WSN nodes have limited resources thus it is very important that data transfer uses as little energy as possible, consumes the energy as evenly as possible and thus extend the time that a network can operate. Nodes close to the sink are most affected because farther nodes out of sink's communication range have to use them to re-transmit their data to the sink. This need introduces a new demand on routing protocols used for convergecast.

A possible solution to the convergecast problem without data aggregation is to construct a data gathering spanning tree, where sensor nodes with sufficient amount of energy and free communication capacity are used to transfer data to the sink. Other nodes that are almost drained, are connected as leafs that only collect their own data and send them to more powered nodes. The construction of an optimal spanning tree in the sense of maximum lifetime (minimum energy consumption in nodes) is NP-complete problem [7]. Thus for larger networks we need to design a heuristic algorithm that constructs a convergecast tree. The distributed nature of the algorithm is beneficial as distributed algorithms are usually more prone to network failures and changes and can react to them quickly.

2   Related Work

There has been an extensive research aimed at resource effective routing, broadcast, multicast and convergecast in an environment of ad-hoc wireless networks and wireless sensor networks in recent years.

The general purpose routing algorithms for example Optimal Link State Routing (OLSR) algorithm [5] and broadcast and multicast algorithms for example Broadcast Incremental Power (BIP) [11] and its distributed version described in [4] can be used for data gathering but they are not specifically optimized for the scenario of data convergecast. The first group of algorithms specifically designed for convergecast apply a perfect data aggregation where all data received in intermediate nodes are combined to a single data unit. Examples could be found in [3], [6], [8] and [9]. Data gathering with perfect data aggregation limits the usage of sensor networks only to certain types of scenarios as not all types of collected data can be perfectly aggregated without the information loss.

The second group of convergecast algorithms contain those without the (perfect) data aggregation. The size of transferred data here is proportional to the number of nodes that need to use the intermediate node for data transfer. Centralized algorithms MITT [7] and similar MLTTA [12] compute *min-max-weight spanning tree*. The main difference between them is that the later one assumes adjustable communication distance and has a slightly different approach to the optimization function which is described rather vaguely. Representatives of distributed convergecast algorithms are Dozer [1] and MeeCast [13] algorithms.

Our algorithm is designed to solve an *min-max-weight spanning tree* optimization problem similar to MITT [7] or MLTTA [12]. The main contribution of this paper is the distributed approach to the solution of the optimization problem.

3   Network Model and Problem Statement

We assume that all links in the network are bidirectional and reliable (mainly for the sake of an easy comparison with MITT and MLTTA), and that there is always at least one route that connects every sensor node to the sink. For each pair of nodes that can communicate directly we can estimate an euclidean distance between them (e.g. signal/noise ration) and for the sake of a better performance we also expect that all nodes can adjust their transmission range [12]. Each node has also its unique identifier (e.g. MAC address). In the rest of this chapter we formalize network model and present the problem of constructing a *min-max-weight spanning tree* for maximizing a network lifetime.

3.1   Network Model

We use an undirected graph $G = (V, E)$ to represent a sensor network in the paper. For each sensor node $v_i \in V$ we define a function $Er(v_i) \in \langle 0, e_{\max} \rangle$ that returns an actual residual energy of a sensor node in v_i's battery. For each $e_{ij} = (v_i, v_j) \in E$ we define variable d_{ij} that represents an estimated euclidean distance between vertices v_i and v_j. We define n-(hop)neighborhood $neigh(G, v_i, n)$ of node v_i in network G as a set of all nodes with a hop distance at most n. 1-neighborhood of node v_i contains all nodes in communication distance of v_i.

For an arbitrary tree T rooted in s and the node $v_i \in T$ we call each vertex on the path from v_i to the root/sink s to be a **predecessor of** v_i and the first predecessor of v_i to be a **parent of** v_i. The **descendant of** v_i is defined to be each vertex from a sub-tree rooted in v_i . The size of the sub-tree is denoted $S(T, v_i)$ and the **child of** v_i is each descendant c_j of v_i for which there is an edge $(v_i, c_j) \in T$. **Tree neighbours** of v_i are its parent and children. They define the current communication distance $d(T, v_i)$ to be the minimal range that allows v_i to reach all its tree neighbours. We use d to denote the minimal range for the currently processed node.

$ST(G, s)$ is the set of all spanning trees on graph G rooted in the sink node s. The **lifetime** $L(T, v_i)$ for each network node v_i and tree T is the maximal number of data gathering rounds the node v_i can process. It decreases with increasing $S(T, v_i)$ and increases with increasing $Er(v_i)$ (more detailed description of estimate in section 4). The **lifetime** of **tree** T's defined to be $L(T) = min(L(T, v_i))$ over all $v_i \in V$.

3.2   Problem Statement

A natural goal would be to find the best *min-max-weight spanning tree* $T_{\max}$ rooted in sink s that satisfies $L(T_{\max}) = \max_{T_j \in ST(G,s)} \min_{v_i \in V}(L(T_j, v_i))$ (tree with maximum lifetime). The solution to the problem is NP-complete [7], so we rather propose algorithm to find the sub-optimal tree that approximates the lifetime of $T_{\max}$.

4   Metric

For the construction of a *min-max-weight spanning tree* we need to know an estimate of the energy consumed by the sensor data transfer in each node v_i during one data gathering round.

$$Ec(T, v_i) = Ec_{rx}(T, v_i) + Ec_{tx}(T, v_i)$$

The energy consumption is composed of two parts. Ec_{rx}, energy necessary for the reception of all data units from v_i's descendants and Ec_{tx}, energy necessary for the transmission of received data and v_i's own data to the parent node.

$$Ec_{rx}(T, v_i) = Ec_{elect} l S(T, v_i)$$
$$Ec_{tx}(T, v_i) = Ec_{elect} l (S(T, v_i) + 1) + Ec_{amp} l d^2 (S(T, v_i) + 1) \tag{1}$$

Ec_{elect} and Ec_{amp} are constants, l is the length of data in bits.

Lifetime $L(T, v_i)$ estimate for the node v_i in the current tree T is:

$$L(T, v_i) = \left\lfloor \frac{Er(v_i)}{Ec(T, v_i)} \right\rfloor . \tag{2}$$

Recall that the tree lifetime is defined to be a minimal lifetime over all T's nodes. The global maximum is expensive to compute in terms of messages necessary to gather in nodes (a.i. consumed energy). Thus we substitute the global minimum with a local minimum computed from n-neighborhood as follows:

$$L_{min}(T, v_i) = \min \left(L(T, v_i), \min_{v_j \in neigh(G, v_i, n)} L(T, v_j) \right) .$$

$Ec_{min}(T, v_i)$ is the amount of energy necessary for node v_i to gather data during the estimated tree T's lifetime. The energy necessary to support a single descendant $\varphi(T, v_i)$ depends on $L_{min}(T, v_i)$ as well:

$$Ec_{min}(T, v_i) = L_{min}(T, v_i) \times Ec(T, v_i)$$
$$\varphi(T, v_i) = L_{min}(T, v_i)(2Ec_{elect} l + Ec_{amp} l d^2)$$

To optimize the tree lifetime we need to decrease the energy consumption in nodes with the minimal lifetime. This can be achieved in two ways. The first most significant method lies in removing some of v_i's descendants. To achieve this we need to calculate the number of descendants that must be disconnected. Consequently the nodes that accept the removed nodes as their new descendants must be found. The second method is due to the adjustable transmission range used in Ec_{tx}. If the parent is the farthest tree neighbor of v_i than v_i can try to connect a closer parent to decrease the necessary communication distance d.

We first introduce **nodes' classification** to find nodes that have to be optimized. We classify nodes to three classes distinguished by colors:

Red nodes are bottleneck nodes having a large load or a low energy. They need to get rid of at least one child node to become a gray node. These nodes satisfy the following inequality.

$$Er(v_i) < L_{\min}(T, v_i)Ec(T, v_i) + \varphi(T, v_i)$$

Gray nodes are sub-bottlenecks. These are nodes that would become bottlenecks if they accept at least one more descendant which means that they satisfy:

$$L_{\min}(T, v_i)Ec(T, v_i) + \varphi(T, v_i) \leq Er(v_i) < L_{\min}(T, v_i)Ec(T, v_i) + \varphi(T, v_i) + \frac{\varphi(T, v_i)}{Er(v_i)} \ .$$

The factor $\frac{\varphi(T, v_i)}{Er(v_i)}$ is used to control the size of the gray area. The main purpose of introducing this factor is that the nodes with the short lifetime accept less descendants compared to nodes with the maximal lifetime.

Green nodes are those having a low load or a high energy. They can accept at least one more descendant without becoming red.

Consequently we need to compute the number of descendants a green node can accept without becoming red and the number of nodes a red node must remove to become gray. For this purpose we define a **capacity** for each node v_i.

$$C(T, v_i) = \left\lfloor \frac{(L(T, v_i) - L_{\min}(T, v_i))Ec(T, v_i) - \varphi(T, v_i)}{\varphi(T, v_i)} \right\rfloor$$

For the sink node the capacity is defined as $|V|$. For red nodes v_i it is defined as $min(C(T, v_i), -1)$. Gray nodes have zero capacity and green nodes' capacity equals to $C(T, v_i)$. From these values we compute the final value of the capacity $Cf(T, v_i)$:

$$Cf(T, v_i) = min(C(T, v_i), C(T, u_1) \ldots C(T, u_n))$$

It is defined as the minimum of $C(T, v_i)$ and $C(T, u_j)$ over all v_i's predecessors $u_j \in neigh(G, v_i, n)$.

Final configuration we want to achieve is the state where no bottleneck node can neither disconnect any of its descendants nor it can decrease its transmission distance without causing some other node/s to become a bottleneck.

5   Algorithm

First we present a short overview of the main parts of WSTDO algorithm in this section. Then we define and describe variables stored in each node followed by a detailed description of the main algorithm parts and messages the algorithm use.

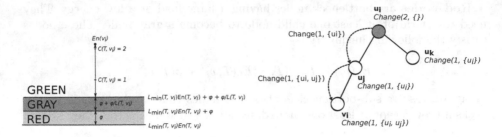

Fig. 1. Nodes' classification and capacity (left) and a scheme of recursive calls to *Change* procedure and sending of *Change* message (right)

5.1 Overview

The algorithm consists of four main parts: the initialization, the tree maintenance, the tree optimization and the data transfer.

During the initialization (see 5.2) an initial tree is constructed, local variables are initializes and the information on n-neighborhood nodes are gathered. The initialization is executed first and the other parts of the algorithm are started and performed in parallel after the initialization has finished.

Tree maintenance (see 5.3) procedure is used to collect and process up-to-date nodes' state and topology changes in each nodes' n-neighborhood. It also serves to detect and repair possible cycles that can be formed during the tree optimization.

The main part of the algorithm is the tree optimization (see 5.4) where nodes' loads are locally optimized according to their classification and capacity. Up-to-date state of each node is then transmitted to its n-neighborhood.

The last part is the data transfer protocol. We expect a periodic data sampling in all nodes and the consecutive data transfer to the sink over the constructed tree.

All parts of the algorithm are written as executed in the local node v_i. Each node v_i maintains several **variables that describe v_i's state** and also the state of its neighbors. The most important variables are $lifetime = L(T, v_i)$, *state* with possible values *red*, *gray*, *green*, and *capacity* $= Cf(T, v_i)$. The variable *disconnectTries* is a set that contains a number of possible attempts to disconnect each child of v_i. For child c_j $triesMultiplyer * (S(T, c_j) + 1)$ is defined. Each time v_i sends a disconnect message to a child it decrements the value of *disconnectTries* for the child. This variable makes sure that each bottleneck stops trying to disconnect its children after a finite number of tries so that the algorithm can reach a stable state.

Variables *lifetime*, *capacity* and *parent* are broadcasted to n-neighborhood in *NodeState* message. They are also stored locally as a part of v_i's state of n-neighborhood nodes in *neighbours* set. Node v_i obtains these values from *NodeState* and *NodeLoad* messages described in Tree maintenance subsection (see 5.3).

5.2   Initialization

The result of the initialization is the initial tree, the values of v_i's local variables set to their initial values and the initial state of v_i is broadcasted to nodes in its n-neighborhood. Symmetrically the same information is obtained by v_i from its n-neighborhood.

At the beginning of this phase we use modified Echo algorithm [2] to construct an initial tree and to initialize the values of local variables *parent*, *children* and $S(T, v_i)$ for each v_i. Then nodes exchange their *id* and value of $L(T, v_i)$ with its 1-hop neighbors.

After node v_i receives values of $L(T, u_i)$ from all its 1-hop neighbors u_i, it computes initial values of variables *state* and *capacity* and then it broadcasts these values in a *NodeState* message to n-neighborhood and ends initialization phase.

5.3   Tree Maintenance

This part of the algorithm is used to maintain v_i's information on its neighbors in n-neighborhood and also to detect and remove cycles damaging the tree topology.

Two types of messages are used to transfer the information. The *NodeState* message is broadcasted periodically in regular intervals to nodes in n-neighborhood. It contains a state information of the sending node. After the initialization only incremental changes of node's state are sent. If there is no change or only the lifetime of the node differs, the message is not sent. Neighbors are able to estimate v_i's lifetime from the last value that was received minus the number of data gathering periods since it was last received.

The *NodeLoad* message is sent from node u_i to its parent as a unicast message if a child node connects/disconnects to/from u_i and it is recursively forwarded to the sink. It is used to inform predecessors of the node u_i on the changed number of the descendants. The value of *possibleTries*$[c_j]$ is updated to *triesMultiplyer* $* (S(T, c_j) + 1)$. The message contains *id* of a newly connected child for a purpose of a cycle detection. If v_i receives its own *id* in the message it has detected a cycle and it is forced to connect to a different parent. v_i also stores an information on previous connection attempts that caused a cycle for the further cycle prevention.

5.4   Tree Optimization

The tree optimization is performed periodically starting from the end of the initialization phase. Its main purpose is to find bottlenecks in the convergecast tree and remove them. It starts by pseudo-code in Algorithm 1.

The **procedure Change** has two parameters. Parameter *excesiveLoad* determines the number of descendants v_i has to remove from its sub-tree. Parameter *predecessors* is an ordered set that contains n-neighborhood predecessors that launched recursive calls of the *Change* procedure (see Figure 1).

Algorithm 1. Tree Optimization

---

update node state
if v_i becomes *red* **and** has no red predecessor **then**   ▷ v_i is bottleneck
   $x = -Cf(T, v_i)$
   call *Change(x, Ø)*
else
   broadcast *NodeState* message neighbors
end if

Algorithm 2. *Change(excesiveLoad, predecessors)*

                                                    ▷ Optimize node's load
$swichStatus = v_i \rightarrow Switch(predecessors)$ ▷ try to switch to different parent first
if $switchStatus == true$ **then**   ▷ switched to different parent
   update node state
   broadcast *NodeState*
else   ▷ else select victims r_k from $v_i \rightarrow children$ to disconnect
   add all children c_j with $v_i \rightarrow possibleTries[c_j] > 0$ to *removeCandidates*
   remove all r_k not known from n-neighborhood from *predecessors* and add v_i
   sort *removeCandidates* set from farthest to closest
   while $excessiveLoad \geq 0$ **do**
      select and remove first r_k from *removeCandidates*
      $disconnectLoad = S(T, r_k) + 1$
      if $disconectLoad < excesiveLoad$ **then** $disconectLoad = excesiveLoad$
      end if
      $v_i \rightarrow possibleTries[r_k] = v_i \rightarrow possibleTries[r_k] - 1$
      $excessiveLoad = excessiveLoad - disconectLoad$
      send Change($disconnectLoad, predecessors$) to r_k
   end while
   update node state and broadcast *NodeState*
end if

The *Change* procedure is called either directly after a node detects its *red* state or upon a receipt of a *Change* message. Change message serves to inform a child node of v_i that it must re-connect to a different parent and thus decrease load of v_i. It contains the *predecessors* set of nodes that recursively called the *Change* procedure before the *Change* message arrived to v_i. *Predecessors* set contains only nodes known to v_i's parent from n-neighborhood. The *Change* message also contains the number of descendants the child node of v_i has to remove. The pseudo-code of the *Change* procedure is in Algorithm 2.

The **function Switch** is called as a part of the *Change* procedure when v_i tries to find a new parent. There are two possible reasons to switch to a different parent. The first case occurs when v_i's state becomes red. It can be detected by the empty *predecessors* set. In this case it is sufficient for v_i to decrease its current communication distance d if possible.

The second case of calling *Switch* occurs when v_i is selected as a victim for the disconnection from its parent. In this case it is not necessary to connect

---

Algorithm 3. *Switch(predecessors)*

                                                  ▷ Try to switch to another parent

$candidates = \emptyset$
for all n_j neighbours of v_i node **do**
   if *IsPossibleCandidate(n_j)* **then**
      $candidates = candidates \cup \{v_i\}$
   end if
end for
if $candidates == \emptyset$ **then return** $false$
end if
sort $candidates$ set from nearest to farthest
for all $c_k \in candidates$ **do**
   try to connect to c_k
   if not sucessfull for any c_k **then return** $false$
   end if
end for
set $v_i \rightarrow parent$ to selected c_k
broadcast $NodeState$
return $true$

---

to a closer parent; it is sufficient to switch to another parent to decrease the load of v_i's parent. There are other criteria for the selection of a neighbor as a candidate c_k for a new parent. Namely c_k must not turn red after accepting v_i as a child, must not be a descendant of v_i known from n-neighbourhood, must not have caused a cycle in previous attempts and must not have the same predecessor as v_i known from n-neighbourhood. All these criteria are checked in the *IsPossibleCandidate(candidate)* function.

6 Performance Evaluation

We compare our algorithm to MITT [7] and MLTTA [12] algorithms in this section. Both MITT and MLTTA solve a very similar spanning tree optimization problem as our algorithm does. The main difference is that these algorithms are centralized and ours is distributed. Because MITT and MLTTA algorithms are targeted to wireless sensor networks environment we assume that they have to be executed in the sink node. It implies that the network topology and the residual energy must be transferred to the sink and then the resulting spanning tree topology must be disseminated back to network nodes.

We do not assume any link and/or node failures during simulations thus for the sake of a fair comparison we collect the network topology only once at the beginning of the network lifetime. On the other hand the residual energy is collected before each MITT or MLTTA execution. Both algorithms are executed after each data gathering round. A similar execution scenarios for centralized algorithms in wireless sensor networks are suggested in [7] and [13].

We use ideal Time Division Multiple Access (TDMA) described for example in [13] as MAC layer model for our simulations for a closer comparison to MITT

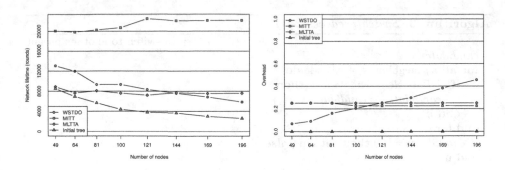

Fig. 2. Number of data gathering rounds (left) and overhead (right) for different number of nodes in network with sink on position (50, 50) and initial energy from interval $\langle 5, 10 \rangle$ J

and MLTTA and an assessment of a raw optimization ability of our algorithm. Oracle is used in the model to schedule nodes communication so that there are no transmission collisions, idle listening (nodes transmit/receive message or sleep) or overhearing. All channels are reliable with no data loss or damage during the communication.

We use data message size of 560 bits throughout the simulations, the maximal transmission range of nodes is set to be 25m as in [7]. Energy parameters are set like this: $Ec_{\text{elect}} = 50\text{nJ/bit}$. $Ec_{\text{amp}} = 100\text{nJ/bit}$ for MITT and 100pJ/bit/m^2 for MLTTA and WSTDO. MITT and MLTTA use $k_1 = 100$ and $k_2 = 1000$. In MLTTA we use same optimization part as it is in MITT because the optimization part of MLTTA algorithm is described very vaguely. Multiplication constant *triesMultiplyer* is set to 50.

Simulations show that the knowledge of 2-neighborhood is sufficient for an optimal performance of our algorithm. 1-neighborhood does not support enough neighborhood knowledge. 3-neighborhood on the other hand causes a huge message overhead to maintain the up-to-date state information and it is also highly demanding in storage space (considering the limited memory of sensor nodes).

6.1   Simulations Results

The important result is that WSTDO always converges to the final configuration.

For networks with 100 nodes or less and initial energy $\langle 5, 10 \rangle$ J the overhead of our algorithm is smaller than it is for both centralized algorithms. This is caused by the fact that each node has only few neighbors and thus it only drains energy from few nodes during the broadcast of its state.

As the main result we compare the network lifetime reached by all three algorithms. Figure 2 shows that our algorithm can perform approximately half of the data gathering rounds compared to MITT algorithm for sparse networks with 100 or less nodes and one third for denser networks. Comparison to MLTTA shows that our algorithm performs the same or better.

The main reasons for the worse performance of our algorithm is that both centralized algorithms perform optimization using the complete knowledge of the network topology and the energy status of all nodes. The local optimization in the sink also means that each topology change during the computation in the sink does not cost any messages to perform. Thus the energy metric can be more fine grained and even the slightest change in the network topology that saves only a small portion of energy can be considered in the optimization. MITT and MLTTA algorithms change network topology only once between data gathering rounds after the optimization computation was finished. On the other hand our distributed algorithm works continuously.

The results of simulations can be influenced by an estimate of the overhead in centralized algorithms used for the information collection and the distribution in/from the sink node. The overhead is not considered in centralized algorithms but it must be considered for a better assessment of the performance in a more real sensor network deployment scenario and the comparison to our algorithm.

Centralized algorithms also leverage from the fact that they collect network topology only once. The topology collection before each algorithm execution would be necessary only if we assume network failures and changes. This would decrease their performance significantly while the dynamic topology processing in our distributed algorithm does not influence its performance.

7   Conclusion

In this paper we proposed the distributed convergecast algorithm targeted the to sensor networks environment. It is based on local optimization of the spanning tree structure according to the load of tree nodes.

We compared our algorithm to two centralized spanning tree optimization algorithms MITT and MLTTA. Simulations show that our algorithms performance is between one half and one third of the MITT's performance. It has better results for sparse networks. Our algorithm compensates this fact by its ability to work effectively in the scenario with link failures and topology changes. Convergecast tree can be repaired locally and the optimization is performed paralelly in disjoint sub-trees.

Centralized algorithms do not consider the energy consumption needed for network status gathering in the sink node and for optimization results distribution to network nodes. We made the modest assumptions about this energy consumption for the sake of comparing the centralized algorithms to our distributed one. The need for an estimate of centralized algorithms overhead makes the comparison of distributed algorithm to centralized ones difficult.

8   Future Work

Our distributed algorithm will be tested and optimized (including parameter settings) in more real MAC layer model using COOJA simulator [10]. Overall design of WSTDO allows its deployment in real MAC layer with only few modifications. During tests our algorithm should be compared to Dozer and MeeCast as these algorithms are currently the state-of-the-art for data gathering without aggregating. We will also prove the convergence of WSTDO to the final configuration, its approximation factor and its complexity.

References

1. Burri, N., von Rickenbach, P., Wattenhofer, R.: Dozer: ultra-low power data gathering in sensor networks. In: Proceedings of the 6th International Conference on Information Processing in Sensor Networks, IPSN 2007, pp. 450–459. ACM, New York (2007)
2. Chang, E.J.H.: Echo algorithms: Depth parallel operations on general graphs. IEEE Trans. Software Eng. 8(4), 391–401 (1982)
3. Hariharan, S., Shroff, N.B.: On optimal energy efficient convergecasting in unreliable sensor networks with applications to target tracking. In: Proceedings of the MobiHoc 2011, pp. 24:1–24:10. ACM, New York (2011)
4. Ingelrest, F., Simplot-Ryl, D.: Localized broadcast incremental power protocol for wireless ad hoc networks. Wirel. Netw. 14(3), 309–319 (2008)
5. Jacquet, P., Muhlethaler, P., Clausen, T., Laouiti, A., Qayyum, A., Viennot, L.: Optimized link state routing protocol for ad hoc networks. In: Proceedings of the IEEE International Conference IEEE INMIC 2001, pp. 62–68 (2001)
6. Levin, L., Segal, M., Shpungin, H.: Energy efficient data gathering in multi-hop hierarchical wireless ad hoc networks. In: Proceedings of the 7th International Workshop on Foundations of Mobile Computing, FOMC 2011, pp. 62–69. ACM, New York (2011)
7. Liang, J., Wang, J., Cao, J., Chen, J., Lu, M.: An efficient algorithm for constructing maximum lifetime tree for data gathering without aggregation in wireless sensor networks. In: Proceedings of the 29th Conference on Information Communications, INFOCOM 2010, pp. 506–510. IEEE Press, Piscataway (2010)
8. Luo, D., Zhu, X., Wu, X., Chen, G.: Maximizing lifetime for the shortest path aggregation tree in wireless sensor networks. In: INFOCOM, pp. 1566–1574. IEEE (2011)
9. Onodera, K., Miyazaki, T.: An autonomous multicast-tree creation algorithm for wireless sensor networks. In: Proceedings of the Future Generation Communication and Networking, FGCN 2007, vol. 01, pp. 268–273. IEEE Computer Society, Washington, DC (2007)
10. Osterlind, F., Dunkels, A., Eriksson, J., Finne, N., Voigt, T.: Cross-level sensor network simulation with cooja. In: Proceedings of the 2006 31st IEEE Conference on Local Computer Networks, pp. 641–648 (November 2006)
11. Wieselthier, J., Nguyen, G., Ephremides, A.: On the construction of energy-efficient broadcast and multicast trees in wireless networks. In: INFOCOM 2000, vol. 2, pp. 585–594 (2000)
12. Yuan, J., Zhou, H., Chen, H.: Constructing maximum-lifetime data gathering tree without data aggregation for sensor networks. In: Lee, R. (ed.) Computer and Information Science 2011. SCI, vol. 364, pp. 47–57. Springer, Heidelberg (2011)
13. Zeng, W., Arora, A., Shroff, N.: Maximizing energy efficiency for convergecast via joint duty cycle and route optimization. In: Proceedings of the 29th Conference on Information Communications, INFOCOM 2010, pp. 16–20. IEEE Press, Piscataway (2010)

Distributed Maintenance of Anytime Available Spanning Trees in Dynamic Networks

Arnaud Casteigts[1], Serge Chaumette[1], Frédéric Guinand[2], and Yoann Pigné[2]

[1] LaBRI, University of Bordeaux, France
{arnaud.casteigts,serge.chaumette}@labri.fr
[2] LITIS, University of Le Havre, France
{frederic.guinand,yoann.pigne}@univ-lehavre.fr

Abstract. We address the problem of building and maintaining a forest of spanning trees in highly dynamic networks, in which topological events can occur at any time and any rate, and no stable periods can be assumed. In these harsh environments, we strive to preserve some properties such as cycle-freeness or existence of a unique root in each fragment regardless of the events, so as to keep these fragments functioning uninterruptedly to a possible extent. Our algorithm operates at a coarse-grain level, using atomic pairwise interactions akin to population protocol or graph relabeling systems. The algorithm relies on a perpetual alternation of *topology-induced splittings* and *computation-induced mergings* of a forest of trees. Each tree in the forest hosts exactly one token (also called root) that performs a random walk *inside* the tree, switching parent-child relationships as it crosses edges. When two tokens are located on both sides of a same edge, their trees are merged upon this edge and one token disappears. Whenever an edge that belongs to a tree disappears, its child endpoint regenerates a new token instantly. The main features of this approach is that both *merging* and *splitting* are purely localized phenomenons. This paper presents the algorithm and establishes its correctness in arbitrary dynamic networks. We also discuss aspects related to the implementation of this general principle in fine-grain models, as well as embryonic elements of analysis. The characterization of the algorithm performance is left open, both analytically and experimentally.

1 Introduction

Spanning trees are essential components in communication networks. The availability of such structures simplifies a large number of tasks, among which broadcasting, routing, or termination detection. From the standpoint of distributed computing, constructing a spanning tree implies the collaboration of neighboring nodes in order to establish selective relationships that inter-connect the whole network without cycle.

The problem is very different in essence in static and dynamic networks. In a static network, there is generally a distinction between the construction of a tree

J. Cichoń, M. Gębala, and M. Klonowski (Eds.): ADHOC-NOW 2013, LNCS 7960, pp. 99–110, 2013.
© Springer-Verlag Berlin Heidelberg 2013

and its effective use, both taking place at different times. In truly dynamic networks (e.g. vehicular networks), the set of communication links evolves rapidly and continuously. As a result, the trees need to be updated on a constant basis and while they are used. Early works addressing the spanning tree problem in dynamic graphs (see e.g. [4,10,6] and the references therein) applied strong restrictions on the dynamicity; namely, these works assumed the network stabilizes eventually, or recurrently offers stable periods during which the tree can entirely be recomputed. These assumptions are certainly appropriate in the case of occasional failures or reconfigurations of the topology. But they are not reasonable in highly dynamic scenarios like mobile ad hoc networks.

We are interested in understanding what can still be done in the harshest dynamic context. In particular, we consider networks in which no stability period is ever expected; no information is available about future topological events; no restrictions apply to the rate of these events; and no contemporaneous end-to-end connectivity is assumed (that is, we address *delay-tolerant networks* [8]). On the other hand, we allow ourselves to reason at a high level of abstraction, using a coarse-grain interaction model akin to recent *population protocol* models [3]. While we find the problem in this model interesting in its own right, we still hope and believe the principles highlighted here can help subsequent effort to make it work in finer-grain (e.g. message passing) models.

The algorithm relies on a perpetual alternation of *topology-induced splittings* and *computation-induced mergings* of a forest of spanning trees. Each tree in the forest hosts exactly one token (also called root) that performs a random walk *inside* the tree, switching parent-child relationships as it crosses edges. When two tokens are located on both sides of a same edge, their trees are merged upon this edge and one token disappears. Whenever an edge that belongs to a tree disappears, its child endpoint regenerates a new token instantly. The main features of this approach is that both *merging* and *splitting* are purely localized phenomenons.

After reviewing some relevant work in Section 2, we define the network model and assumptions, as well as the computational model in Section 3. The algorithm is then presented in detail and proved correct in Section 4. This presentation is followed by a discussion regarding some important implementation choices (e.g. priority between different rules of interaction). In Section 5, we provide preliminary results on the analysis of the algorithm, which we regard as a coalescing particle system involving random walks in trees. We conclude in Section 6 with some perspectives.

2   Related Work

The problem of building distributed spanning trees in communication networks, and more generally in graphs, has been extensively studied during the last three decades and a large literature exists on the topic. It is noteworthy that the problem was studied by different communities (self-stabilization, stochastic processes, distributed computing) using different paradigms and terminologies (*e.g.*

token, mobile agent, random walk, legal state, stabilization time, coalescing time, tree, forest, etc.). We review below the most relevant concepts and approaches to solve this problem.

Self-stabilization: A system that reaches a *legal* state starting from an *arbitrary* state is called *self-stabilizing*. After a fault in the system, the time required to reach the legal state is called the *stabilization time*. In the context of spanning trees in dynamic networks, topological changes are the faults, and having the entire network covered by a single tree, or in case of partitioned networks one tree per connected component, is the legal state. One approach to transform a non-self-stabilizing algorithm into a self-stabilizing one, is to *reset* the states of the nodes when a fault occurs, so that a new execution of the algorithm is initiated. This approach has been considered by most self-stabilizing algorithms proposed so far for the spanning tree problem, and an optimal-time solution was proposed in [4] (as a coarse-grain graph algorithm, more recently transposed into the message passing model in [6]). We refer the reader to [10] for a more general survey on self-stabilizing spanning tree algorithms. In these works, the algorithms assume that no additional fault occur during the stabilization period, which is not acceptable in highly dynamic networks.

Random Walk: A random walk is a sequence of nodes such that each node in the sequence (except the starting node) is randomly selected among the neighbors of its predecessor. Random walks have been used to solve several problems in distributed systems, such as leader election, voting, or spanning trees [7]. The idea of using random walks to compute spanning trees was first proposed by Aldous in [2], where a single random walk is considered. Anytime, the set of all covered nodes, along with the edges from which they were visited the first time, defines a random tree that spans the nodes already visited.

Mobile Agents: Mobile agents are entities that can travel across the network, and perform tasks on the underlying nodes. These agents may or may not carry their own memory, and adopt a variety of strategies to move within the network. In [5], distributed random walks of mobile agents (called *tokens* in the paper) are used. More precisely, colored tokens are annexing territories while walking within the network. Each token builds a tree (a subtree of the global spanning tree). When two tokens meet or when a token visits a node that have already been visited, the two trees are merged into one. This operation is performed by a *wave propagation*, which is a broadcast-based process that occurs along the edges of the trees. The network is assumed connected and no topological changes are allowed during the construction of the tree. Unique identifiers are also required. A related approach was proposed in [1], where mobile colored agents (equivalent to tokens) construct subtrees that are progressively merged into a final spanning tree. Whenever one agent enters the region of another, the agent that have the larger color progressively takes control of the nodes and eventually destroys the other agent. The advantage of this gradual process is that it avoids the wave propagation. However, unique identifiers are still required to generate the colors and some global information (an upper bound in the cover time of the random walk) is needed to regenerate an agent. Finally, the approach does not tolerate frequent topological events.

In comparison to these approaches, the one we propose does not require stable periods or unique identifiers (nor any global information). This is, to the best of our knowledge, the first attempt in this direction.

3   Network Model and Assumptions

We represent the network as an evolving graph $\mathcal{G} = \{G_1, G_2, ...\}$, all elements of which correspond to snapshots of the topology, and the transitions between them bijectively reflect the occurrence of one, or several simultaneous topological events (appearance or disappearance of edges). More elaborate variants of evolving graphs can be found in the original paper [9]. However, this basic variant is suitable enough for our purpose.

At a given moment, the network is therefore represented by an undirected simple graph $G_i = (V, E_i)$, where the set of nodes V is assumed to be constant, while the set of edges varies without restriction from one G_i to the next. The temporal span of each G_i is arbitrary and in particular, it is not bounded (whether from above or below). We do not require the existence of unique identifiers for the nodes, but we assume they are able to distinguish between their incident edges and assign a local value to them (thus, an edge typically has two values, one on each side). Note that in practice, especially in a wireless network, this feature would require unique identifiers to be implemented. It is however a weaker assumption from a theoretical standpoint. Further, it is more natural to think of our algorithm without identifiers.

3.1   Computational Model

We consider a coarse-grain interaction model akin to *population protocols* [3] or *graph relabeling systems* [11]. In these models a computation step is an atomic pairwise interaction. Precisely, a computation step takes as input the state of a pair of nodes (together with their common edge), and modifies these states according to some rule. For example, the rule $\underset{0}{\overset{inside}{\bullet}} \underset{0}{\overset{outside}{\bullet}} \longrightarrow \underset{2}{\overset{inside}{\bullet}} \underset{1}{\overset{inside}{\bullet}}$ may represent the construction of a rooted spanning tree in a static network from some distinguished `inside` node. We assume in general that two interactions can occur in parallel so long as they are disjoint (they do not imply a common node). The way interactions are selected, that is, the *scheduling*, is typically not a part of the algorithm (e.g. it can be adversarial with some constraints, or probabilistic, or result from some finer-grain interaction). The general properties we establish on our algorithm are insensitive to these concerns. Note that the guard of a rule (left part) may represent two nodes in a same state. In this case, despite the absence of unique identifiers, symmetry is broken by the application of the rule – however, the choice of what role is played by each node is not controlled by the algorithm (it is up to the scheduler).

Dealing with a dynamic graph (the usual population protocols deal with static graphs), we consider another type of operation in addition to pairwise interaction. This operation, triggered by topological events, consists in updating the state of a node immediately after one of its edges disappears. As such, an algorithm can associate reactive operations to the loss of a link.

4   The Spanning Forest Algorithm

Informally, the algorithm is based on three operations on tokens: *circulation*, *merging*, and *regeneration*, which aim at maintaining exactly one token per tree. Initially, every node forms a tree of its own and is the root of that tree (it has the token). When two token owners interact over a common edge, their tokens are merged into one and their common edge is added to the tree (*merging* rule r_1, see Figure 1 below). The parent-child relation is set accordingly. The rest of the time, each token performs a random walk along the edges of its own tree (*circulation* rule r_2, see Figure 2 below) in search of new merging opportunities; parent-child relations are flipped as the circulation proceeds, so that a node can always tell, locally, which edge leads to the token. Whenever an edge of the tree disappears, the node on the child side regenerates a token (*regeneration* rule r_a, see Figure 3 below), which re-enables its orphan tree to keep running the process.

4.1   State Space and Initialization

At any time, the state of the system is fully described by two functions: one function for the state of the nodes $\lambda : V \to \{T, N\}$, where T means this node has a token, while N means it does not; and one function for the state of the edges *locally to both endpoints* $\lambda : V \times E_i \to \{0, 1, 2\}$, where E_i is the current set of edges. The domain of both functions being different and non-ambiguous from the context, we authorize a unique symbol λ to denote them. State 0 for an edge means it does not belong to a tree. States 1 or 2 mean it does, and the local direction is from child to parent (state 1) or from parent to child (state 2). Hence, an edge whose state is 1 at one end, must be in state 2 at the other end. Notice that one bit of information is enough to encode the state of a node, and two bits, locally at each node, are sufficient for an edge.

Initialization: Given the first graph $G_0 = (V, E_0)$, we set $\lambda(v) = T$ for all $v \in V$. We also set $\lambda(v, e) = 0$ and $\lambda(u, e) = 0$ for all $e = (u, v) \in E_0$. In words, every node initially holds a token and none of the edges belong to a tree.

4.2   State Transitions

The evolution of the process is determined by two sources of events: topological events (i.e., appearance or disappearance of an edge) and computational events (i.e., pairwise interaction). We specify both separately. Keep in mind the principle presented here is intended to be extremely general, and several important questions, like priority among rules or the role played by each node in the rule, are deliberately set aside at this point. (They are discussed shortly after.)

Transitions Induced by Pairwise Interaction

Merging Rule: Given two nodes u and v involved in an interaction over an edge $e = (u,v)$, the operation is specified as follows. If $\lambda(u) = T$ and $\lambda(v) = T$, then set $\lambda(v) = N$, $\lambda(v,e) = 1$, and $\lambda(u,e) = 2$. This rule, called *merging* rule (r_1), can be represented graphically as shown in Figure 1.

$$r_1 : \overset{T}{\bullet}\!\!-\!\!-\!\!-\!\!-\!\!\overset{T}{\bullet} \longrightarrow \overset{T}{\bullet}\!\!\!\!\leftarrow\!\!-\!\!-\!\!-\!\!\overset{N}{\bullet}$$

Fig. 1. Merging rule (graphical representation)

Circulation Rule: Given two nodes u and v involved in an interaction over an edge $e = (u,v)$, the operation is specified as follows. If $\lambda(u) = T$ and $\lambda(v) = N$ and $\lambda(u,e) = 2$, then set $\lambda(u) = N$, $\lambda(v) = T$, $\lambda(v,e) = 2$, and $\lambda(u,e) = 1$. This *circulation* rule (r_2) can be represented graphically as shown on Figure 2.

$$r_2 : \overset{T}{\bullet}\!\!\!\!\leftarrow\!\!-\!\!-\!\!-\!\!\overset{N}{\bullet} \longrightarrow \overset{N}{\bullet}\!\!-\!\!-\!\!-\!\!-\!\!\!\!\overset{T}{\bullet}$$

Fig. 2. Circulation rule (graphical representation)

Transitions Induced by Topological Events

Given two consecutive graphs G_i and G_{i+1} in $\mathcal{G}$, the transition from one to the other induces the following updates on the states of the system.

Appearance of an Edge: For all $e = (u,v) \in E_{i+1}\backslash E_i$, both $\lambda(u,e)$ and $\lambda(v,e)$ are set to 0. In words, new edges are initialized with state 0 on both sides.

Disappearance of an Edge: For all $e = (u,v) \in E_i\backslash E_{i+1}$, if $\lambda(u,e) = 1$, then set $\lambda(u) = T$; else if $\lambda(v,e) = 1$, then set $\lambda(v) = T$. In words, if a node loses the edge leading to its parent, it regenerates a token immediately. This rule, called *regeneration* rule (r_a), can be represented graphically as shown on Figure 3.

$$r_a : \overset{N}{\bullet}\!\!\overset{\text{off}}{\longrightarrow} \longrightarrow \overset{T}{\bullet}$$

Fig. 3. Regeneration rule (graphical representation)

An example execution sequence of the algorithm is provided on Figure 4.

4.3   Correctness

In this Section we establish some properties of the spanning forest algorithm, namely, that there is always exactly one root (token) in every tree, and no cycle can possibly occur.

Lemma 1. *At any time, there is at least one token per tree.*

Proof. The lemma holds initially, when every node is the root of its own tree. Now observe that both *merging* and *circulation* operations perserve this property. Indeed, the application of r_1 merges two trees but suppresses one token, while r_2 just moves a token within the underlying tree. We can thus focus on the disappearance of edges. Whenever an edge e disappears, either e did not belong to a tree or it did. If it did not, nothing has to be done. If it did, then this tree is now split into two trees, one of which is left token-less. By rule r_a, whose application is immediate, a token is regenerated on the orphan side of that edge (edge state 1). If several such edges had disappeared simultaneously, the same mechanism would have occurred relative to each fragment. □

Lemma 2. *At any time, there is at most one token per tree.*

Proof (By contradiction). The only rule leading to the creation of a token is r_a. Since the lemma holds initially, the presence of more than one token in a tree must result from one of these events:

1. Rule r_a was applied despite the existence of another token in the tree.
2. Rule r_a was applied several times simultaneously in the tree.

In the first case, the contradiction stems from the fact that r_a is applied on the child endpoint of a lost edge. By construction, the token is thus on the other side and the local subtree is token free. In the second case, the contradiction is slightly less direct. Let v and v' be two nodes of a same tree, both of which have applied r_a simultaneously. Three cases are possible regarding the relative position of v and v' in the tree:

1. (a) v is an ancestor of v'. This is impossible because the application of r_a by v' results from the disappearance of its parent edge.
   (b) v' is an ancestor of v. Same argument for v.
   (c) v and v' have a common ancestor. This is again impossible because the application of r_a results from the disappearance of a parent edge, therefore neither v nor v' can have an ancestor at all. □

Theorem 1. *At any time, there is exactly one token per tree.*

Proof. By Lemmas 1 and 2. □

Theorem 2. *At any time, the trees are cycle-free.*

Proof. The property holds initially. The only way an edge can be added to a tree is by means of applying r_1, which involves two tokens. By Lemma 2, there is at most one token per tree, thus at most one application of r_1 can occur at a time for a given tree, and the two tokens must belong to different trees. □

4.4   Discussion

The algorithmic principle introduced here is very general. In particular, the correctness of the properties we have considered so far does not depend on the order in which the edges are selected for interaction, nor whether some interactions should be favored over others (e.g. r_1 over r_2).

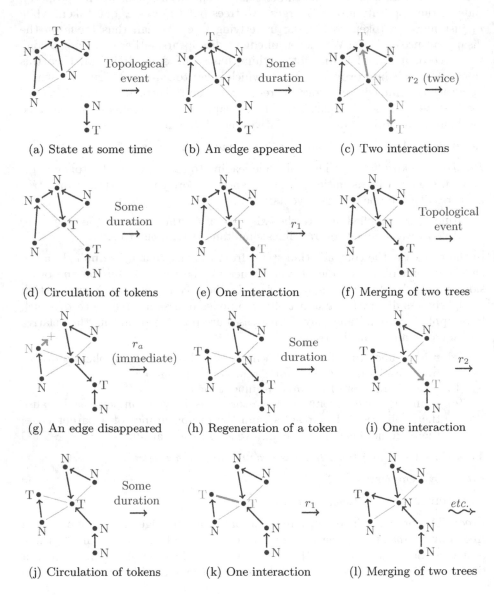

(a) State at some time   (b) An edge appeared   (c) Two interactions

(d) Circulation of tokens   (e) One interaction   (f) Merging of two trees

(g) An edge disappeared   (h) Regeneration of a token   (i) One interaction

(j) Circulation of tokens   (k) One interaction   (l) Merging of two trees

Fig. 4. A possible sequence of execution of the spanning forest algorithm

On the other hand, these aspects can have a tremendous impact on the ability of the trees to merge with each other and converge towards a single tree per connected component (remind that the network is expected to be partitioned in general).

Priority among Rules: In general, given two neighbor nodes at a given time, there might be more than one eligible rule. This was not the case with this algorithm, since r_1 and r_2 have two incompatible guards (preconditions). However, the matter is worth being discussed. Priority among the rules could be understood in a *weak* sense, enforcing the fact that a rule should not be applied by *these two* nodes if they are able to apply another rule first. Another, much *stronger* sense of priority consists in forbidding a node to apply a given rule as long as another rule is applicable *with any* of its neighbors.

Clearly, in the case of the spanning forest algorithm, merging should be preferred over circulation whenever possible. Enforcing strong priority would thus come to forbid the application of r_2 whenever r_1 can be applied. This behavior is expected to produce larger trees, but at the cost of a strong constraint on the scheduler (probing the state of an entire neighborhood prior to interaction). Without speculating on finer-grain implementations of our principle – which is not the object of this paper – we believe a strong priority mechanism remains somewhat natural in a wireless environment, where nodes routinely broadcast their state to all neighbors, in particular if we assume a synchronous communication model such as $\mathcal{LOCAL}$ or $\mathcal{CONGEST}$ [12].

Role Played by Both Nodes in an Interaction: The reader may have noticed that, in the definition of the circulation rule r_2, the guard of the rule is not tested on both sides. That is, u implicitely plays the role of the left node, and v that of the right node. As far as the present work is concerned, we do not want to impose a preferred way to solve this question, as it does not affect correctness. As a suggestion, the scheduler may select edges in a directed way (with a left node, and a right node), or the second direction systematically when an edge is selected and the rule is not applicable in the first direction.

High-Level View of the Process: Assuming the token has equal probability to move to each neighbor (in the tree), we can regard the circulation as a random walk in the tree. Further, if we assume *strong* priority enforcement between r_1 and r_2, the circulation and merging processes turn into a specific variant of coalescing random walks [7]. This point of view is the one we consider in the next section.

5   Preliminary Analysis

In this section, we study the question of how frequent the mergings are. We only provide preliminary results and some thoughts about the complete analysis of this process (which is far beyond the scope of this paper). Hence, we characterize the number of token moves expected in a stationary regime, before a merging occurs between two given trees in a static context. This value is given as a function of their size and the number of edges connecting them (called *bridges*).

5.1 Random Walks in Trees

For the sake of analysis (and with loss of generality), we look at the process of merging and circulating tokens as a system of particles that perform random walks *in trees* and coalesce whenever they *meet*. Here, the concept of meeting between two particles is defined with a special meaning. Indeed, in most coalescing particle systems, two particles are said to meet if they happen to be located at a same node, whereas in our case, they meet if they are located at both endpoints of a same edge (remind that the tokens cannot travel beyond their trees).

5.2 Bridges

Given two different trees $\mathcal{T}_1$ and $\mathcal{T}_2$, there may be some edges whose endpoints lie in $\mathcal{T}_1$ on one side, and $\mathcal{T}_2$ on the other side – we call such edges *bridges*. Figure 5 shows an example of two trees that share four bridges.

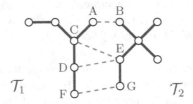

Fig. 5. Example of two trees sharing four bridges (dashed lines)

As discussed in Paragraph 4.4, the enforcement of a *strong* notion of priority between merging and circulation allows one to assume that if two tokens are located on a same bridge, then merging occurs. (This is at least true in the case of two trees, which is the one addressed here.) Hence, the probability that merging occurs is that of having both tokens located at a same bridge.

Let us denote by $Bridges(\mathcal{T}_1, \mathcal{T}_2)$ the set of edges (u, v) such that $u \in E_{\mathcal{T}_1}$ and $v \in E_{\mathcal{T}_2}$. The probability that $\mathcal{T}_1$ and $\mathcal{T}_2$ merge at a given time is equal to:

$$P_{merge(\mathcal{T}_1, \mathcal{T}_2)} = \sum_{(u,v) \in Bridges(\mathcal{T}_1, \mathcal{T}_2)} P[\lambda(u)=T \wedge \lambda(v)=T]. \tag{1}$$

5.3 Probability of Being Located at a Node

In a stationary regime, the probability for a token to be located at a given node v in a graph G (tree or not) is a well-known result in random walk theory, which only depends on the ratio between the degree of v, $d_G(v)$, and the sum of all degrees in G.

In a tree $\mathcal{T}$, the probability a node v hosts the token is thus

$$P(\lambda(v) = T) = \frac{d_{\mathcal{T}}(v)}{2|E_{\mathcal{T}}|} \tag{2}$$

where $|E_{\mathcal{T}}|$ is the size of $\mathcal{T}$. Keep in mind this value corresponds to the stationary regime (when the probabilities no more depend on the initial configuration).

5.4   Expected Merging Time in the Stationary Regime

We are interested in the mean number of steps (token moves) required to merge the two trees, assuming the walks are in a stationary regime. Moreover, as trees are bipartite graphs, if both tokens move synchronously it may happen, depending on their initial position, that they never meet. Thus, we assume here that the moves are asynchronous (i.e., one at a time). Equations 1 and 2 allow us to state that the probability for two trees $\mathcal{T}_1$ and $\mathcal{T}_2$ to merge at any step is

$$P_{merge}(\mathcal{T}_1, \mathcal{T}_2) = \sum_{\{(u,v) \in Bridges(\mathcal{T}_1, \mathcal{T}_2)\}} \frac{d_{\mathcal{T}_1}(u)}{2|E_{\mathcal{T}_1}|} \times \frac{d_{\mathcal{T}_2}(v)}{2|E_{\mathcal{T}_2}|} \tag{3}$$

which in turn gives the *expected merging time* in number of steps, as $E_{merge}(\mathcal{T}_1, \mathcal{T}_2) = P_{merge}(\mathcal{T}_1, \mathcal{T}_2)^{-1}$.

However limited, a quick look at these results teaches us some preliminary facts. First, the merging time of two trees of size n in which the nodes degrees d are fairly distributed is in $O(\frac{n^2}{nbBridges \cdot d^2})$. The d^2 term could actually be omitted if we consider that degrees are bounded by some constant (a fair assumption in most wireless networks). Whence a time of $O(\frac{n^2}{nbBridges})$ steps. Whether this time is linear or quadratic in the sizes of the trees depends on the number of bridges (e.g. merging time is linear if the number of bridges is in $O(n)$; it is quadratic if that number is constant; etc.). That in turn, depends on the networking scenario which is considered, and in particular, what *mobility model* is used.

A deeper look is required to understand the behavior of this process. We expect it to be quite difficult to analyze in the general case. Not only the algorithm involves much more than two trees in general, but it is intended to run over highly dynamic topologies, where splittings and mergings occur concurrently. In fact, the metric of interest might be different than convergence time, since the merging process is never expected to converge. A better metric here could be the average number of trees per connected component in a stationary regime.

6   Conclusion

This paper proposed a new mechanism for building and maintaining a forest of spanning trees in highly dynamic networks. The originality of the approach is that the construction is a perpetual ongoing process that takes place at the same time as the trees are used. The principle is very general and relies on token circulation techniques that turns *splittings* and *mergings* of the trees into purely localized phenomenons. After presenting the algorithm using a coarse grain interaction model, we provided some preliminary observations on the analysis of the corresponding process, regarded for the occasion as a system of coalescing random walks. A deeper analysis of this process is still far from reach, and we expect it to be technically challenging in the general case. As the process is never expected to converge, completion time is not the most relevant metric here (however its characterization in a static and connected context might already be very insightful). Characterizing the average number of trees per connected component in the stationary regime seems to be the relevant metric.

Besides analysis, an avenue of research is to transpose the algorithm into finer-grain communication models. We believe this can be done, at least in synchronous message passing models. Finally, de-randomizing the way tokens circulate (e.g. using Propp machine-like mechanisms) may lower the cover time, and possibly, speed-up the merging process. These questions are open.

References

1. Abbas, S., Mosbah, M., Zemmari, A.: Distributed computation of a spanning tree in a dynamic graph by mobile agents. In: Proc. of IEEE Int. Conference on Engineering of Intelligent Systems (ICEIS), pp. 1–6 (2006)
2. Aldous, D.J.: The random walk construction of uniform spanning trees and uniform labelled trees. SIAM J. Discret. Math. 3(4), 450–465 (1990)
3. Angluin, D., Aspnes, J., Diamadi, Z., Fischer, M., Peralta, R.: Computation in networks of passively mobile finite-state sensors. Distributed Computing 18(4), 235–253 (2006)
4. Awerbuch, B., Kutten, S., Mansour, Y., Patt-Shamir, B., Varghese, G.: Time optimal self-stabilizing synchronization. In: Proc. of the 25th ACM Symp. on Theory of Computing (STOC), New York, USA, pp. 652–661 (1993)
5. Baala, H., Flauzac, O., Gaber, J., Bui, M., El-Ghazawi, T.: A self-stabilizing distributed algorithm for spanning tree construction in wireless ad hoc networks. Journal of Parallel and Distributed Computing 63, 97–104 (2003)
6. Burman, J., Kutten, S.: Time optimal asynchronous self-stabilizing spanning tree. In: Pelc, A. (ed.) DISC 2007. LNCS, vol. 4731, pp. 92–107. Springer, Heidelberg (2007)
7. Cooper, C., Elsässer, R., Ono, H., Radzik, T.: Coalescing random walks and voting on graphs. In: Proc. of the 31st ACM Symp. on Principles of Distributed Computing (PODC), pp. 47–56 (2012)
8. Fall, K.: A delay-tolerant network architecture for challenged internets. In: Proc. of Int. Conf. on Applications, Technologies, Architectures, and Protocols for Computer Communications (SIGCOMM), pp. 27–34 (2003)
9. Ferreira, A.: Building a reference combinatorial model for MANETs. IEEE Network 18(5), 24–29 (2004)
10. Gaertner, F.C.: A Survey of Self-Stabilizing Spanning-Tree Construction Algorithms. Technical report, EPFL (2003)
11. Litovsky, I., Métivier, Y., Sopena, E.: Graph relabelling systems and distributed algorithms. In: Handbook of Graph Grammars and Computing by Graph Transformation, vol. III, pp. 1–56. World Scientific Publishing (1999)
12. Peleg, D.: Distributed computing: a locality-sensitive approach. Society for Industrial and Applied Mathematics (2000)

Evaluating and Bounding Operations Performance in Heterogeneous Sensor and Actuator Networks with Wireless Components

José Cecílio, João Costa, Pedro Martins, Nickerson Ferreira, and Pedro Furtado

University of Coimbra,
Coimbra, Portugal
{jcecilio,jpcosta,pmom,nickerson,pnf}@dei.uc.pt

Abstract. When wireless sensor networks are introduced in industrial settings, they will be part of a much larger heterogeneous sensor and actuation network that includes those sub-networks together with Ethernet cabled Programmable Logic Controllers (PLCs) and control stations. Performance issues arise in such systems. We propose mechanisms to measure, verify, control and debug expected operation time bounds in such heterogeneous sensor and actuator networks.

Keywords: WSAN, Heterogeneity, Operations Timing Guarantees.

1 Introduction

A sensor and actuation infrastructure is typically a heterogeneous environment, consisting of software running on computer-boards, control stations, and multiple networked sensors and actuators. It can be composed of multiple parts: wireless sensor sub-networks will do the sensing and actuation over parts of the plant, and they will be part of a much larger heterogeneous network that includes those sub-networks together with Ethernet cabled computer-boards and control stations.

Network and control engineers deploying and putting to work such a system will need to control and verify whether time bounds are within acceptable values. There are many options involved, and there is the need for both planning and testing. For instance, with the appropriate Time Division Multiple Access (TDMA) schedule, it is possible to implement simple closed-loops within a single wireless sensor network, with guaranteed behavior. On the other hand, it is also possible to have situations where a closed-loop is required with sensing happening in a wireless sensor network, decision logic in a computer, and actuation happening in another wireless sensor network. In this case the control loop will traverse multiple, most probably non-real-time hardware and software systems, nevertheless the control loop will still need to be under verifiable expected time bounds.

In this paper we define measures and metrics for surveillance of expectable time bounds and an approach for debugging, using tools and mechanisms to explore and report problems. This surveillance can be used in any distributed system to verify

J. Cichoń, M. Gębala, and M. Klonowski (Eds.): ADHOC-NOW 2013, LNCS 7960, pp. 111–122, 2013.

performance compliance. Assuming that we have monitoring or closed-loop tasks with timing requirements, this allows users to constantly monitor timing conformity.

We will also show experimental results concerning bounds and debugging tool. We create a simulation environment where we introduce some random delays in the messages, to demonstrate how the debugging tool works and its usability. These practical experimental results will show the values obtained in that environment, proving that the network and control engineers will have the adequate measures, metrics and tools to plan and debug adequately.

The rest of the paper is organized as follows: section 2 reviews related work; section 3 describes the heterogeneous monitoring and control architecture. It explains the architecture components considered in this work. Section 4 defines measures and metrics used by our approach to evaluate and bound operations performance in heterogeneous sensor networks. Section 5 describes the addition of debugging modules to the heterogeneous monitoring and control architecture. The debugging node component and operation performance monitor component are described. It is also described how the performance information is collected and processed. An example of operation performance monitor UI is presented, which allows users to evaluate the performance. In section 6 we define the experimental setup and section 7 shows results obtained and the conclusions. Lastly, section 8 concludes the paper.

2   Related Work

Our proposal is an approach to verify and control expected operation time bounds in a heterogeneous system such as Sensor and Actuator Networks (SANs) with wireless sensor sub-networks. Related work includes monitoring tools, studies on performance, latency and delay analysis.

Monitoring is crucial to control operations performance of a sensor and actuator network, and tools are needed to measure specific parameters of performance. These methods are used to monitor all necessary parameters that assure that all functionalities are working as expected.

There already exist some tools to monitor network performance in wireless sensor networks. Sympathy [1] is a monitoring system in which sensor nodes are supplied with specific monitoring software, and periodically send specific parameters to a sink node. The mechanism developed is aimed at detecting and debugging failures in sensor networks and is specifically designed to be used with data gathering applications. All evaluation is done at the sink. When a failure occurs, Sympathy triggers failure localization and reporting.

DiMo [2] presents a distributed and scalable solution for monitoring the nodes and the topology, along with a redundant topology for increased robustness. The aim is to operate in safety-critical wireless sensor networks, where all sensor nodes must be up and functional. This tool provides two functionalities, network topology maintenance and network health status monitoring. Monitoring is done by using observer nodes that check the reception of heartbeats within a certain monitoring time. If not, the observer sends a node missing message to the sink. There is always one observer for each node. The sink is always aware of which nodes are being observed in the network, and therefore always knows which nodes are up and running.

Our work is related to these ones in what concerns network performance monitoring. However our approach is oriented towards high-level operations, where we propose operation time failure detection mechanisms.

There are also some works addressing latency and delays for WSNs [3, 4, 5, 6]. These works have considered the extension of the Network Calculus methodology [7] to WSNs. End-to-end delay bounds for real-time flows in WSNs have been studied in [8]. The authors propose closed-form recurrent expressions for computing the worst-case end-to-end delays, buffering and bandwidth requirements across any source-destination path in the cluster-tree assuming error free channel. They propose and describe a system model, an analytical methodology and software tool that permits the worst-case dimensioning and analysis of cluster-tree WSNs.

There also exist approaches to monitor latencies and delays in distributed control systems based on wired components. The authors of [9] and [10] show two studies on modeling and analyzing latency and delay stability of network control systems. The authors of [11] modeled end-to-end time delay dynamics for the internet using system identification tools. The study in [12] presents an analytical performance evaluation of the switched Ethernet with multiple levels from timing diagram analysis, and experimental evaluation from an experimental testbed with a networked control system. These works assume a wired network. However, a distributed control system may include wireless and wired parts. Our proposed approach includes application-level, end-to-end message losses, message delivery delays, commands delays, specification of bounds to provide high degree of performance in all components of the SAN (wired and wireless).

3   Heterogeneous Monitoring and Control Architecture

With the evolution and increased adoption of wireless sensor technology and networks, and their easier and much cheaper deployment, there is a current solution to partially replace or complement the existing infrastructure. When deployed in industrial settings, it will build a heterogeneous sensor and actuation network. The global network can be composed by wireless sensor networks (WSN), PLC, computers and control stations (Fig. 1).

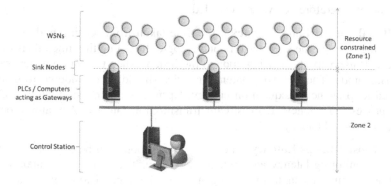

Fig. 1. General Architecture

One way to try to provide operation timing guarantees is to deploy WSN sub-networks with real-time specific algorithms that would include at least completely pre-planned synchronous time-division mechanisms. However, sensor and actuation infrastructures are typically heterogeneous systems. The system includes specific software parts, some possibly offering real-time, while others do not. In that context, deployment of sensor and actuator networks requires a high level of reliability control concerning measures such as latencies, delays and message losses. It is important to ensure that monitoring and control loops will still be under specified time bounds. For instance, control engineers can specify operation time bounds to deliver the data to or from supervision controllers.

4   Measures and Metrics

Operation timing issues in terms of monitor and closed-loops control can be controlled with the help of two measures, which we denote as Latency and Delay of Periodic Events.

Latency consists of the time required to travel between a source and a destination. Sources and destinations may be any site in the distributed system. For instance, the latency can be measured from a WSN leaf node to a sink node, or from a WSN sensing node to a computer, control station or backend application.

Delay of Periodic Events consists of the extra time taken to receive a message with respect to the predefined periodic reception instant.

Given the above time measures, we define metrics for sensing and control. The metrics allow us to quantify timing behavior of monitoring and closed-loops.

Monitoring Latency – the time taken to deliver a value from sensing node to the control station, for display or alarm computation. The following latency metrics are therefore all considered: Acquisition latency, Transmission latency, Control Station processing latency, End-to-end latency.

Monitoring Delay – the amount of extra time from the moment when a periodic operation was expected to receive some data to the instant when it actually received. When users create a monitoring task, they must specify a sensing rate. The control station expects to receive the data at that rate, but delays may happen in the way to the control station, therefore delays are recorded.

Event-Based Closed-Loop Latency – the time taken from sensing node to actuator node passing through the supervision control logic. It will be the time taken since the value (event) happens at a sensing node to the instant when the action is performed at the actuator node. The following latency metrics should be considered to determine the closed-loop latency: Acquisition latency, Upstream transmission latency, Control Station processing latency, Downstream transmission latency, Actuator processing latency, End-to-end latency.

Periodic Closed-Loops Latency – periodic closed-loops can be associated with two latencies: Monitoring latency and Actuation latency. The Monitoring latency can be defined as the time taken from sensing node to the supervision control logic. The Actuation latency corresponds to the time taken to reach an actuator. We also define an end-to-end latency as the time from the instant when a specific value is sensed and the moment when an actuation is done which incorporates a decision based on that value. The following latency metrics should be considered to determine the closed-loop latency for synchronous or periodic closed-loops: Acquisition latency; Upstream transmission latency; Wait for the actuation instant latency; Control Station processing latency; Downstream transmission latency; Actuator processing latency; End-to-end latency;

Closed-Loop Delays – In periodic closed-loop operations, actuation is expected within a specific period. However, operation delays may occur in the control station and/or command transmission. The closed-loop delay is the excess time.

In event-based closed-loops, there can be monitoring delays. This means that a sample expected every x time units may be delayed.

5   Addition of Debugging Modules to Monitoring and Control Architecture

In the previous sections we defined measures and metrics useful to evaluate operation performance. In this section we will discuss how to add debugging modules to the monitoring and control architecture.

The architecture can be divided into two main parts: nodes and control station. To add debugging functionalities, we need to add a Debugging Module (DM) to nodes and a Performance Monitor Module (PMM) to the control station. The debugging module collects information from operations in nodes, then formats and forwards information to the Performance Monitor module. The Performance Monitor gathers the status information coming from nodes, stores it in a database and processes it according to bounds defined by the user.

In the next sections we describe how the Debugging module and Performance Monitor module work.

5.1   The Debugging Module

The Debugging module (DM) stores all information concerning node operation (e.g. execution times, battery level) and messages (e.g. messages received, messages transmitted, transmission fails, transmission latencies). This information is stored inside the node. It can be stored either in main memory, flash memory or other storage device.

DM is an optional module that can be activated or deactivated. It generates a debugging report, either by request or periodically, with a configurable period.

DM has two modes of operation:

- Network debugging – the DM runs in all nodes and keeps the header information of messages, where it adds timestamps corresponding to arrive and departure instants. After, this information is sent periodically or by request to the Performance Monitor (described in the next sub-section), which is able to calculate metrics. This operation mode may be deactivated in constrained devices, because it consumes resources such as memory and processing time.

- High-level operation debugging – instead of collecting, storing and sending all information to the Performance Monitor, the DM can be configured to only add specific timestamps to messages along the path to the control station.

Assuming a monitoring operation in a distributed control system with WSN sub-networks, where data messages are sent through a gateway, the DM can be configured to add timestamps in the source node, sink node, gateway and control station. Fig. 2 illustrates nodes, gateways and a control station in that context.

The approach assumes that WSN nodes are clock synchronized. However, they may not be synchronized with the rest of the distributed control system. Gateways, computers and control stations are also assumed clock synchronized (e.g. the NTP protocol can be used).

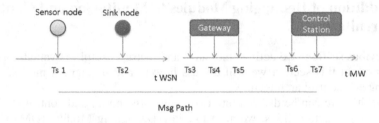

Fig. 2. Message path – example

In Fig. 2, the DM starts by adding a generation timestamp (source timestamp) in the sensor node ($Ts1$). When this message is received by the sink node, it adds a new timestamp ($Ts2$) and indicates to the gateway that a message is available to be written in the serial interface. Upon receiving this indication, the gateway keeps a timestamp that will be added to the message ($Ts3$), and the serial transmission starts. After concluding the serial transmission, the gateway takes note of the current timestamp ($Ts4$) and adds $Ts3$ and $Ts4$ to the message.

Upon concluding this process and after applying any necessary processing to the message, the gateway adds another timestamp ($Ts5$) and transmits it to the control station. When the message is received by the control station, it adds a timestamp ($Ts6$), processes the message and adds a new timestamp ($TS7$), which indicates the instant of message processing at the control station was concluded. After that, at the control station, the Performance Monitor module (described in the next section) receives the message and, based on the timestamps that come in the message, it is able to calculate metrics.

If there is only one computer node and the control station, there will only be $Ts1$, $Ts6$ and $Ts7$.

5.2   The Performance Monitor Module and UI

In this sub-section we describe the Performance Monitor module (PMM), which debugs operations performance in the heterogeneous distributed system. The PMM stores events (data messages, debug messages), latencies and delays into a database. It collects all events when they arrive, computes metric values, classifies events with respect to bounds, and stores the information in the database. Bounds should be configured for the relevant metrics.

Assuming the example shown in Fig. 2, PMM collects the timestamps and processes them it to determine partial and end-to-end latencies.

The following partial latencies are calculated:

- WSN upstream latency ($Ts2 - Ts1$)
- WSN to Gateway interface latency ($Ts4 - Ts3$)
- Middleware latency ($Ts6 - Ts5$)
- Control station latency ($Ts7 - Ts6$)
- End-to-end (($Ts2 - Ts1$) + ($Ts4 - Ts3$) + ($Ts6 - Ts5$) + ($Ts7 - Ts6$))

After concluding all computations, PMM stores the following information in the database: Source node id, Destination node id, Type of message, MsgSeqId, [Timestamps], partial latencies, end-to-end latency. This information is stored for each message, when the second operation mode of debugging is running. When the first operation mode of the debugging component is running, a full report with link-by-link information and end-to-end information is also stored.

The PMM user interface shows operations performance data, and alerts users when there is a problem detected by metric exceeds bounds. Statistical information is also shown and is updated for each event that arrives or for each timeout that occurs.

Fig. 3 shows a screenshot of PMM. We can see how many events (data messages) arrived in-time, out-of-time (with respect to defined bounds), and the corresponding statistical information. This interface also shows a pie chart to give an overall view of the performance.

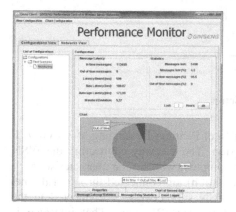

Fig. 3. PMM user interface   **Fig. 4.** PMM user interface – event logger

Fig. 4 shows the event logger of PMM. This logger shows the details on failed events. A list of failures is shown and the user can select one of each and see all details, including the latency in each part of the distributed control system.

When a problem is reported by the PMM, the user can explore the event properties (e.g. delayed messages, high latencies) and find where the problem occurs. If a problem is found and the debugging report is not available at the PMM, nodes are requested to send their debugging report. If a node is dead, the debugging report is not retrieved and the problem source may be due to the dead node. Otherwise, if all reports are retrieved, the PMM is able to detect the message path and check where it was discarded or where it took longer than expected.

PMM allows users to select one message and see all details, including the latency in each part of the distributed control system.

6   Experimental Setup

In this experimental section we use a testbed prototype to show that our approach is useful to monitor and debug operations performance in the whole SAN (WSN nodes, gateways, control stations and end-user applications).

In the testbed there is a WSN sub-network, which is a planned network designed to provide performance control, gateways (computer nodes) and control stations. Fig. 5 shows a sketch of our setup.

The WSN sub-network includes 12 TelosB nodes organized hierarchically in a 1-1-2 tree and one sink node (composed by one TelosB node and one computer that acts as gateway of the sub-network). The setup also includes a control station that receives the sensor samples, monitors operations performance using bounds, and collects data for testing the debugging approach.

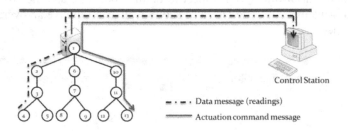

Fig. 5. Setup

The control station is a computer with an Intel Pentium D, running at 3.4 GHz. It has 2 GB of RAM and an Ethernet connection. The gateway connecting the WSN sub-network to the cabled network is another computer with similar characteristics.

All computer nodes (gateway and control station) are connected through Ethernet cables and GigaBit network adapters. The WSN sub-network is connected to the gateway using the serial interface provided by TelosB nodes. That interface is configured to operate at 460800 baud/second.

All computers run Linux OS and have specific components developed using Java to do specific tasks. For instance, the gateway computer has a gateway software component to read data from the serial interface and send it to the control station, and to receive messages destined for the WSN and deliver them through the serial interface. The control station has a software component which implements remote configuration and performs functionalities such as monitoring and closed-loop control.

The WSN sensor nodes run Contiki OS and generate one message per time unit with a specified sensing rate. Each message includes data measures such as temperature and light. GinMAC [13] is used at the MAC layer by the WSN nodes.

7   Testing Bounds and the Performance Monitoring Tool

To exercise the use of bounds and debugging, we created a monitoring operation and introduced a "liar" node which injects 10 ms of delay in the first of every two consecutive messages that travel through it. Using the setup shown in Fig. 5, we will replace node 3 by the "liar" node (Fig. 6).

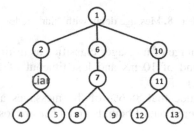

Fig. 6. Setup with "liar" node

Moreover, to simulate some losses in the network, we changed the node 4 configuration to consecutively send one message and discard the next message. This allows us to simulate 50% of message losses.

Fig. 7 and Fig. 8 show the results concerning message delays. Fig. 7 reports values concerning delay without the "liar" node, while Fig. 8 reports delays after replacement of node 3 by the "liar" node.

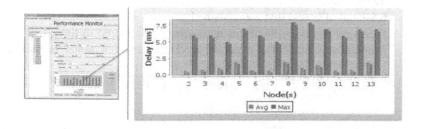

Fig. 7. Message delay without "liar" node

From Fig. 7 we can conclude that consecutive messages arrive at the control station, in average, within 0.5 to 2 ms. This value can grow up to a maximum of 8 ms.

After introducing our "liar" node and running the monitoring operation for 24 hours, we obtained the chart of Fig. 8. This figure shows that the delay of nodes 4 and 5 increased. These nodes send their messages to the control station passing through the "liar" node, which is node 3. In this case, we can see that the delay of two consecutive messages increased, in average, to 12 ms, and up to a maximum of 20 ms.

Using the PM proposed in this paper, we can also define bounds for the message delay. Assuming that each message should arrive at the control station within a maximum delay of 10 ms, we can define a delay bound and analyse the results.

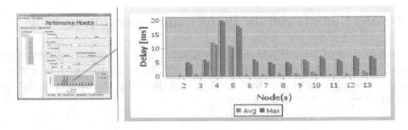

Fig. 8. Message delay with "liar" node

Fig. 9 shows the percentage of messages classified as in-time, out-of-time and lost, according to a delay bound of 10 ms and lost timeout of 1s (the timeout when a message is considered lost).

From Fig. 9 we conclude that 88.6% of the messages are delivered within the bound, but 6.8% are delivered out of bounds and 4.5% are lost. These numbers are as expected:

- Messages lost – there are 12 nodes sending data messages, node 4 fails one in every two messages. That results in $\dfrac{1}{12*2}$ losses, which agrees with the result of 4.5% losses that was obtained.
- Messages out of bound – there are 12 nodes sending data messages, node 4 sends only half of its messages and half of them arrive delayed. Concerning node 5, half of its messages arrive delayed. That results in $\dfrac{1}{12*3}+\dfrac{1}{12*2}$ messages out of bound, which agrees with the result of 6.8% out of bound that was obtained.
- Messages in time – 88.6% of messages are delivered in time, that results from the total number of expected received messages minus the number of losses and out of bound $\left(1-\left(\dfrac{1}{12*2}+\dfrac{1}{12*3}+\dfrac{1}{12*2}\right)\right)$.

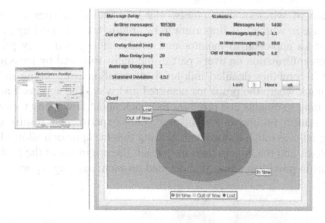

Fig. 9. Message classification according delay bound

The user can also explore event properties (in this case, delays) and find where the problem occurred. For instance, the user interface includes per node evaluation such as the one in Fig. 10, which shows which node(s) is failing.

From Fig. 10 we can conclude that node 4 is responsible for the losses represented in Fig. 9. It is losing 50% of the messages, as expected.

Fig. 10 also shows that the delay bound is not met by nodes 4 and 5. In this case, further debugging allows the user to identify the path of each message and to check where it took longer than expected.

For instance, if we explore the path and delay parts of node 5, we can conclude that messages sent by that node are waiting, in average, 10 ms in the transmission queue of node 3 (our "liar" node), which is greater than the expected average delay value of 2 ms for node 3 sending messages to the control station seen in Fig. 7.

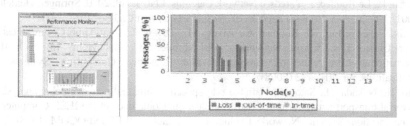

Fig. 10. Message classification according to delay bounds per node

8   Conclusion

In order to make sensor and actuator networks (SANs) more reliable in practical contexts with constraints such as latencies and delay bounds, there is a need for approaches to help a user to correctly evaluate and debug the performance problems in those SANs. This is especially true in heterogeneous systems with non real-time components, where it is very important to determine where and why problems occur.

The introduction of wireless sensor networks into industrial sites has created one heterogeneity case in which it is important to be able to track and debug problems.

We proposed an approach to monitor and debug latency and delay problems. The approach collects information on every part of the paths followed by messages and also on delays, and it combines detailed link-level information with timing information to allow users to track where the problems occurred and why. The approach also classifies messages according to bounds and provides feedback about how many messages arrived at the control station in-time or out-of-time. Our experimental results consisted in injecting timing failures into a testbed and showing that the approach allowed the users to detect the problems and track them until they reach the originator of the problem. Future work will consist in providing automated detailed problem tracking reports.

References

1. Ramanathan, N., Chang, K., Kapur, R., Girod, L., Kohler, E., Estrin, D.: Sympathy for the Sensor Network Debugger. Presented at ACM Sensys (2005)
2. Meier, A., Motani, M., Siquan, H., Künzli, S.: DiMo: Distributed Node Monitoring in Wireless Sensor Networks. Presented at 11th ACM International Conference on Modeling, Analysis and Simulation of Wireless and Mobile Systems (MSWiM 2008), Vancouver (2008)
3. Koubaa, A., Alves, M., Tovar, E.: Modeling and Worst-Case Dimensioning of Cluster-Tree Wireless Sensor Networks. In: 27th IEEE International RealTime Systems Symposium, RTSS 2006, pp. 412–421 (2006)
4. Lenzini, L., Martorini, L., Mingozzi, E., Stea, G.: Tight end-to-end per-flow delay bounds in FIFO multiplexing sink-tree networks. Performance Evaluation 63(9-10), 956–987 (2006)
5. Roedig, U., Gollan, N., Schmitt, J.: Validating the Sensor Network Calculus by Simulations. In: Network 2006 (2006)
6. Schmitt, J., Zdarsky, F., Roedig, U.: Sensor Network Calculus with Multiple Sinks. In: Proceedings of IFIP NETWORKING 2006 Workshop on Performance Control in Wireless Sensor Networks, pp. 6–13 (2006)
7. Thiran, P., Le Boudec, J.-Y. (eds.): Network Calculus. LNCS, vol. 2050. Springer, Heidelberg (2001), http://ica1www.epfl.ch/PS_files/netCalBookv4.pdf
8. Jurcik, P., Severino, R., Koubaa, A., Alves, M., Tovar, E.: Real-Time Communications Over Cluster-Tree Sensor Networks with Mobile Sink Behaviour. In: 14th IEEE International Conference on Embedded and RealTime Computing Systems and Applications, pp. 401–412 (2008)
9. Sato, K., Nakada, H., Sato, Y.: Variable rate speech coding and network delay analysis for universal transport network. In: Seventh Annual Joint Conference of the IEEE Computer and Communcations Societies Networks Evolution or Revolution, IEEE INFOCOM (1988)
10. Wu, J., Deng, F.Q., Gao, J.: Modeling and stability of long random delay networked control systems. In: International Conference on Machine Learning and Cybernetics, vol. 2, pp. 947–952 (2005)
11. Kamrani, E., Mehraban, M.H.: Modeling Internet Delay Dynamics Using System Identification. In: IEEE International Conference on Industrial Technology, pp. 716–721 (2006)
12. Lee, S., Lee, K., Lee, M., Harashima, F.: Integration of mobile vehicles for automated material handling using Profibus and IEEE 802.11 networks. IEEE Transactions on Industrial Electronics (2002)
13. Suriyachai, P., Brown, J., Roedig, U.: Time-Critical Data Delivery in Wireless Sensor Networks. In: Rajaraman, R., Moscibroda, T., Dunkels, A., Scaglione, A. (eds.) DCOSS 2010. LNCS, vol. 6131, pp. 216–229. Springer, Heidelberg (2010)

MidSN – A Middleware for Uniform Configuration and Processing over Heterogeneous Sensor and Actuator Networks

José Cecílio, João Costa, Pedro Martins, Nickerson Ferreira, and Pedro Furtado

University of Coimbra,
Coimbra, Portugal
{jcecilio,jpcosta,pmom,nickerson,pnf}@dei.uc.pt

Abstract. Applications for Sensor and Actuator Networks (SANs) are being spread to areas in which easy configuration by non-programmers will enhance acceptance, and also contexts where reconfiguration is needed during the application lifespan. Wireless SAN (WSAN) may include wired or wireless devices, computers and control stations arranged in a heterogeneous distributed system. Instead of assuming that embedded device nodes (e.g. a TelosB mote) and control station(s) (e.g. computers running Linux) are disparate entities with their own programming and processing model, it should be viewed as a single heterogeneous distributed system, offering more uniformity, simplicity and flexibility. In this paper we propose MidSN - an approach to hide heterogeneity and offer a single common configuration and processing component for all nodes of that heterogeneous system. This advances the current state-of-the-art, by providing a lego-like model whereby a single simple but powerful component is deployed in any node regardless of its underlying differences and the system is able to remotely configure and process data in any node in a most flexible way, since every node (including for instance computer nodes) has the same uniform API, processing and access functionalities.

Keywords: WSAN, Heterogeneity, (Re)Configuration.

1 Introduction

Wireless Sensor Network (WSN) embedded device nodes such as the TelosB or MicaZ have limited amounts of memory, processing power and energy, therefore they typically run small Operating Systems for embedded devices (e.g. TinyOS, Contiki). They are therefore very different platforms from computer nodes, and frequently they do not run an IP communication stack either. The devices communicate wirelessly and one or more is connected to the internet through the USB adapter and a serial read/write driver as interface, with some form of gateway software above that.

In many industrial contexts the software infrastructure for enterprise networks needs data coming from nodes that can be computers or pervasive devices, such as WSNs, and some contexts also need closed loop control over heterogeneous systems. The inclusion of pervasive devices in industrial control and monitor applications

J. Cichoń, M. Gębala, and M. Klonowski (Eds.): ADHOC-NOW 2013, LNCS 7960, pp. 123–135, 2013.

provides flexibility and cost savings when compared to entirely cabled deployments. In a real industrial setup there will typically coexist wired sensors, wireless sensor (WSN) and wired backbone nodes, forming a heterogeneous programmable distributed system.

An important research issue in that context is how to provide interoperability between different nodes and provide a single configuration and data processing model in that kind of distributed system that handles different realizations, where the same operations can run without any custom programming over different hardware or on different components of the system (sensors, sink nodes or computers), controlling different parts of the system. In this context we propose MidSN, a model to enable such vision. By considering a single node component that can be installed in any node, including nodes outside of the WSN, we ensure that all nodes will have at least a uniform configuration interface (API), important remote configuration and processing capabilities without any further programming or gluing together.

As discussed later in the related work section, MidSN improves the currents state-of-the-art in WSN macro-programming and middleware because, on one hand, it provides more node-focused flexibility than global programming approaches such as TinyDB, while on the other hand it relieves users from having to code extensively or at least from having to code separately parts of the system such as computers, WSN nodes and glue between those.

Figure 1 depicts the development steps for different programming paradigms. Figure 1(a) shows that for typical macro-programming approaches and for OS programming dialects such as NesC, it is necessary to develop code for various WSN nodes (e.g. sensors, other sensors, relays, sink) and for computers to interface with the WSN through a serial interface and to do further processing. In Figure 1(b) we show that global programming approaches such as TinyDB require engine-compatible WSN nodes and development of computer code to get and use the data. Figure 1(c) shows that MidSN approach only requires drivers for different nodes hardware to be loaded and executed (includes computer nodes) and the GinConf API calls to configure execution for each node.

The paper is organized as follows: section 2 discusses related work. Section 3 discusses the architecture of MidSN, and then section 4 is on data processing and configuration. Section 5 deals with node referencing, remote configuration and heterogeneity. Section 6 presents experimental results and section 7 concludes the paper.

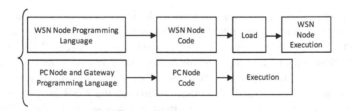

a)   Development steps with most Macro-Programming or OS programming dialects

Fig. 1. Development steps for different programming paradigms

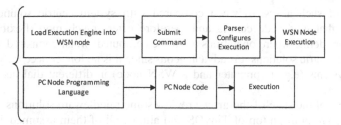

b)   Development steps with Global Programming dialect (TinyDB)

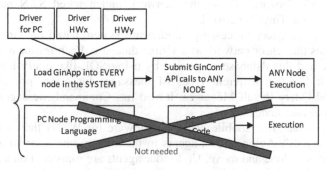

c)   Development steps with MidSN

Fig. 1. (*continued*)

2   Related Work

In this section, we present a comparative study of several state-of-the-art middleware for sensor network.

Motolla [1] reviews some of the most relevant macro-programming approaches in the state-of-the-art. In that review he classifies WSN applications on the basis of purpose (sense-only, sense and react), interaction pattern (one-to-many, many-to-one, many-to-many), space (global, local) and time (periodic, event driven). Then, a taxonomy is built for the WSN macro-programming abstractions. This taxonomy distinguishes programming models and architectures. Within programming model, the classification is based on communication and computation models, programming idiom and distribution model. Approaches such as TinyDB which do not allow programming what individual nodes do are classified as "System Centric", whereas those which allow programming individual nodes are node centric. Node centric solutions are more flexible, since they empower programmers to decide where to process what. However, state-of-the-art node centric solutions are focused on WSN code and require programming computer code separately, by hand, interfacing with a WSN-serial interface and adding any computer-side processing code in a computer programming IDE such as a Java framework. Our proposal bridges this gap in the state-of-the-art, by providing a node centric solution with remote configuration capabilities which can be deployed in any node of a heterogeneous system. This means that a system comprised of one or more WSNs and computers is considered a network of functionally similar, remotely configured nodes. The functional advantage of such a system when compared with system centric solutions such as TinyDB is that it is node centric and therefore offers much more flexibility in that programmers define where operations are to be computed; when compared with state of the art node centric solutions it is advantageous since it is no longer necessary to program different systems (e.g. a computer and a WSN node) in different dialects and to glue them together.

In the rest of the state-of-the-art we review some middleware solutions. Most of the middleware are built on top of TinyOS and almost all of them assumed TinyOS as a requirement and were not designed for abstracting away heterogeneous contexts (e.g. a TinyOS WSN network, a computer network and a wired SAN network), they focused on a single TinyOS network.

TinyDB [3] is a query processing middleware system based on TinyOS. TinyDB approach treats the sensor network as a virtual database. Each sensor node contains a tiny database and this database is queried by using SQL like query language called Tiny SQL. It has two different types of messages for query processing: Query Messages and Query Result Messages. It also has Command Messages for sending commands to sensor nodes.

Agilla [4] is the "first mobile agent middleware for WSNs that is implemented entirely in TinyOS". Agilla allows agents to move from one node to another using two instructions – clone and move. Up to four agents are supported on a single sensor node, running multiple applications on the network simultaneously.

Impala [5] is a middleware that was designed based on an event-based programming model with code modularity, ease of application adaptability and update, fault-tolerance, energy efficiency, and long deployment time in focus.

Borealis [6] is a distributed data stream management system. Sensor networks are interfaced with Borealis nodes by intermediary proxies. Each sensor network provides an adapter in order to provide common information to the Borealis node.

GSN [7] is conceived for a fast and flexible deployment of applications which need integration of data from sensors. The authors use the "Virtual sensors" and Borealis concepts to abstract the sensors from the physical implementations and provide a homogeneous view of sensor data. They build a middleware to operate in PC nodes and gateways, which allows transforming data to a homogeneous format and viewing the whole network as an homogeneous one. In comparison to that approach, our approach can be deployed in any part of the SAN, including WSN nodes.

IrisNet [8] proposes a two-tier architecture consisting of sensing agents (SA), which collect and pre-process sensor data, and organizing agents (OA) which store sensor data in a hierarchical, distributed XML database. This database is modeled after the design of the Internet DNS, and supports XPath queries.

None of these works handles the issue of distributed configuration with heterogeneity that we are solving in this paper. Some of these approaches provide heterogeneity out-of-WSN with communication code and application-level issues developed hand-programmed for each WSN node. Other approaches, such as TinyDB, are unable to run in heterogeneous sensor networks and do not provide homogeneous configuration capabilities, where every computation-capable node including server nodes should be viewed as a similar node for configuration and data processing.

3   MidSN Driver-Based Architecture

MidSN is a conceptual architecture. Its power rests on the simplicity of a single uniform software component that every node that wants to participate in the distributed system implements – GinApp, and a configuration component that talks with any node (or group of nodes) in the distributed system to configure it/(them). In this context, a driver is an implementation of the GinApp software component for a specific hardware. It is developed only once for each specific hardware (e.g. computer, tiny-OS based motes, contiki-based motes, arduino, waspmote).

A practical implementation/deployment is composed of possibly heterogeneous nodes, where each node has a specific hardware and is loaded with the relevant driver for that hardware.

Every node is referenceable and configurable from a configuration component. This allows every node, including a control station outside the WSN, to have the same configuration and processing capabilities readily available without any customized programming – e.g. detection of thresholds and alarms, closed control supervision code, statistic reports to be computed periodically from the signals. This way the programmer configures processing of a complete WSN application, including what to do with the data outside the WSN, without writing any custom stream processing application for any node.

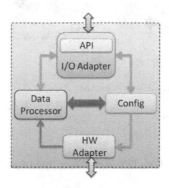

Fig. 2. GinApp Component

The GinApp depicted in Figure 2 has a unique address and the following modules:

- A Node Processor, which implements node operations. Node operations include data processing and any other operations that may be defined for a node. Data Processing concerns computations and actuations over the signal streams that come to the component from other nodes or from sensors.
- A Config module (CM), which configures data processor and I/O Adapter for sensing and actuating. The CM processes commands, modifying processing, acquisition and communication according to issued configuration.

- I/O Adapter, which implements a protocol to exchange data and commands with other GinApp components. Remote configuration is based on API calls through the I/O Adapter. The API defines an external configuration interface offered by the GinApp component.
- HW Adapter, which performs all sensor and actuator actions required to gather data from sensors or actuate. This module must be present in nodes doing actual sensing and actuating.

4   Data Processing Model and Configuration

The processing part of the GinApp component implements a (simple) stream processor that is configured by the configuration part. A complete working system is built based on deploying GinApp components on all nodes, and a system configuration component which will run on a controlling station.

Figure 3 illustrates a complete distributed system with the above component.

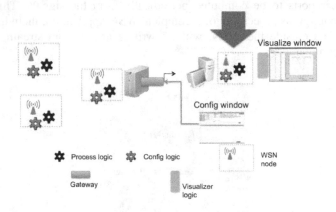

Fig. 3. MidSN-using Distributed System

Every node with the GinApp component has a Config logic and a Process logic. Nodes outside the WSN (e.g. the control station node pointed out by a red arrow in the figure) also have the same component. Every node can be addressed and configured through the same config. interface (implemented as either the Config window for users, a simple command-line console or a programming interface for applications).

In this model sensor signals are inserted into streams, from then on operations specify manipulations over streams. The creation of streams is through a call to an API method as exemplified next:

```
Node.createStream(Zone1SensorNodes,
    "pressureStreamfromZone1SensorNodes",
      1s, {(PRESSURE, VALUE)}
);
```

A control station can receive signals from the sensors and determine an alarm when pressure rises above a specified threshold:

```
Alarm.createAlarm( controlSation1, "ServerPressureAlarm",
    pressureStreamfromZone1SensorNode,
    CONDITION( PRESSURE.VALUE, >, 5)
);
```

Due to the uniform approach of MidSN and remote configuration capabilities, the same command can be submitted for configuration of any node, by simply changing the address field (controlStation1 and Zone1SensorNodes are node referencing variables).

5   Node Referencing, Remote Configuration and Heterogeneity

Communication and remote access relies on a uniform distributed system. However, actual platforms with WSN and control stations are heterogeneous, with embedded devices featuring tiny OS and communication protocols. MidSN defines a gateway component to handle parts of the distributed system that do not support standard IP protocol. This provides support for communication with non-IP embedded devices. A global catalog identifies IPs that must be routed through the gateway, and the gateway is an IP node itself which implements proprietary protocols for non-IP sub-networks.

MidSN defines a scheme for saving the assignment information when assigning the IP addresses to ZigBee ID which is assigned to the sensor node in wireless sensor networks. The assigned IP address is IPv6 global unicast and gateway has to prepare it in advance (similar to DHCP server). The address assignment is managed and processed in gateway. Each IP address which is used in client requests is mapped to a corresponding ZigBee ID. The translation information can be found on the scheme, and it is used by gateway when sending the information of the client request to the sensor node.

The scheme of the IP address assignment consists of identifier (nodeID) of wireless sensor networks, the hostname, the group id (groupID) of the sensor node, and IPv6 address (ipADDR) which was allocated to sensor node.

MidSN maintains a catalog of nodes with hostname, IP and current node configuration data for each node, as well as a catalog of groups of nodes with group name and list of IPs. The catalog is XML-based. The MidSN communication protocol defines formats for control and data messages. There is a control message format for remote configuration of nodes which defines the configuration API method to call and parameters. A configuration application uses those control messages to send configuration commands to any nodes.

Figure 4 is an illustration of the gateway mechanism of MidSN.

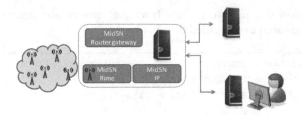

Fig. 4. MidSN Gateway Components

The MidSN router gateway translates between two network protocols – in the figure example this is Rime for Contiki and IP for the internet. Such gateway component was implemented in our testbed prototype: WSN embedded devices (TelosB) run Contiki OS and the Rime network protocol. The gateway was implemented as a module that seats on both a sink TelosB node and the computer to which it is connected. If the WSN supports IP the gateway is replaced by a bridge for IP over the tiny OS.

6   Experimental Evaluation

In this experimental section we use a prototype to show that we are able to configure any node in the whole SAN (WSN nodes, intermediate computers and a control station) using the same remote interface. We also show that the configuration resulted in corresponding behavior modification.

We have deployed MidSN in an industrial environment in context of European FP7 research project - Ginseng on performance-controlled WSNs, in which applicability in actual industrial scenarios is an important aspect. In that setting, MidSN allows users to configure and change what any part of the system is expected to do easily and remotely, concerning data collection, alarms and actuation.

We have used both industrial and lab testbed to test our approach. The testbed is a totally planned network (Ginseng focuses on totally planned networks with performance guarantees) with two sub-networks where each one include computer nodes and TelosB nodes. The first (sub-network 1) includes 16 TelosB nodes organized hierarchically in a 3-2-1 tree and one sink node (computer node). The second one (sub-network 2) includes 6 TelosB nodes organized hierarchically in a 1-2-1 tree. The setup also includes a control station that receives the sensor samples and alarm messages. All computers nodes (sink 1, sink 2 and control station) are connected with Ethernet cable and GigaBit network adapter. Figure 5 shows a sketch of our setup.

In our setup the WSN nodes of the first tree run the Contiki operating system (sub-network 1) with a TDMA network protocol (GinMac [9]) to provide precise schedule-based communication. The TDMA schedule had an epoch time of 640 msecs. The nodes of the second tree run TinyOS (sub-network 2) with an S-MAC protocol implementation [11]. All WSN nodes are time-synchronized and awake for their predefined slot time. The other nodes (computer nodes) are also time-synchronized using NTP protocol.

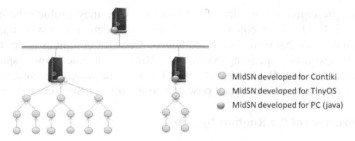

Fig. 5. Node deployment

MidSN was implemented in three programming languages, Contiki-C to be supported by ContikiOS in the first tree, nesC to be deployed in TinyOS and Java to run in computer nodes with linuxOS.

Each sink is connected to the WSN through a bridge, which allows configuring both WSN nodes and sink with any of the pre-defined operations (e.g. getting data from other nodes, firing alarms based on conditions, closed-loop control logic from input signals that come from other nodes).

The evaluation of MidSN will be done in two parts. The first part consists on evaluating the resources needed to run GinApp, since that component will have to be able to run in memory and computation resource constrained motes. The second part will evaluate performance. Furthermore, we describe how MidSN can be used in real-world applications.

6.1   Memory Footprint

The Typically, WSN devices have limited memory and computation capabilities, so the occupied memory is an important issue. GinApp can be deployed on WSN devices or more powerful nodes, such as computers.

In this sub-section we will evaluate the amount of memory needed when GinApp is deployed in a WSN node, such as TelosB node and a computer node with Java. Table 1 shows the amount of memory needed by each component of GinApp when it is ported to ContikiOS, TinyOS and Java.

Table 1. Memory Consumption

Component	TelosB (Contiki-C) [Bytes]	TelosB (nesC) [Bytes]	Computer (Java) [KBytes]
I/O Adapter & API	1260	800	8803
Config. Manager (CM)	880	1002	6000
Data Processor	5104	3029	20890
HW Adapter	544	206	200

The size of CM depends of the flexibility required by the application context, because its size depends on the number of commands that are to be sent and interpreted by the node.

The values presented are the size of the runtime code, they exclude the operating system. From Table 1 we can conclude that GinApp implementation was significantly small to fit all devices that were tested. Implementations for computer nodes need less than 40 KB (without operating systems). These consume more space than implementation for TelosB platform (using Contiki-C or nesC) because it is java-based, but computers resources do not pose any constraints on such code sizes.

6.2   Performance of the Runtime System

To test the MidSN we have written a simple web service client to submit commands at specific time instants and we collected data logs in the control station to analyze timings and modifications to sampled data coming from the SAN. With this data we have built timeline charts to show the results.

Experiments involved configuring nodes with the following alternatives:

1. A pressure reading is obtained every 3 seconds from every sensor node;
2. An alarm is raised in the control station an in a node when a threshold is passed.

Test 1 - Sensor Readings – The following code shows the calls to configuration API methods that were issued by the configuration software to start a sensor collection stream with a 3 seconds sampling rate.

1. Create stream with readings every three seconds:

```
// create stream for reading from sensor
Node.createStream(Node1.2, "pressureStreamfromNode1.2", 3s,
   {(PRESSURE, VALUE)} );

// forward the stream to the control station
Node.createStream(ControlStation, "pressureSensor",
   pressureStreamfromNode1.2 );
```

Figure 6 shows the results of these commands for node 2 of sub-network 1. From the figure it is possible to see that pressure samples started to be collected at a rate of a sample every 3 seconds soon after the sensor receives the command.

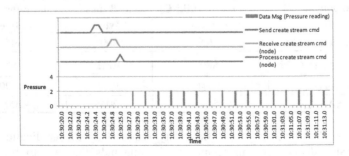

Fig. 6. Test 1: Sensor Readings and Reconfigure

The time taken to deliver a command is an important measure of reliability in performance controlled WSNs, partly enforced by a TDMA protocol. Figure 7 shows the measured maximum, minimum and average delay to deliver a command for TelosB nodes of each sub-network. These networks are identified with a prefix number before each node id (e.g. 1.2 – node two of sub-network 1).

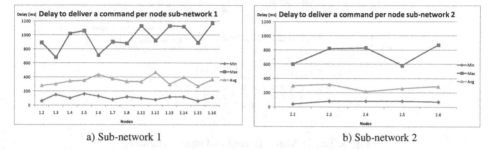

a) Sub-network 1   b) Sub-network 2

Fig. 7. Time taken to deliver a command

The results show that all commands sent to TelosB nodes were delivered in less than 500ms on average, and always below 1200 ms. These times reflect the TDMA schedules: every node transmits in its slot every complete cycle. The delays were smaller in network 2, which is mostly due to its smaller size. We also measured the delay to deliver a command to each computer node, but that one was less than 1ms.

This experimental setup and results show that the middleware enabled configuration and processing over completely heterogeneous nodes using a uniform remote approach, in this case over ContikiOS, TinyOS and Linux on computers.

Test 2 – Alarms Raised at Either Control Station or Sensor Node (1.2) - In this test we configured an alarm on the control station when sensor stream data is above a certain threshold, and another alarm on a sensor node for the same effect but with a different threshold. The fact that both parts of the system contain the same configuration and processing component and are directly referenced by an address variable allows uniform configuration in spite of being very different platforms (a TelosB node and a linux control station). We programmed sub-network 1 to sample pressure every 3 seconds (in order to more easily exercise the alarm thresholds we generated simulated pressure values instead of reading the actual values).

Raise alarm on the control station every time pressure goes above a value of 9:

```
Node.createAlarm( controlSation, "ServerPressureAlarm",
    pressureStreamfromNet1, CONDITION( PRESSURE.VALUE, >, 9)
);
```

An alarm is also to be raised on the sensor node every time pressure goes above a value of 7:

```
Node.createAlarm( Node1.2, "pressureStreamfromNode1.2",
    pressureStreamfromNet1, CONDITION( PRESSURE.VALUE, >, 7)
);
```

The test ran for 30 minutes and we logged all alarm occurrences and sample readings. After the test we collected those alarms and samples from the log and created the chart shown in figure 8 (showing only two minutes).

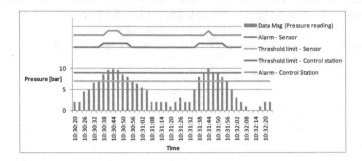

Fig. 8. Test 2: Alarms (sensor and control station)

The chart shows that the configuration worked well and the alarms are raised when the thresholds are passed: above 7 bars the sensor node alarm was triggered – shown in the figure by the red alarm line – and above 9 bars the control station alarm was also raised.

7   Conclusion

In this paper we proposed a middleware model for uniform configuration and operation over heterogeneous networks comprising WSNs and nodes outside the WSNs (control stations). The model advances the state-of-the-art since it views the whole system as a distributed system and any computing device as a node (inside or outside of the WSN, regardless of hardware or operating system) with the same configuration API, remote configuration capabilities and operation interface. This results in a lego-like system with great configuration flexibility and uniform API. We have described the modules and details of the component. Then we used our experimental testbed and defined a set of tests that show the system is able to configure both sensor nodes and control stations very easily and using exactly the same calls. From our test runs we extracted logs and displayed a timeline that proves correct configuration of both sensor nodes and control station using the approach. We also show results on delay statistics for the data.

References

1. Mottola, L.: Programming Wireless Sensor Networks: From Physical to Logical Neighborhoods, PhD Thesis (2008)
2. Levis, P., Culler, D.E.: Maté: a Tiny Virtual Machine for Sensor Networks. Architectural Support for Programming Languages and Operating Systems (2002)
3. Madden, S., Franklin, M., Hellerstein, J., Hong, W.: TinyDB: An acquisitional query processing system for sensor networks. ACM Trans. Database Syst. (2005)

4. Fok, C.-L., Roman, G.-C., Lu, C.: Agilla: A mobile agent middleware for self-adaptive wireless sensor networks. ACM Transactions on Autonomous and Adaptive Systems (2009)
5. Liu, T., Martonosi, M.: Impala: a middleware system for managing autonomic, parallel sensor systems. In: Proceedings of the Ninth ACM SIGPLAN Symposium on Principles and Practice of Parallel Programming, PPoPP 2003 (2003)
6. Abadi, D., Ahmad, Y., Balakrishnan, H., Balazinska, M., Cetintemel, U., Cherniack, M., Hwang, J., Jannotti, J., Lindner, W., Madden, S., Rasin, A., Stonebraker, M., Tatbul, N., Xing, Y., Zdonik, S.: The Design of the Borealis Stream Processing Engine. In: Proc. of the Second Biennial Conference on Innovative Data Systems Research, CIDR (2005)
7. Aberer, K., Hauswirth, M., Salehi, A.: The global sensor networks middleware for efficient and flexible deployment and interconnection of sensor networks, T. Report LSIR-REPORT-2006-006, Ecole Polytechnique Fédérale de Lausanne, EPFL (2006)
8. Gibbons, P.B., Karp, B., Ke, Y., Nath, S., Seshan, S.: IrisNet: An Architecture for a World-Wide SensorWeb. IEEE Pervasive Computing (2003)
9. Suriyachai, P., Brown, J., Roedig, U.: Poster Abstract: A MAC Protocol for Industrial Process Automation and Control. In: Proc. of EWSN 2010 (2010)
10. Dunkels, A., Schmidt, O., Voigt, T., Ali, M.: Protothreads: Simplifying Event-Driven Programming of Memory-Constrained Embedded Systems. In: Proc. of the 4th Int. Conf. on Emb. Netw. Sensor Syst. (2006)
11. Ye, W., Heidemann, J., Estrin, D.: An Energy-Efficient MAC Protocol for Wireless Sensor Networks. In: IEEE INFOCOM (June 2002)

Exploring and Making Safe Dangerous Networks Using Mobile Entities⋆

Mattia D'Emidio[1], Daniele Frigioni[1], and Alfredo Navarra[2]

[1] Department of Information Engineering, Computer Science and Mathematics,
University of L'Aquila, Via Gronchi 18, I–67100, L'Aquila, Italy
{mattia.demidio,daniele.frigioni}@univaq.it
[2] Department of Mathematics and Computer Science, University of Perugia,
Via Vanvitelli, 1, I-06123, Perugia, Italy
alfredo.navarra@unipg.it

Abstract. We consider synchronous and mobile entities that have to explore and make safe a network with faulty nodes, called *black-holes*, that destroy any entering entity. We are interested in the scenario where the destruction of an entity by means of a black-hole also affects all the entities within a fixed range r (in terms of hops), and we ask for the minimum number of synchronized steps needed to remove all the black-holes from that network. Clearly, if there are b black-holes in the network, then $k \geq b$ entities are necessary.

First, we show that the problem is NP-hard even for $b = k = 1$; second, we provide an asymptotical optimal solution for the case of $r = 0$ and a general lower bound for the case of $r > 0$; third, we propose two different strategies plus a refined heuristic for the case of $r = 1$, and we prove they are all asymptotically optimal; finally, we provide an experimental study to show the practical performance of the proposed strategies.

1 Introduction

The exploration task of unknown graphs by means of mobile entities has been widely considered during the last years. The increasing interest to the problem comes from the variety of applications that it meets. In robotics, it might be very useful to let a robot or a team of robots exploring dangerous or impervious zones. In networking, software agents might automatically discover nodes of a network and perform updates and/or refuse their connections. In this paper, we are interested in the exploration of a network with faulty nodes, i.e. nodes that destroy any entering entity. Such nodes are called *black-holes*, and the exploration of such kind of networks is usually referred as *black-hole search*. According to the assumed initial settings of the network, the knowledge, and the capabilities of the involved entities, many results have been provided.

⋆ Research partially supported by the Research Grant 2010N5K7EB 'PRIN 2010' ARS TechnoMedia (Algoritmica per le Reti Sociali Tecno-mediate)' from the Italian Ministry of University and Research, and by "Fondazione Cassa di Risparmio della Provincia dell'Aquila", project ARISE.

J. Cichoń, M. Gębala, and M. Klonowski (Eds.): ADHOC-NOW 2013, LNCS 7960, pp. 136–147, 2013.
© Springer-Verlag Berlin Heidelberg 2013

Related Works. Pure exploration strategies, without dealing with black-holes, have been widely addressed (see for instance [13,17] and references therein). In that case, the requirement is usually to perform the exploration as fast as possible. When black-holes are considered, along with the time (or equivalently the number of edge traversals) required for a full exploration, the main goal resides in minimizing the number of entities that may fall into some black-hole. A full exploration in this case means that, at the end, all the edges which do not lead to any black-hole must be "marked" as safe edges.

In this research area, many different models have been investigated. The system might be either synchronous [6,16] or asynchronous [8,9]. The input graph might be undirected [9,16] or directed [5,18]. It can be known in advance [10,16] or just a bound on the number of nodes is provided [18]. It can refer to a specific topology like rings [2,9], trees [6], hypercubes [7], tori [19]. Entities may communicate only when they meet [4], or by means of white-boards associated to the nodes of the graph [2], or simply by opportunely disposing available pebbles [11]. The objective function may ask for the minimum number of entities, the minimum number of steps performed by all the entities, the minimum number of synchronous steps.

Recently, a hybrid model in between exploration and black-hole search has been introduced in [4], named the *Explore and Repair* problem. The idea is to perform the exploration of the graph as fast as possible with the constraint that if an entity meets with a black-hole, it disappears along with the black-hole. So, the objective is to provide an exploration strategy that ensures to make safe the whole graph in the fastest way. In [4], it is assumed that if more than one entity enter a black-hole, only one gets destroyed while the others can continue the exploration. However, it is possible to think about scenarios where entities are mobile robots exploring an impervious area disseminated of land-mines, and if more than one robot meet concurrently with an explosion, then all of them get involved. Furthermore, the explosion may affect also robots within a fixed range.

Contribution of the Paper. In this paper, we consider this last scenario so that given an integer number r, all entities within distance r from an entity met with a black-hole instantaneously disappear from the network along with the black-hole. Clearly, if there are b black-holes, then $k \geq b$ entities are necessary to remove all the black-holes from the network. It might happen that the network is not explored completely, but an exploration algorithm must guarantee that all the black-holes have been removed.

The results of this paper can be summarized as follows: we first show that the problem is NP-hard even for $b = k = 1$; second, we consider the case of $r = 0$ and show that a simple variation of the strategy proposed in [4] can be applied to give an asymptotically optimal solution; third, we consider the case of $r > 0$, and provide a general lower bound; fourth, we consider the case of $r = 1$ and propose two different exploration strategies plus a refined heuristic, and we prove they are all asymptotically optimal; finally, we provide an experimental study to show the practical performance of the proposed strategies among different graph classes by varying on their size, on the number of black-holes, and on the number of available entities.

2   Definitions and Notation

We represent the exploration area as an undirected graph $G = (V, E)$ where V is a finite set of n nodes and E is a finite set of m edges. An edge in E between two nodes $u, v \in V$ is denoted as $\{u, v\}$. Given $v \in V$, $N(v)$ denotes the set of neighbors of v, and $\deg(v) = |N(v)|$ denotes the degree of v. A path P in G between nodes u and v is denoted as $P = (u, ..., v)$. The length of P, denoted as $l(P)$ is equal to the number of edges in P. A *shortest path* between nodes u and v is a path from u to v with the minimum length. The *distance* $d(u, v)$ from u to v is the length of a shortest path from u to v. The diameter D of G is the maximum distance between two nodes in G. Given G and a source node $s \in V$, a Breadth First Search (BFS) tree of G, denoted from now on as T, is the result of a Breadth First Search algorithm on G, starting at s. Given T and a node v of T, the *level* of v equals the distance between s and v. A tree is *ordered* if and only if an ordering is specified for the children of each vertex.

Given a graph $G = (V, E)$, we consider a synchronous model of repairable faults, also referred to as *black-holes*, whose number and locations are unknown. A set of k entities are initially placed at a common node $s \in V$ assumed to be safe, and referred to as the *home-base* of G. Synchronously, entities move from s through G by traversing one edge per time step. If an entity enters a black-hole residing at node v, it is assumed to sacrifice itself while repairing the black-hole. If more than one entity enters the same black-hole concurrently, they all die. If an entity enters a node where initially there was a black-hole that subsequently has been repaired, then the entity does not notice any differences. In fact, after the repair, which is completed instantaneously, the node acts normally and other entities can pass through it as if the black-hole never existed.

We consider the general case where the repair of a black-hole affects also its neighborhood. We consider a parameter $r \in \mathbb{N}$, named *radius*, which represents the distance (in terms of number of edges) from a black-hole. All nodes but s in such a range are affected by the repairing operation. The home-base s is always assumed to be safe, no matter its distance from black-holes. This means that entities not in s within range r from a black-hole get destroyed during the repair of such a black-hole.[1]

An exploration strategy is *correct* if it ensures to repair *all* the b black-holes in a network by means of $k \geq b$ entities within a finite number of time steps. It is *optimal* if it requires the minimum number of time steps. We refer to this problem as the r-ER problem, that is, the *Explore and Repair* problem of [4] with radius r. The input of the r-ER problem is an undirected graph $G = (V, E)$, a home-base node $s \in V$, and a positive integer k representing the number of available entities. Entities are numbered from 1 to k (the entities' identifiers), they are synchronized, and they can communicate only when they meet at a node of G.

---

[1] The variant where also other black-holes within range r are affected during the repairing operation like in a "chain explosion" is not considered here.

During the exploration, entities may decrease in number as some of them can meet black-holes. On the one hand, since $k \geq b$, entities suffice to repair all the black-holes. On the other hand, while an entity is acting to repair a black-hole, it is a must to avoid that other entities get involved (in particular if $k = b$). The entities move through the graph according to some exploration algorithm. More precisely, if an entity is at the beginning of the current step at a node v, then its next move is determined by the exploration algorithm based on the topology of the network, the entity's identity, the entity's state (the state of its memory), the states of all other entities which are at the same time step at node v. The entity's ability to access the states of all other entities currently at the same node models the communication among the entities when they meet.

3   Complexity and Algorithms

In this section, we first show that the r-ER problem is NP-hard even for $b = k = 1$; second, we consider the case of $r = 0$ and show that a simple variation of the strategy proposed in [4] can be applied to give an asymptotically optimal solution; third, we provide a general lower bound for the case of $r > 0$; finally, we consider the case of $r = 1$ and propose two different exploration strategies plus a refined heuristic, and we show they are all asymptotically optimal.

Theorem 1. *The r-ER problem is NP-hard for $b = 1$ and $k = 1$.*

Proof. It is enough to observe that even the fastest way to explore a graph in order to detect where the unique black-hole resides must involve all the nodes of the input graph G. Hence, $n - 1$ time steps are required for the full exploration by means of the unique entity. This corresponds to the problem of finding a Hamiltonian Path in G, if possible. Since the Hamiltonian Path problem is known to be NP-complete [12], the claim holds for the r-ER problem.   □

The r-ER problem is subject to a simple lower bound concerning the time steps required by any exploration algorithm with $k \geq b$. In fact, the input instance might need an entity to move from the home-base s to the farthest possible node, whose distance from s is the diameter D of G. Moreover, if k entities are available, then each of them can explore a portion of the graph equal to $\frac{n}{k}$ nodes, concurrently. This implies a lower bound of $\Omega(\frac{n}{k} + D)$ time steps.

In the model proposed in [4], where each black-hole can affect only one entity, the lower bound has been increased to $\Omega(\frac{n}{k-b} + \frac{\log f}{\log \log f} D)$, with $k > b$ and $f = \min\{\frac{n}{k-b}, \frac{n}{D}\}$. The authors also provided an asymptotical optimal exploration algorithm that meets this bound. Clearly, the same lower bound holds also for the r-ER problem, which requires further constraints with respect to the model of [4]. The upper bound is not extendable straightforwardly as it depends on the radius r.

3.1   The Case of $r = 0$

When $r = 0$, the only difference with respect to the model presented in [4] is just to avoid that more than one entity at the same time enters a node with a black-hole as otherwise all of them would disappear. Assume a bunch of entities has to move from a node u to a node v concurrently (in one time step). The same movement can be obtained in two time steps as follows: first move only one entity (say the one with the smallest identifier) from u to v; then, in the subsequent time step, move all the remaining entities from u to v, while the first entity, if survived, waits for them at v. In this way, if node v is occupied by a black-hole, only one entity will be affected. The case where a node is entered concurrently from different neighbors by means of two sub-teams of entities can also be avoided by preliminary computations as all the exploration paths for each entity can be predetermined before letting entities go into the network. Hence, we can state the next theorem by simply exploiting the exploration algorithm proposed in [4] with the only modification of applying the above considerations.

Lemma 1. *The 0-ER problem can be solved in $\Theta(\frac{n}{k-b} + \frac{\log f}{\log \log f}D)$ time steps, with $k > b$.*

3.2   The Case of $r > 0$

In general, when $r > 0$, the exploration strategy of [4] enriched as described in Section 3.1 is no longer sufficient to guarantee that all black-holes of the graph will be repaired. In fact, when the first entity of a punch of entities moves from a node u to a node v, if v is a black-hole, then this entity and all the entities waiting on u and on any other neighbor of v will die. Hence, we need to develop a completely new strategy for the case of $r > 0$. The next theorem provides a lower bound to the number of time steps required. Its proof can be found in the extended version of the paper.

Theorem 2. *For any $r > 0$ and $b \leq k = o(n)$, the r-ER problem requires $\Omega(n)$ time steps in the worst-case.*

3.3   The Case of $r = 1$

In what follows, we propose two exploration strategies, named the *snake* strategy and the *scout* strategy, respectively, and show that they are both asymptotically optimal. Then, we introduce a variant of the *scout* strategy named *parallel-scout* which is able to take advantage of parallelism.

In all strategies, we assume that an ordered BFS tree T of the input graph G, rooted at the home-base s, has been pre-computed and that all the entities know it. Given an edge $\{u, v\}$ in T, such that the level of u is smaller than the level of v, then an entity is said to move in the *down* (*up*, respectively) direction if it moves from u to v (v to u, respectively). An exploration strategy terminates if all the black-holes have been repaired within a finite number of time steps.

Given T, we call *exploration walk* the path obtained by visiting T by a Depth First Search algorithm. Clearly, this visit traverses each edge of the ordered tree T twice, once in the *down* direction and once in the *up* direction. Then, it follows that the *exploration walk* has length equal to $2(n-1)$.

The *Snake* Strategy. In this strategy, we distinguish three different types of entities: *head*, *normal* and *backward*. At the beginning, all the entities are of *normal* type and reside in the home-base s. Then, on the basis of the identifiers of the entities, a *head* entity is elected. By now, no *backward* entity exists. The *snake* strategy allows at each time step, the existence of at most one *head* entity. The entities move synchronously through the *exploration walk* starting from s, according to their type. In general, as the strategy name suggests, the entities proceed in line in a *snake*-like visit as follows: the *head* entity starts moving from s, edge by edge in the *down* direction through the *exploration walk*, and every two time steps a *normal* entity leaves s along the same walk of the *head*. This implies that there is always an empty node between each pair of consecutive entities composing the *snake*. The order of the *normal* entities is based on the identifiers. Usually, entities do not meet, except in the following cases:

(i)- the *head* visits a child of s. If survived, the head is back to s in two time-steps, hence meeting all the other entities. If not survived, a new head is elected among the remaining entities.

(ii)- the *head* visits a leaf node of T without meeting with a black-hole, at a certain time step t. In this case, at step $t+1$ the *head* meets the first *normal* entity following it. This *normal* entity changes its type to *backward* and both the entities start traversing the *exploration walk* in the *up* direction one edge per time step. During this traversing the two entities can meet other *normal* entities which turn of *backward* type as well and join this group. When the entities meet, they communicate in order to know which part of T has already been visited. This visit continues until the entities reach a node v of T with at least one unvisited child. At this point, the exploration continues as follows: the *head* visits the first unvisited child of v, while the *backward* entities move to the parent of v (unless $v \equiv s$) in the *up* direction, where they meet a *normal* entity (if any), and stay there. In this case, the *normal* entity keeps its own type. Now, every two time steps, a *normal* entity meets the *backward* entities group and is made aware of the visited part of T making the *normal* entities able to follow the *head* according to the knowledge of the unvisited part of T. Note that, also in this phase, the *normal* entities keep their own type. When, at a certain step, no *normal* entity visits the parent of v, every two time steps a *backward* entity turns back of *normal* type, leaves this node and continues the visit as usual, following the *head* along the unvisited part of T. The case of $v \equiv s$ is similar to start the exploration with $k' \leq k$ entities on T' subtree of T with $b' \leq b$ black-holes.

(iii)- a *normal* entity visits a leaf node of T without meeting with a black-hole. This can happen if the *head* entity and possibly some *normal* entities have met with black-holes. In this case, the leading *normal* entity immediately changes its type to *head* and the visit proceeds as previously described.

(iv)- the surviving entities (if any) have gone throughout all the *exploration walk* and are back in s, that is, the exploration has terminated.

Note that, while a group of *backward* entities is waiting at a certain node z which is parent of a node v in T, another group of *backward* entities may reach z from v. In this case, the two groups join and all the *backward* entities start behaving like the *backward* entities coming from v, that is, the most recent ones.

Lemma 2. *The* snake *strategy solves the 1-ER problem in* $O(n)$ *time steps.*

Proof. By assuming $b = 0$, the *head* traverses the *exploration walk* in exactly $2(n - 1)$ synchronous steps. When $b > 0$, the leading entity of the *snake* can be either a *head* or a *normal* entity. Every time the leading entity meets with a black-hole, two synchronous steps are required to have the new leading entity in the same position where its predecessor died, and hence to keep continuing with the traversing of the *exploration walk*. Therefore, the *snake* strategy requires $2(n - 1) + 2b$ overall steps for solving the 1-ER problem. As b cannot be larger than n, it follows that $O(n)$ time steps are required.   □

The *Scout* Strategy. In this strategy, we distinguish two different types of entities: *scout* and *team*. At the beginning, all the entities are of *team* type and reside in s. Then, on the basis of the identifiers, a *scout* entity is elected. The *scout* strategy allows at each time step, the existence of at most one *scout* entity. The entities move synchronously through the *exploration walk* starting from s, according to their type as follows. An entity of *scout* type traverses two edges in the *down* direction and one edge in the *up* direction (*scout round*). An entity of *team* type waits two synchronous steps and then traverses an edge in the *down* direction (*team round*).

Usually, at the third time step of a round, the *scout* entity and the *team* entities meet at a node v of G. This implies that the *scout* has visited only safe nodes and the exploration can continue from v. Otherwise, if the *team* entities do not meet the *scout* at node v, this means that the *scout* has met with a black-hole and has died by repairing it. At this point, a new *scout* is elected among the *team* entities in v and the exploration continues from v.

The only exception to the above behavior is when the *scout* visits a child node v of s. In this case, the *scout* and the *team* entities meet at the second step of a round. However, since all entities know how T must be explored, the *team* entities can infer that the *scout* can come back after only two synchronous steps, and hence the exploration can continue as usual.

When the *team* entities reach the parent node of a leaf in T, then all the entities start traversing the *exploration walk* in the *up* direction, an edge per step, until they meet in a node v of T with at least one unvisited child or they are back to s. In the first case, the *scout* (elected now if needed) moves to the first unvisited child of v and the *team* entities move to the parent node of v. At this point, the entities start over with their normal behavior. In the second case, the exploration keeps continuing on an unexplored child of s, if any. If all the children of s have been already explored then the exploration is terminated.

Lemma 3. *The* scout *strategy solves the 1-ER problem in $O(n)$ time steps.*

Proof. By assuming $b = 0$, in general each node is made safe within three time steps in order to let know the *team* entities about it. Moreover, the *team* entities make one further step for each reached node in the *up* direction. This means that in general, any node different from the home-base s requires four time steps by the whole process. Exceptions are represented by leaves that are never reached by the *team* entities. Moreover, the children of the home-base s are made safe by the *scout* in two steps each, making known also the *team* entities. Finally, whenever the *team* and the *scout* walk together in the *up* direction until reaching a node x with an unexplored child y, the *team* goes towards the parent of x while the *scout* explores y. In the subsequent step, both the *team* entities and the *scout* (if survived) meet at x, hence y requires only two steps to be explored.

Let us denote as F the set of leaves in T, as $I = V \setminus \{F \cup \{s\}\}$, and as $deg_T(v)$ the degree of a node v in T. By summing up overall the contributions, the *scout* strategy solves the 1-ER problem in a number of time steps equal to:

$$4(n-1) - |F| - deg_T(s) - \sum_{x \in I} (deg_T(x) - 2) =$$
$$4(n-1) - \sum_{x \in F} deg_T(x) - deg_T(s) - \sum_{x \in I} deg_T(x) + \sum_{x \in I} 2 =$$
$$4(n-1) - \sum_{x \in V} deg_T(x) + 2|I| = 2(n-1) + 2|I|.$$

When $b > 0$, the same result holds by observing that only *scout* entities can meet with black-holes while the *team* is always maintained safe along its walk. As $|I|$ cannot be larger than n, it follows that $O(n)$ time steps are required.   □

By Theorem 2 and Lemmata 2 and 3, the following Theorem can be stated.

Theorem 3. *Both* snake *and* scout *strategies solve the 1-ER problem in $\Theta(n)$ time steps.*

Although the *snake* and *scout* strategies have been shown to be asymptotically optimal, none of them exploit possible parallel explorations. The *scout* strategy is independent on b, while the *snake* strategy is independent on the input graph topology, and hence on the structure of T. For instance, when the input graph is a path with one extreme as the home-base, the *scout* strategy requires $2(n-1) + 2|I| = 4(n-1) - 2$ steps, and this might be close to the double of what is required by the *snake* strategy.

Nevertheless, the *scout* strategy reveals an important peculiarity that may help in introducing some parallel moves, hence reducing the number of required time steps. In fact, each time the scout entity explores one new node, it starts moving after having met all the other entities. In this way, all the entities are always aware of the current part of the graph already explored.

The *Parallel-Scout* Strategy. In the general step of the *scout* strategy, when the scout entity e wants to explore a node z, it moves from node x, where also all the other entities reside, to a neighbor y which is the parent of z in the BFS tree T. From y, e can move towards z without damaging any other entity since they are all waiting at distance 2 in x. If y has many children, then all the ones which are not directly connected to z in G could be explored by other entities in parallel with e without incurring in multiple destructions if a black-hole is met. In order to implement this variant of the *scout* strategy called *parallel-scout*, we need to introduce the following further step: in the general step described above, the entities compute an independent set in the subgraph of G induced by the children of node y in T. Then, if enough entities are available, all the nodes belonging to this independent set can be explored concurrently in one time step, and the results made known to all the entities in the subsequent step when they all meet again if survived. If not survived, still all the other entities can imply that some black-holes have been repaired and that the computed independent set is now safe. The acquired information will be used in the subsequent steps of the exploration in order to move safely over nodes already explored, and hence saving in the number of required time steps.

Of course the *parallel-scout* strategy does not give advantages in terms of worst-case asymptotic bounds. However, it can give a practical gain with respect to the basic *scout* strategy in terms of the effective number of time steps needed, which can make *parallel-scout* very competitive in practice. We will give experimental evidence of all the above observations in the next section.

4   Experiments

In this section we report the results of our experimental study, which has been performed on a workstation equipped with a Quad-core 3.60 GHz Intel Xeon X5687 processor with 24 GB of main memory. The programs have been compiled with GNU g++ compiler 4.4.3 under Linux (Kernel 2.6.32). The experiments consist of simulations within the OMNeT++ 4.0p1 environment [20].

Executed Tests. We implemented and tested the *snake*, *scout* and *parallel-scout* strategies by means of OMNeT++. Concerning *parallel-scout*, we implemented the $(\Delta + 2)/3$-approximation algorithm of [14] to find the required independent set, where Δ is the maximum degree of the nodes in the network.

We measured the number of time steps required by each algorithm to explore and repair a network. As input to our simulations we used both real-world and artificial instances of the problem. In detail, we used real-world power-law topologies of the *CAIDA dataset* [15], which are very sparse, random power-law graphs generated by the *Barabási-Albert* algorithm [1], which are slightly denser than the CAIDA ones, and *Erdős-Rényi* random graphs [3] with a number of edges equal to 20% that of the complete graph with the same number of nodes, which makes these graphs denser than the CAIDA and *Barabási-Albert* ones.

Concerning CAIDA instances, we parsed the files provided by CAIDA to obtain a weighted undirected graph, which consists of almost 35000 nodes. We extracted six different subgraphs of this graph, with n ranging from 50 to 1050, with a step of 200. We generated random graphs of the same sizes by using also the *Barabási–Albert* and the *Erdős-Rényi* algorithm.

We tested the algorithms on such instances with different values of b. In particular, for each graph with n nodes, we have chosen b black holes by using a uniform probability distribution, with b ranging from 10%n to 50%n with a

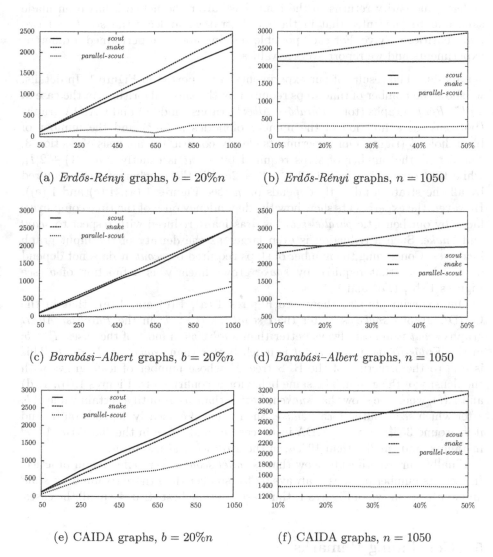

(a) *Erdős-Rényi* graphs, $b = 20\%n$ (b) *Erdős-Rényi* graphs, $n = 1050$

(c) *Barabási–Albert* graphs, $b = 20\%n$ (d) *Barabási–Albert* graphs, $n = 1050$

(e) CAIDA graphs, $b = 20\%n$ (f) CAIDA graphs, $n = 1050$

Fig. 1. Number of time steps required by *scout*, *snake* and *parallel-scout* on *Erdős-Rényi* graphs (a) and (b), *Barabási–Albert* graphs (c) and (d), and CAIDA graphs (e) and (f). In Figures (a), (c) and (e), the x-axis represents the number of nodes n ranging from 50 to 1050, with step 200, and b is fixed to $20\%n$. In Figures (b), (d) and (f), the x-axis represents the number of black-holes b ranging from $10\%n$ to $50\%n$, with step $10\%n$, and n is fixed to 1050.

step of $10\%n$. Without loss of generality, the number of entities k is always set to $b+3$. This is done only to compare the strategies in a common setting where at least one entity returns to the home-base after the network has been made safe, and to guarantee that in the *scout* strategy at least the *scout* and two *team* entities survive. For each possible test instance, we performed 5 different experiments and we report average values.

Analysis. The results of our experiments are reported in Figure 1. In details, we show the number of time steps required by the three algorithms in the case of *Erdős-Rényi* graphs (top), *Barabási–Albert* graphs (middle) and CAIDA graphs (bottom), as a function of the number of nodes n (left) and the number of black holes b (right). Our experiments clearly confirm the analysis of Section 3, that is: (i) the number of steps required by *scout* is exactly $2(n-1)+2|I|$, while that of *snake* is exactly $2(n-1)+2b$; (ii) the number of steps required by all the strategies directly depends on n (see Figures 1.(a), 1.(c) and 1.(e)). However, the experiments show how the dependency on n of the time complexity faced by our heuristic *parallel-scout* is drastically reduced with respect to *scout* and *snake*. Such a reduction is even larger as the density of the input graph increases. Concerning the number of steps required by *scout* it does not depend on b, whereas that required by *snake* grows linearly as a function of b (see Figures 1.(b), 1.(d) and 1.(f)).

Our experiments also show (see Figures 1.(a), 1.(c) and 1.(e)) that in the CAIDA graphs *snake* is better than *scout*, as $|I| > b$; in the *Barabási–Albert* graphs *scout* is almost always better than *snake*, as in most of the cases $|I| < b$; in the *Erdős-Rényi* graphs *scout* is much better than *snake*, as $|I| << b$. This is due to the structure of the BFS tree T, whose number of leafs grows with the density of the graph. The same behavior is confirmed by Figures 1.(b), 1.(d) and 1.(f), where we show that *snake* is better than *scout* until a certain value of b, after which *scout* outperforms *snake*. Such value of b linearly depends on $|I|$ and it is around $35\%n$ in the CAIDA instances, around $20\%n$ in the *Barabási–Albert* instances, and smaller than $10\%n$, in the *Erdős-Rényi* instances.

Finally, our experiments show that *parallel-scout* is very effective in practice. In fact, its number of steps is always by far smaller than those required by *scout* and *snake*, as shown in Figures 1, thus resulting the best strategy in all the cases.

5   Concluding Remarks

In this paper, we have studied the problem of using mobile entities to explore and make safe a graph with black-holes. In particular, we have considered the scenario where the destruction of an entity by means of a black-hole also affects all the entities within a fixed range r. A possible future research direction is the extension of both the *scout* and *snake* strategies to the cases of $r > 1$. While for the *snake* strategy the number of time steps required for the exploration would be $2(n-1)+2rb$, for the *scout* and its *parallel-scout* variant the analysis is not so straightforward.

References

1. Albert, R., Barabási, A.-L.: Emergence of scaling in random networks. Science 286, 509–512 (1999)
2. Balamohan, B., Flocchini, P., Miri, A., Santoro, N.: Time optimal algorithms for black hole search in rings. Discrete Mathematics, Algorithms and Applications 3(4), 457–472 (2011)
3. Bollobás, B.: Random Graphs. Cambridge University Press (2001)
4. Cooper, C., Klasing, R., Radzik, T.: Locating and repairing faults in a network with mobile agents. Theor. Comput. Sci. 411(14-15), 1638–1647 (2010)
5. Czyzowicz, J., Dobrev, S., Královič, R., Miklík, S., Pardubská, D.: Black hole search in directed graphs. In: Kutten, S., Žerovnik, J. (eds.) SIROCCO 2009. LNCS, vol. 5869, pp. 182–194. Springer, Heidelberg (2010)
6. Czyzowicz, J., Kowalski, D., Markou, E., Pelc, A.: Searching for a black hole in synchronous tree networks. Combinatorics, Probability and Computing 16, 595–619 (2007)
7. Dobrev, S., Flocchini, P., Kralovic, R., Prencipe, G., Ruzicka, P., Santoro, N.: Black hole search in common interconnection networks. Networks 47(2), 61–71 (2006)
8. Dobrev, S., Flocchini, P., Prencipe, G., Santoro, N.: Searching for a black hole in arbitrary networks: optimal mobile agents protocols. Distributed Computing 19(1), 1–18 (2006)
9. Dobrev, S., Flocchini, P., Prencipe, G., Santoro, N.: Mobile search for a black hole in an anonymous ring. Algorithmica 48(1), 67–902 (2007)
10. Dobrev, S., Flocchini, P., Santoro, N.: Improved bounds for optimal black hole search with a network map. In: Kralovic, R., Sýkora, O. (eds.) SIROCCO 2004. LNCS, vol. 3104, pp. 111–122. Springer, Heidelberg (2004)
11. Flocchini, P., Ilcinkas, D., Santoro, N.: Ping pong in dangerous graphs: Optimal black hole search with pebbles. Algorithmica 62(3-4), 1006–1033 (2012)
12. Garey, M.R., Johnson, D.S.: Computers and Intractability, A Guide to the Theory of NP-Completeness. W.H. Freeman and Company, New York (1979)
13. Gąsieniec, L., Klasing, R., Martin, R.A., Navarra, A., Zhang, X.: Fast periodic graph exploration with constant memory. Journal of Computer and System Sciences 74(5), 802–822 (2008)
14. Halldórsson, M.M., Radhakrishnan, J.: Greed is good: Approximating independent sets in sparse and bounded-degree graphs. Algorithmica 18, 145–163 (1997)
15. Hyun, Y., Huffaker, B., Andersen, D., Aben, E., Shannon, C., Luckie, M., Claffy, K.: The CAIDA IPv4 routed/24 topology dataset, http://www.caida.org/data/active/ipv4_routed_24_topology_dataset.xml
16. Klasing, R., Markou, E., Radzik, T., Sarracco, F.: Approximation bounds for black hole search problems. Networks 52(4), 216–226 (2008)
17. Kosowski, A., Navarra, A.: Graph decomposition for improving memoryless periodic exploration. Algorithmica 63(1-2), 26–38 (2012)
18. Kosowski, A., Navarra, A., Pinotti, C.: Synchronous black hole search in directed graphs. Theoretical Computer Science 412(41), 5752–5759 (2011)
19. Markou, E., Paquette, M.: Black hole search and exploration in unoriented tori with synchronous scattered finite automata. In: Baldoni, R., Flocchini, P., Binoy, R. (eds.) OPODIS 2012. LNCS, vol. 7702, pp. 239–253. Springer, Heidelberg (2012)
20. OMNeT++. Discrete event simulation environment, http://www.omnetpp.org

Approximating Maximum Disjoint Coverage in Wireless Sensor Networks

Shagufta Henna and Thomas Erlebach

Bahria University, Department of Computer Science,
Islamabad, Pakistan
University of Leicester, Department of Computer Science,
Leicester, United Kingdom
shagufta_henna@yahoo.com, t.erlebach@le.ac.uk

Abstract. Due to limited battery life and fault tolerance issues posed by Wireless Sensor Networks (WSNs), efficient methods which ensure reliable coverage are highly desirable. One solution is to use disjoint set covers to cover the targets. We formulate a problem called MDC which addresses the maximum coverage by using disjoint set covers S_1 and S_2. We prove that MDC is $\mathcal{NP}$-complete and propose a $\sqrt{n}$-approximation algorithm for the MDC problem to cover n targets.

Keywords: Disjoint Coverage, Disjoint Coverage $\mathcal{NP}$-complete, $\sqrt{n}$-approximation algorithm.

1 Introduction

Wireless sensor networks support a variety of applications, such as environmental monitoring and battle surveillance. More often WSNs comprise of thousands of sensors randomly deployed in some particular area to cover some particular targets. Due to random deployment of sensor nodes in a particular area, the only better way to achieve adequate target coverage is to use more sensors than the optimal number. If a target is in the sensing range of a sensor, we say that the sensor provides coverage to that particular target. Sensors which can cover targets can be divided into sets, called set covers, where each set cover can monitor the specified targets. The coverage problem is one of the important research issues in WSNs, and reflects how well a set of targets is monitored by a set of deployed sensors. Coverage problems can be classified into area coverage and target coverage. However, our work in this paper is related with the target coverage problem only.

Sensors are small devices with limited battery for which it may not be possible to replace or recharge them. Further, sensors are prone to software and hardware failures. Sometimes harsh weather or physical environment may contribute to the failure of sensors. It is therefore critical to provide a fault-tolerant coverage that may still continuously monitor the critical targets despite some sensor failures. Coverage problems in sensor networks can be categorized into single coverage and multiple coverage. In single coverage, a target is monitored by at least a

J. Cichoń, M. Gȩbala, and M. Klonowski (Eds.): ADHOC-NOW 2013, LNCS 7960, pp. 148–159, 2013.

single sensor, whereas in multiple coverage, a target is monitored by k different sensors [1].

The problem of coverage in WSNs has been studied in various applications. In [2,3] coverage problems have been discussed to achieve an objective related with the quality of service of a sensor network. Coverage problems have also been discussed in various studies to maintain connectivity. In [4], improved coverage has been discussed for multi-hop ad hoc networks considering the constraint of limited path length. In order to achieve fault-tolerant coverage, the initial studies focus on the problem of finding a maximum number of set covers to cover some targets. Cardei and Du [5] prove that the problem to find maximum set covers to cover some targets is a $\mathcal{NP}$-complete problem, where a sensor may participate in more than one set covers. Cardei et al. [6] have proposed a breadth first search algorithm for the computation of connected set covers from a Base Station (BS) to particular targets. In the same work, they propose a distributed minimum spanning tree algorithm to address the same problem. Jaggi et al. [7] propose a set cover algorithm to maximize the network lifetime. In their problem, they try to maximize the number of disjoint set covers. In [8] the target coverage problem maps disjoint sensor sets to disjoint set covers. These disjoint set covers monitor all the targets. They give a lower bound of 2 for any polynomial time approximation algorithm for disjoint set covers such that every set cover can monitor all the targets. In [9] a heuristic known as constrained least coverage is proposed to find a maximum number of disjoint set covers.

Maximum Set Cover (MSC) [5] for complete target coverage computes non-disjoint set covers where each set cover can cover all the targets. The main objective of MSC is to determine a number of set covers where each set cover covers all the targets such that the network lifetime is maximized by alternating among these set covers. The MSC problem is a well known $\mathcal{NP}$-complete problem. However, this solution does not guarantee fault tolerance because a covering sensor may participate in more than one set covers, and therefore may deplete energy. Cheng et al. [10] have discussed that the MSC problem and similar problems which aim to achieve complete coverage by using non-disjoint set covers assume unlimited number of covering sensors to cover targets. These techniques do not consider the bandwidth constraints. They propose the use of disjoint set covers to solve this problem. In their work they compute disjoint set covers such that a set can cover no more than an assigned number of targets and their main objective is to maximize the number of disjoint set covers. Another variation of the MSC problem is MSC with disjointness constraints [11] for complete target coverage by using disjoint set covers and the main objective is to maximize the number of disjoint set covers. In [12] Abrams et al. discuss a variation of the k-set cover problem. In their problem, they relax the coverage constraint where each node may cover only partial targets, and their main objective is to increase the number of set covers to cover some targets. In order to solve this problem, they have proposed three algorithms. The first algorithm computes k-set covers with a fraction of $1 - 1/e$ of the optimum solution. The second algorithm is based on a greedy approach and gives a solution with $\frac{1}{2}$-approximation ratio. The third algorithm computes a solution with $(1 - 1/e)$-approximation ratio.

In this paper, we consider a variation of the target coverage problem of computing two disjoint set covers S_1 and S_2 such that the first set cover S_1 achieves complete target coverage, whereas the second set cover S_2 can achieve maximum coverage. In other words, our problem relates to both MSC and MSC with disjointness constraints. In particular for the first set cover S_1 our problem is based on MSC, whereas the second set cover tries to achieve the maximum coverage while holding the disjointness constraint. Our problem called Maximum Disjoint Coverage (MDC) computes two set covers S_1 and S_2 such that S_1 achieves complete target coverage and the coverage of the second disjoint set cover S_2 is maximized. In our work, first we reduce the NOT-ALL-EQUAL-3SAT problem to MDC problem. Further, we present an approximation algorithm called Disjoint Set Covers for Maximum Disjoint Coverage (DSC-MDC) to compute two disjoint set covers S_1 and S_2 for the MDC problem. We also show that DSC-MDC achieves approximation ratio $\sqrt{n}$, where n denotes the number of targets.

The remainder of the paper is organized as follows. In Section 2, we present some preliminaries that are necessary for our work in this paper. In Section 3, we formulate the MDC problem, and prove its $\mathcal{NP}$-completeness. In Section 4, we present a $\sqrt{n}$-approximation algorithm DSC-MDC to compute two disjoint set covers for MDC problem. In Section 5, we present an approximation analysis of the algorithm presented in Section 4. Finally Section 6 concludes the paper.

2   Preliminaries

In this section we discuss some of the preliminaries related with MDC problem.

2.1   Set Cover Problem

Given a universal set of elements U and a collection C of subsets of U, the set cover problem is to choose a minimum number of sets from C such that the union of all the sets covers all elements in U.

2.2   Target Coverage Problem (TCP)

Given T targets with known locations, and n sensors with known energy constraints deployed in a wireless sensor network, The Target Coverage Problem (TCP) is to schedule the activity of sensors S such that all the targets are continuously monitored and the overall network lifetime is maximized [5]. For every target $t_j \in T$, there is at least one sensor $s_i \in S$ that covers t_j, and each $s_i \in S$ may cover several targets. In order to maximize the network lifetime, activity among the sensors can be scheduled as follows:

– Based on the information of the sensor nodes S, BS uses some scheduling algorithm and broadcasts this schedule information to the sensor nodes.

– According to the schedule information received from the BS, sensor nodes S follow sleep or active intervals.

The main objective of TCP is to maximize network lifetime and at the same time continuously observe all the targets, so one viable solution is to compute a number of set covers to cover the targets. Each set cover $S_i \subset S$ covers all the targets T. The BS can schedule the activity among these set covers to adjust their sleep or active intervals in order to maximize the network lifetime.

Figure 1 shows an example of a sensor network, where base station BS has to cover targets t_0, and t_1. Assume that each of the sensors s_0, s_1, r_0, and r_1 has a battery life of one time unit. If BS uses all the sensors S to cover all the targets, it may result in network lifetime of one time unit. However, BS may select two set covers $S_1 = \{s_0, s_1\}$ and $S_2 = \{r_0, r_1\}$ to cover all the targets. BS can schedule the activity among S_1 and S_2 such that at time $t = 1$ set cover S_1 can be activated and at time interval $t = 2$ S_2 can be activated resulting in a network lifetime of two time units.

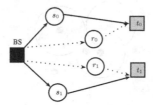

Fig. 1. Two disjoint set covers to cover targets selected by the BS

2.3 Maximum Set Covers (MSC)

The target coverage problem can be formally modelled as the combinatorial optimization problem called MSC. Given a collection S of subsets of a finite set T, compute set covers $S_1, \ldots, S_p$ and weights $w_1, \ldots, w_p$ in $[0, 1]$ such that the total weight $w_1 + \ldots + w_p$ is maximized, and for each sensor $s \in S$, s appears among set covers $S_1, \ldots, S_p$ with weight w_i such that $\sum_{i:s \in S_i} w_i \leq 1$, where 1 denotes the lifetime of each sensor [5]. An example illustrating the MSC problem is shown in Figure 2.

The MSC problem can be further illustrated with the help of Figure 2, where the three set covers computed are: $S_1 = \{s_1, s_4\}$, $S_2 = \{s_2, s_3\}$, and $S_3 = \{s_1, s_2\}$. Network lifetime can be maximized by allowing different sets to be operational at different time intervals. In this case S_1 can be active for $w_1 = 0.5$ time, S_2 for $w_2 = 0.5$ time, and S_3 for $w_3 = 0.5$ time resulting in a total network lifetime of 1.5.

2.4 Disjoint Set Covers (DSC)

Given a collection S of subsets of a finite set T, the objective of the DSC problem is to compute the maximum number of disjoint set covers for T [8]. Each set cover

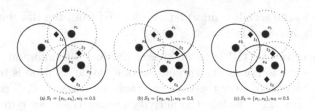

Fig. 2. Three set covers S_1, S_2, and S_3 to cover all the targets

$S_i \subset S$ must be such that every element $t_j \in T$ can be covered by at least one element of S_i, and for any two set covers S_i and S_k, $S_i \cap S_k = \emptyset$. In DSC, all the set covers should be disjoint, therefore one sensor can contribute to one set cover only. As shown in Figure 3, it is possible to cover all the targets by using two disjoint set covers $S_1 = \{s_1, s_3\}$ and $S_2 = \{s_2, s_4, s_5\}$, according to DSC problem. Both S_1 and S_2 can be activated alternatively resulting in a total network lifetime of 2 time units.

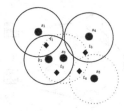

Fig. 3. Randomly deployed sensors with overlapping sensing ranges to cover targets

3   Maximum Disjoint Coverage (MDC) Problem

In this section, we define the MDC problem. It can be seen as a variation of the minimum set cover problem [13] and we prove its $\mathcal{NP}$-completeness.

3.1   Problem Description

Let us assume that there is a set of n sensors $s_i, i = 1, \ldots, n$ to cover m targets t_k, $k = 1, \ldots, m$. The goal is to divide the sensors into two disjoint set covers S_1 and S_2, such that S_1 completely covers all the targets and S_2 covers a maximum number of targets. A target is covered by a sensor if it lies within the sensing range of that sensor. Next we formally define the MDC problem which can be modelled as the combinatorial optimization problem of the target coverage problem.

Definition 1. *MDC Problem:* *Given a collection S of subsets of a finite set T, find two disjoint set covers S_1 and S_2 for T. Both set covers are subsets of S, i.e., $S_1 \subset S$ and $S_2 \subset S$, such that every element of T is covered by at least one member of S_1, and a maximum number of elements of T is covered by members of S_2, and for the set covers S_1 and S_2, $S_1 \cap S_2 = \emptyset$.*

The decision version of the MDC problem called Disjoint Coverage (DC) is stated as follows:

Disjoint Coverage (DC): Given a set of targets T and a collection S of subsets of T, determine whether S can be partitioned into two disjoint set covers that cover all elements of T.

Given a collection S of subsets of a finite set T, where S denotes the set of sensors and T denotes the set of targets, and each sensor monitors a subset of targets, an instance of the DC problem can be represented as a bipartite graph, where the set of sensors S represents the set of S-vertices and the set of targets T represents the set of T-vertices in the bipartite graph. For every target represented as vertex $t_j \in T$-vertices, there is at least one sensor represented as vertex $s_i \in S$-vertices, where $i = 1, \ldots, m$ and where $j = 1, \ldots, n$. The coverage of target t_j by some sensor s_i is represented by an edge between $t_j \in T$-vertices and $s_i \in S$-vertices in the bipartite graph. The DC problem is to compute two disjoint set covers S_1 and S_2 from S-vertices, such that both S_1 and S_2 cover all the elements of T-vertices in the bipartite graph. An instance of the DC is represented as a bipartite graph G with S-vertices and T-vertices in Figure 4.

Figure 4(a) shows an example of a sensor network where 5 sensors $s_1, \ldots, s_5$ are deployed to cover 8 targets $t_1, \ldots, t_8$. Figure 4(b) shows the corresponding bipartite graph for the sensor to target coverage for the sensor network in Figure 4(a). One solution to the MDC problem in Figure 4 is two disjoint set covers using $S_2 = \{s_1\}$ to cover 7 targets $t_1, \ldots, t_7$ and $S_1 = \{s_2, s_3, s_4, s_5\}$ to cover all the 8 targets $t_1, \ldots, t_8$.

3.2   DC Is $\mathcal{NP}$-Complete

It can be proved that DC$\in \mathcal{NP}$ because a non-deterministic solution can partition the collection C into two disjoint sub-collections and it is verifiable in polynomial time if both sub-collections S_1 and S_2 cover T completely.

We reduce the NOT-ALL-EQUAL-3SAT problem, which is known to be $\mathcal{NP}$-complete, to DC in polynomial time. The NOT-ALL-EQUAL-3SAT problem appears in Garey and Johnson [13] and is defined below:

Definition 2. *NOT-ALL-EQUAL-3SAT problem:* *Given a set U of variables, and a collection C of clauses over U, such that each clause $c_i \in C$ has size $|c_i| = 3$, the decision problem is to find out, if there is a truth assignment of U such that each clause $c_i \in C$ has at least one literal true and at least one literal false.*

Theorem 1. *DC is $\mathcal{NP}$-complete.*

Proof. Let $U = \{x_1, x_2, x_3, \ldots, x_n\}$ be a given set of variables. Let $C = c_1 \wedge c_2 \wedge \ldots \wedge c_m$ be a collection of clauses given over U, where each clause $c_i \in C$ takes variables from U, i.e., $c_i = (u_{i1} \vee u_{i2} \vee u_{i3})$, where each u_{ij} denotes x_h, or $\bar{x}_h$ for

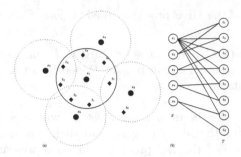

Fig. 4. (a). Example of a sensor network with 8 targets and 5 covering sensors (b). Corresponding bipartite graph for (a)

some variables in U where $h = 1, \ldots, n$. We show how to construct a bipartite graph G in polynomial time such that U has a NOT-ALL-EQUAL-3SAT truth assignment for clauses C, if and only if G has two disjoint set covers that cover all elements of T.

We first define graph H (as it can be seen in Figure 5) as a bipartite graph with the sensors S_H in one set of the bipartition, and T_H in the other, where $S_H = \{x_i, \bar{x}_i\}$ and $T_H = \{y_i\}$, where T_H denotes the set of targets. The only vertices from the H sub-graph which will be adjacent to other vertices of G will be from the set S_H, i.e., x_i and $\bar{x}_i$. So, every copy of sub-graph H in graph G will be represented by x_i and $\bar{x}_i$.

We can construct a graph G for the set C of clauses as illustrated in Figure 6 and explained below. In the construction of the graph G we have n copies of the sub-graph H shown in Figure 5. In the example the copies are H_1, H_2, and H_3, where H_1 is represented by vertices x_1 and $\bar{x}_1$, H_2 by vertices x_2 and $\bar{x}_2$, and H_3 by vertices x_3 and $\bar{x}_3$. We also add one vertex for each clause in C to G. These vertices are called clause vertices. In graph G each vertex c_i representing a clause $c_i = (u_{i1} \vee u_{i2} \vee u_{i3})$ is connected to x_{i1} if $u_{i1} = x_{i1}$, to $\bar{x}_{i1}$ if $u_{i1} = \bar{x}_{i1}$, to x_{i2} if $u_{i2} = x_{i2}$, to $\bar{x}_{i2}$ if $u_{i2} = \bar{x}_{i2}$, to x_{i3} if $u_{i3} = x_{i3}$, to $\bar{x}_{i3}$ if $u_{i3} = \bar{x}_{i3}$. The vertices x_h and $\bar{x}_h$ for $h = 1, \ldots, n$ are the S-vertices of G and the remaining vertices are the T-vertices as discussed in Section 3.1. Figure 6 shows the construction of graph G for three clauses $c_1 = \{\bar{x}_1 \vee x_2 \vee x_3\}$, $c_2 = \{x_1 \vee \bar{x}_2 \vee \bar{x}_3\}$, and $c_3 = \{x_1 \vee \bar{x}_2 \vee x_3\}$.

Given a satisfying assignment to the NOT-ALL-EQUAL-3-SAT instance, let D_1 = variables x_h in U that are true, and $D_2 = U - D_1$ constitute variables x_h in U that are false. We can construct two disjoint set covers S_1 and S_2 as follows: for each x_h in G, if $x_h \in D_1$ then place x_h in S_1, and place $\bar{x}_h$ into S_2, and if $x_h \in D_2$, then place x_h in S_2, and $\bar{x}_h$ into S_1. For example in Figure 6 for $C = \{\bar{x}_1 \vee x_2 \vee x_3\} \wedge \{x_1 \vee \bar{x}_2 \vee \bar{x}_3\} \wedge \{x_1 \vee \bar{x}_2 \vee x_3\}$, one NOT-ALL-EQUAL-3-SAT truth assignment is to let $D_1 = \{x_1, x_2, x_3\}$, then for each $x_k \in D_1$, place the literal in S_1, and place $\bar{x}_k$ in S_2. So, we have $S_1 = \{x_1, x_2, x_3\}$ and $S_2 = \{\bar{x}_1, \bar{x}_2, \bar{x}_3\}$ two disjoint set covers because both sets S_1 and S_2 cannot contain x_h and $\bar{x}_h$ in the same set cover. Therefore both S_1 and S_2 disjointly cover all T-vertices of G.

Conversely, suppose graph G has disjoint set covers S_1 and S_2. For every variable x_i, S_1 must contain one literal and S_2 its opposite. If both literals, i.e., x_i and $\bar{x}_i$ are in S_1 then y_i cannot be covered by S_2 because y_i is covered by x_i and $\bar{x}_i$ only and therefore S_2 cannot be a set cover. We may define a truth assignment by assigning true values to literal x_h if and only if its corresponding vertex in S_1, i.e., for each $x_h \in U$, if $x_h \in S_1$, set x_h to true, and if $x_h \in S_2$ then set x_h to false. Then every clause c_i is satisfied since S_1 covers the clause vertices and S_2 covers the clause vertices thus providing the disjoint coverage DC which is also a NOT-ALL-EQUAL-3SAT assignment. Finally, the reduction from NOT-ALL-EQUAL-3SAT to DC is polynomial-time computable.   □

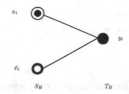

Fig. 5. Graph H, building block for graph G

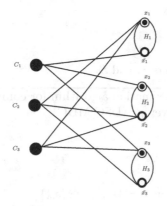

Fig. 6. Graph G for $C = \{\bar{x}_1 \vee x_2 \vee x_3\} \wedge \{x_1 \vee \bar{x}_2 \vee \bar{x}_3\} \wedge \{x_1 \vee \bar{x}_2 \vee x_3\}$

4   Approximation Algorithm for Maximum Disjoint Coverage (DSC-MDC)

In this section, we present an approximation algorithm Disjoint Set Covers for Maximum Disjoint Coverage (DSC-MDC) for the MDC problem. Given a collection S of subsets of a finite set T we want to determine two disjoint set covers S_1 and S_2 such that S_1 covers T completely and S_2 covers a maximum number of elements of T. DSC-MDC takes as an input a collection of subsets $S = \{s_1, s_2, \ldots, s_m\}$ and $T = \{t_1, t_2, \ldots, t_n\}$ where every set s_i for $1 \leq i \leq m$ denotes a set of elements in T.

DSC-MDC uses a greedy approach and selects a subset s_i from S which can cover a maximum number of elements of T. DSC-MDC evaluates if the elements of s_i can still be covered by other subsets in S, if it is true it adds s_i into set cover S_2. The algorithm DSC-MDC repeats until it has added all the possible subsets from S into set cover S_2 while ensuring that all elements of T can be still covered by subsets of S. From the remaining subsets in S, i.e., $S \setminus S_2$, a simple greedy algorithm is used to compute a disjoint set cover S_1 which can cover all the elements of set T. Finally, DSC-MDC returns two disjoint set covers S_1 and S_2 where S_1 covers all the elements of T and S_2 covers as many as possible. The algorithm is shown in pseudo-code in Algorithm 1 and Algorithm 2.

Data: Collection of subsets $S = \{s_1, s_2, \ldots, s_m\}$
Result: Two Disjoint Set Covers S_1, and S_2
$X \Leftarrow S$
$S_2 \Leftarrow \emptyset$
while $X \neq \emptyset$ **do**
    Let $s_i \in X$ be a set that increases the coverage of S_2 by as much as possible
    if *all elements of s_i can still be covered by some other sets in S* **then**
        $S_2 \Leftarrow S_2 \cup \{s_i\}$
        $S \Leftarrow S \setminus \{s_i\}$
    end
    $X \Leftarrow X \setminus \{s_i\}$
end
$S_1 \Leftarrow Greedy(S)$
Output Disjoint Set Covers S_1 and S_2

Algorithm 1: $\sqrt{n}$-approximation algorithm for computing disjoint set covers for maximum disjoint coverage (DSC-MDC)

Data: Collection of subsets $S = \{s_1, s_2, \ldots, s_m\}$
Result: Set Cover S_1
$S_1 \Leftarrow \emptyset$
while S_1 *does not cover all targets* **do**
    Let $s_i \in S$ be a set that increases the coverage of S_1 by as much as possible
    $S_1 \Leftarrow S_1 \cup \{s_i\}$
    $S \Leftarrow S \setminus \{s_i\}$
end
Return S_1

Algorithm 2: A greedy algorithm to compute set cover S_1 $(Greedy(S))$

5   Approximation Analysis

Theorem 2. *The approximation ratio of DSC-MDC is at most $\sqrt{n}$ where n is the number of elements of T.*

Proof. Let us have a collection of subsets $S = \{s_1, s_2, \ldots, s_m\}$ such that all the subsets in S can cover all n elements of T. Let us say set s_i is selected by DSC-MDC to add it into S_2 and s_i covers k_i elements in the ith iteration, where $1 \leq i \leq A$, where A is the number of sets the Algorithm 1 adds to S_2. In A iterations the total number of elements covered by S_2 denoted by $Coverage_{DSC\_MDC}$ is given as follows:

$$|Coverage_{DSC\_MDC}| = k_1 + k_2 + k_3 + \ldots + k_A \tag{1}$$

Let $\mathcal{OPT}$ denote an optimal solution to compute S_2 to cover targets T. We describe two cases below to show that in the worst case compared to the optimum $\mathcal{OPT}$ DSC-MDC is a $\sqrt{n}$-approximation algorithm.

Case 1:
Let us assume that in the A iterations, the total number of elements covered by S_2 with DSC-MDC is greater than or equal to $\sqrt{n}$. $\mathcal{OPT}$ can cover at most n elements. So we can compare the coverage of elements of T using S_2 by both DSC-MDC and $\mathcal{OPT}$ as follows:

$$k_1 + k_2 + k_3 + \ldots + k_A \geq \sqrt{n}$$
$$C(\mathcal{OPT}) \leq n$$
$$\Rightarrow \frac{C(\mathcal{OPT})}{k_1 + k_2 + k_3 + \ldots + k_A} \leq \sqrt{n}$$
$$\Rightarrow \sqrt{n} - Approximation$$

Case 2:
Let us assume that the total number of elements of S covered by DSC-MDC using S_2 in A iterations is less than or equal to $\sqrt{n}$. Let us say for each iteration, DSC-MDC covers k_i elements and the other sets covering these elements are the last available sets in S, each covering $k_i - 1$ other elements. It means for these k_i elements, DSC-MDC loses at most $k_i(k_i - 1)$ elements in the ith iteration, where $1 \leq i \leq A$. So, the total loss of elements for DSC-MDC for A iterations, denoted by $|Loss_{DSC\_MDC}|$, is at most $k_1(k_1 - 1) + k_2(k_2 - 1) + k_3(k_3 - 1) + \ldots + k_A(k_A - 1)$, i.e.,

$$|Loss_{DSC\_MDC}| \leq \sum_{i=1}^{A} k_i(k_i - 1) \tag{2}$$

On the other hand, every element covered by $\mathcal{OPT}$ is either covered by DSC-MDC or belongs to $Loss_{DSC\_MDC}$. So the total number of elements covered by $\mathcal{OPT}$ for A iterations is given as follows:

$$|\mathcal{OPT}| \leq |Loss_{DSC\_MDC}| + |Coverage_{DSC\_MDC}|$$

$$\leq \sum_{i=1}^{A} k_i(k_i - 1) + \sum_{i=1}^{A} k_i$$

We can compare the number of elements of T covered by DSC-MDC using S_2 to the total elements of T covered by S_2 computed by $\mathcal{OPT}$, i.e., $C(\mathcal{OPT})$ as follows:

$$k_1 + k_2 + k_3 + \ldots + k_A \leq \sqrt{n}$$
$$C(\mathcal{OPT}) \leq k_1^2 + k_2^2 + k_3^2 + \ldots + k_A^2$$
$$\leq (k_1 + k_2 + k_3 + \ldots + k_A)^2$$
$$\leq (k_1 + k_2 + k_3 + \ldots + k_A).\sqrt{n}$$
$$\Rightarrow \frac{C(\mathcal{OPT})}{k_1 + k_2 + k_3 + \ldots + k_A} \leq \sqrt{n}$$
$$\Rightarrow \sqrt{n} - Approximation$$

From both cases, we can conclude that DSC-MDC is a $\sqrt{n}$-approximation algorithm.   □

6   Conclusion

Due to several limitations including limited battery life and fault tolerance issues posed by wireless sensor networks, efficient methods which can ensure reliable coverage of targets are highly desirable. One solution to provide reliable coverage to targets is to organize the sensors in set covers. These set covers can monitor the targets completely. However, one sensor may participate in more than one set covers to monitor the targets which is not a very energy efficient and fault-tolerant solution. Another solution is to divide the sensor into disjoint set covers which can completely cover the targets. However, sometimes it is not possible to achieve complete target coverage while keeping the disjointness constraint. In our work we formulate a problem called MDC which is a variation of the target coverage problem. The MDC problem is to use two set covers S_1 and S_2 to maximize the target coverage while holding the disjointness constraint. We proved that the decision version of MDC problem called DC is $\mathcal{NP}$-complete and proposed a $\sqrt{n}$-approximation algorithm DSC-MDC for the MDC problem. Our algorithm computes disjoint set covers S_1 and S_2 in such a way that computation of S_2 maximizes the target coverage whereas S_1 gives the complete coverage.

References

1. Gallais, A., Carle, J.: An adaptive localized algorithm for multiple sensor area coverage. In: Proc. of the Int'l Conf on Advanced Networking and Applications, pp. 525–532 (2007)

2. Srivastava, V., Motani, M.: Worst and best-case coverage in sensor networks. IEEE Transanctions on Mobile Computing 3(4), 84–92 (2004)
3. Li, X.-Y.: Coverage in wireless ad hoc sensor networks. IEEE Transanctions on Computers 52(6), 753–763 (2003)
4. Haas, Z.: On the relaying capability of the reconfigurable wireless networks. In: Proc. of the Int'l Conf. on Vehicular Technology, pp. 1148–1152 (1997)
5. Cardei, M., Thai, M., Li, Y., Wu, W.: Energy-efficient target coverage in wireless sensor networks. In: Proc. of the IEEE Int'l Conf. on Computer Communications, pp. 1976–1984 (2005)
6. Zorbas, D., Glynos, D., Kotzanikolaou, P., Douligeris, C.: Solving coverage problems in wireless sensor networks using coversets. Ad Hoc Networks 8(4), 400–415 (2010)
7. Jaggi, N., Abouzeid, A.A.: Energy-efficient connected coverage in wireless sensor networks. In: Proc. of the Int'l Conf. on Asian International Mobile Computing, pp. 4–7 (2006)
8. Cardei, M., Du, D.Z.: Improving wireless sensor network lifetime through power aware organization. Wireless Networks 11(3), 333–340 (2005)
9. Slijepcevic, S., Potkonjak, M.: Power efficient organization of wireless sensor networks. In: Proc. of the Int'l Conf. on Communications, pp. 472–476 (2005)
10. Cheng, M.X., Ruan, L., Wu, W.: Achieving minimum coverage breach under bandwidth constraints in wireless sensor networks. In: Proc. of the IEEE Int'l Conf. on Computer Communications, pp. 2638–2645 (2005)
11. Cheng, M.X., Ruan, L., Wu, W.: Coverage breach problems in bandwidth-constrained sensor networks. ACM Transactions on Sensor Networks 3(2), 1–23 (2007)
12. Abrams, Z., Goel, A., Plotkin, S.: Set k-cover algorithms for energy efficient monitoring in wireless sensor networks. In: Proc. of the Symposium on Information Processing in Sensor Networks, pp. 26–27 (2004)
13. Garey, M.R., Johnson, D.S.: Computers and Intractability: A Guide to the Theory of NP-Completeness. W. H. Freeman (1979)

Least Channel Variation Multi-channel MAC (LCV-MMAC)

Shagufta Henna and Thomas Erlebach

Bahria University, Department of Computer Science,
Islamabad, Pakistan
University of Leicester, Department of Computer Science,
Leicester, United Kingdom
shagufta_henna@yahoo.com, t.erlebach@le.ac.uk

Abstract. The use of multi-channel MAC protocols improves the capacity of wireless networks. Efficient multi-channel MAC protocols aim to utilize multiple channels effectively. Our proposed multi-channel MAC protocol called LCV-MMAC effectively utilizes the multiple channels by handling the control channel saturation. LCV-MMAC demonstrates significantly better throughput and fairness compared to DCA, MMAC, and AMCP in different network scenarios.

Keywords: Multi-channel MAC, Control channel saturation, Channel busyness ratio.

1 Introduction

The half-duplex nature of transceivers, limited bandwidth, interference and topology of wireless networks are the fundamental constraints on the capacity of wireless networks. There exist different solutions to enable spatial reuse by mitigating interference, including different Media Access Control (MAC) methods, transmission power control protocols, and use of directional antennas instead of omnidirectional antennas. The capacity of wireless networks by using multi-channel communication is well studied by Kyasanur et al. [1]. Their results show that it is possible to improve the capacity of wireless networks even if the number of radios is smaller than the number of channels.

Recently, several multi-packet transmission/reception techniques [2] have been proposed to increase the capacity of wireless networks with an increase in the number of nodes. However, these techniques are expensive to deploy as they require multiple radios at each node. A low cost, yet attractive solution is to use multiple channels offered by the IEEE 802.11 PHY. Its Distributed Coordination Function (DCF) originally uses a single common channel for transmissions, however it has the capability to support multiple channels. Significant capacity improvement is possible by using multiple channels for simultaneous transmissions. Another potential benefit of using multiple channels is fairness.

The use of multiple channels for transmissions has raised several challenges, for example, multi-channel hidden terminal problem, channel switching delay,

J. Cichoń, M. Gębala, and M. Klonowski (Eds.): ADHOC-NOW 2013, LNCS 7960, pp. 160–171, 2013.

and control channel saturation problem. Dedicated control channel techniques simplify channel coordination by eliminating the need for synchronization, however the control channel may become the bottleneck for the performance of the network. A better trade off which can solve the coordination problem, and can mitigate the control channel bottleneck is desirable. Further, dynamic channel assignment techniques may induce significant channel switching delay, which if not controlled properly may result in significant increase in latency.

In this paper, we propose a protocol called Least Channel Variation Multi-channel MAC (LCV-MMAC) based on IEEE 802.11 MAC. The novel part of this protocol is the channel assignment technique, where a mechanism to avoid unnecessary channel assignment and thus channel switching is used. The main highlights of our work in this paper are the following:

- We devise a multi-channel MAC protocol LCV-MMAC which uses an efficient channel selection technique. In particular LCV-MMAC mitigates the control channel saturation and improves the aggregated throughput, with a reasonable fairness in different network scenarios.
- We explore the properties of LCV-MMAC through extensive simulations with the help of ns-2, and compare it with popular existing multi-channel MAC protocols. Experimental results validate that LCV-MMAC achieves better aggregated throughput, and fairness index than other multi-channel MAC protocols in highly congested single hop and multi-hop network scenarios.

The remainder of the paper is organized as follows: Section 2 presents the related work. In Section 3, we discuss the motivation for the design of LCV-MMAC. Section 4 introduce the technique for control channel detection, and in Section 5 we introduce LCV-MMAC. In Section 6, we evaluate the performance of LCV-MMAC in different network scenarios. Section 7 concludes the paper.

2   Related Work

Vaidya and Jungmin [3] propose Multi-channel MAC (MMAC) which is a well known time multiplexed multi-channel MAC protocol. MMAC periodically transmits beacons with a beacon period of $100ms$, with an Ad hoc Traffic Indication Message (ATIM) window of $20ms$ and a DATA window of $80ms$. During the ATIM window, all nodes listen on the control channel and contend for the free channels. Data transmissions take place on all the available data channels during the DATA window. MMAC uses RTS, CTS, and Reservation (RES) packets to negotiate the data channels during the ATIM window to handle the hidden terminal problem. Periodic exchange of RTS/CTS, ATIM, ATIM Acknowledgement (ATIM-ACK), and ATIM Reservation (ATIM-RES) packets during the ATIM window negatively impact the efficiency of the control channel resulting in reduced scheduling capacity.

DCA [4] is a multi-radio multi-channel MAC protocol and requires two radio interfaces. One interface is permanently tuned to the control channel to facilitate the channel negotiation process, and the second one is dynamically tuned to a data channel. The use of a dedicated control channel eliminates the need for synchronization. However, one of its major limitations is the control channel saturation, where the control channel may become a bottleneck for the network performance [5].

Asynchronous Multi-channel Coordination Protocol (AMCP) [6] uses a dedicated control channel and n data channels with a half duplex transceiver. Nodes exchange control messages on the control channel in order to negotiate and reserve the data channel. Each node maintains a table, where each table entry records a *channel*, a *bit* indicating its availability and a *timer* indicating the duration of the time a channel is being used. Further, AMCP uses a *prefer* variable to decrease the probability of collisions which in turn increases the aggregated throughput and fairness among the flows. High contention or queues at the control channel may result in underutilization of data channels.

Bi-directional Multi-channel MAC protocol (Bi-MMAC) [7] is an extension to MMAC with bidirectional flow of data. Bi-MMAC uses the RTS/CTS handshake mechanism complemented with Channel Reservation Notification (CRN). A sender explicitly sends CRN to its neighbours about the channel reservation and its duration. Similar to MMAC, Bi-MMAC uses K channels where one channel is used as the control, and the remaining $K - 1$ as the data channels. However, Bi-MMAC differs from other multi-channel MAC protocols by facilitating the DATA frame exchange in both directions. The exchange of RTS/CTS/CRN on the control channel may result in a bottleneck in heavily loaded network scenarios. Other popular techniques include multi-radio multi-channel MAC approaches [8] with static and dynamic channel selection and target improved network throughput by reducing the number of collisions.

All the above multi-channel MAC protocols suffer from control channel saturation or channel switching problems. Both these problems degrade the performance of these protocols significantly. In this paper, we introduce LCV-MMAC protocol which uses a limited number of channels and a half-duplex transceiver. LCV-MMAC mitigates the control channel saturation and channel switching delay improving the aggregate throughput. LCV-MMAC is simple and does not need any network wide periodic synchronization. It avoids frequent channel switching and channel contention, if there is no significant performance gain. Simulations in ns-2 show that LCV-MMAC improves the aggregated throughput significantly by incorporating both the channel switching gain, and mitigating the control channel saturation.

3   Motivation for Design of LCV-MMAC

The main objective of channel assignment is to switch a limited number of transceivers to available channels in order to transmit data. An efficient multichannel technique requires channel usage information for channel assignment in order to avoid collisions. An efficient multi-channel MAC protocol should be designed by considering two issues: acquiring channel usage information and channel switching delay. For the second problem, a per-packet channel assignment technique is not preferred, instead a channel assignment technique which is valid for long time should be preferred. In order to fully exploit the potential of channel diversity, proper channel selection and control channel saturation problems need to be addressed. Below, we describe control channel saturation problem.

Channel Saturation Problem

Several existing multi-channel MAC approaches assume that the channel switching delay is low and have small switching penalty [3]. However, practical measurements on IEEE 802.11 MAC show that the channel switching penalty is rather high which adds delay. In order to co-ordinate and reserve the channel, both the sender and receiver need to exchange the control messages on the control channel. The control channel is able to accommodate a limited number of data channels. Frequent exchange of control messages without any significant increase in the capacity results in high control channel saturation. An efficient channel assignment technique which can avoid the channel switching and channel contention during control channel saturation may result in efficient use of control and data channels, and therefore may improve the aggregated throughput. We propose a multi-channel MAC protocol called LCV-MMAC which utilizes an effective channel assignment technique and avoids unnecessary channel assignment and control channel contention when the control channel is highly saturated.

4   Detection of Control Channel Saturation

In this section, we characterize how to measure the control channel status, and based on the control channel status we propose a novel channel assignment technique.

4.1   Channel Busyness Ratio: An Accurate Measure of Channel Saturation

In the current IEEE 802.11 MAC standard which is a CSMA-based MAC protocol with the capability to use both the physical, and virtual carrier sensing, the function which can measure the status of the channel, i.e., busy or idle is already available. The channel busyness ratio R_{busy} [9] is defined as the ratio of time intervals a channel is busy due to collisions or transmission to the total time. The channel busyness ratio provides a good early sign of control channel saturation. Let T_{suc}, and T_{col} be the time periods associated with the successful transmission, and a transmission resulting in collision, respectively. Then with the RTS/CTS enabled [9]:

$$T_{suc} = rts + ccts + crn + data + ack + 3sifs + difs \qquad (1)$$

$$T_{col} = rts + cts\_timeout + difs = rts + eifs \tag{2}$$

Where rts denotes time to send an RTS, and $ccts$ denotes time to receive a successful CTS packet from the receiver. The time to send a control channel reservation notification to inform neighbouring nodes about the channel reservation is denoted by crn in Equation 1. The $data$ denotes the average length of the data packet in seconds for successful transmission. Both $sifs$, and $difs$ denote short inter-frame spacing, and distributed inter-frame spacing. When a node experiences a collision it adjusts its NAV with an extended inter-frame spacing $eifs$ period as shown in Equation 2. The channel busyness ratio R_{busy} of the control channel can be computed as follows:

$$R_{busy} = \frac{T_{suc} + T_{col}}{T_{tot}} \tag{3}$$

Equation 3 defines the R_{busy} as the ratio of the total lengths of busy periods due to collisions or successful transmissions to the total time T_{tot} during a time interval. The channel busyness ratio provides a good sign of early control channel saturation, we can use the observed channel busyness ratio at a node for efficient channel assignment. In order to avoid control channel saturation, we take R_{busy} as an inception of control channel saturation and compare it with a predefined threshold Th_{busy}. For MANETs, a payload size of $1000-1500$ bytes is commonly used so according to [10], setting Th_{busy} to 92% is a good way to detect that the control channel has entered into saturation.

5   Protocol Description of LCV-MMAC

5.1   Structures and Variables

LCV-MMAC uses one control channel and N data channels. Each node is equipped with a half-duplex transceiver, and therefore can listen or transmit at a time only. Other data structures and variables used are defined below.

- *Channel Table:* Each node maintains a channel table, where each entry records a data channel, the neighbour who is using it, and the timer when the channel will be released by the neighbour. Each table entry also has an availability called avail bit which indicates that the channel is available or not. Similar concepts have been used in [6].
- *Data Channel Usage Counter (U_i):* A channel usage counter U_i is computed by counting the number of times a particular channel i is used by the neighbouring nodes.
- *$prefer_c$:* The variable $prefer_c$ indicates the preferable channel selected by the node.

5.2   Basic Protocol Operation

Channel selection is not needed for every transmission. If a neighbour already knows the receiver's channel, it selects the same channel for transmission. Further, every node detects the control channel saturation by using the channel busyness ratio R_{busy} as discussed in Section 4. If the control channel is saturated, a node continues with the last known data channel to transmit to the receiver without any channel switching. It is likely that the receiver stays switched on the same channel. If a node has no knowledge about the receiver's last known channel and the control channel is saturated or not saturated, the procedure below is followed for channel selection.

Channel Selection. Node x iterates through the channel table, and looks for the available channels in the table. x compares the current data channel usage U_c of channel c with the data channel usage U_i of all the available channels, where i denotes the ith available channel. If no data channel with lower channel usage than U_c is found, the node prefers to use the current data channel c as the preferable channel $prefer_c$ and does not switch the transceiver to any other channel. If a data channel U_i with lower channel usage than U_c is found, the node selects U_i as the preferable channel $prefer_c$. If the control channel saturation is detected by using channel busyness ratio R_{busy} as discussed in Section 4, the transceiver stays switched on the last known data channel with the particular receiver. Further, in case of control channel saturation, the node will not contend the control channel for channel negotiation, and will directly go on to the Data Transmission stage.

Channel Contention. x contends the control channel only if the node wants to switch to another channel as discussed in Section 5.2. If no control channel saturation is detected then $prefer_c$ is the preferable channel selected by x node. x inserts the $prefer_c$ to its RTS packet and contends on the control channel using the CSMA/CA mechanism.

Channel Negotiation/Reservation. When a node receives the RTS packet, it checks the status of channel $prefer_c$ in its channel table. If channel $prefer_c$ is available, the node replies to x with a Confirming CTS (CCTS) packet containing $prefer_c$. Then, it switches to data channel $prefer_c$ and waits for a DATA packet. However if data channel $prefer_c$ is not available, the node replies to the sender x with a Rejecting CTS (RCTS) and includes its available data channels in it, and stays switched on the control channel.

If sender x receives a CCTS, it broadcasts a CRN on the control channel. If a RCTS is received, the sender randomly selects a common channel among its and the receiver's preferable channels. If a match is found, the sender contends the control channel for RTS/CTS. If no match is found, the sender selects another preferable channel $prefer_c$ with minimum channel usage, and appends it in the RTS to begin a new contention cycle on the control channel. The sender retries to send RTS up to a maximum number of retries, and afterwards discards the packet.

Data Transmission. A node responds with an acknowledgement ACK packet after receiving a DATA packet on the $prefer_c$ channel, and then switches back to the control channel. Node x after receiving the ACK packet also switches to the control channel. However, if the control channel is still saturated, x will continue on the same data channel instead of switching to the control channel.

Deferral of Transmission for the Neighbouring Nodes. Upon overhearing a CCTS, neighbouring nodes update their channel table. Further these nodes adjust NAV according to the channel reservation duration from the CCTS. Initially the neighbouring nodes of sender x which are hidden for the receiver adjust the initial value of NAV according to the channel reservation duration from the RTS, and defer the transmission. However, after receiving a CRN from the sender, they update their NAV based on the channel reservation duration from CRN, and defer their transmission accordingly on that channel. Other than the channel listened from CRN/CCTS, neighbouring nodes can compete for other preferable channels according to the channel usage information. If neighbouring nodes overhear RCTS, no deferring rule is applied and nodes can contend for their preferable channels.

Updating Channel Usage Information/Channel Preferences. Neighbouring nodes update their channel usage information according to the information received from the overheard RTS/CCTS/CRN packets by switching to the control channel.

6   Performance Evaluation

We evaluate the performance of LCV-MMAC by using ns-2, and compare it with MMAC, AMCP, DCA, and single channel 802.11 in different network topologies namely grid, chain, and random. We run all experiments for a simulation time of 300 seconds. We have used the AODV routing protocol for all the experiments. Unless stated otherwise, the distance between the nodes is $250m$. We have used 4 channels for LCV-MMAC, 1 for the control and the other 3 as the data channels. Other simulation parameters are listed in the Table 1.

6.1   Chain Topology

In chain topology, the distance between two nodes is $250m$. At any instant of time, two TCP connections traverse the chain. We repeat each experiment for 10 random runs. Fig 1 below shows that for all the protocols, the aggregated throughput is good for a 4-node topology. Beyond 6 nodes, the dropoff in the aggregated throughput is gradual. For a 4-node network, LCV-MMAC has comparable aggregated throughput with DCA, and better than MMAC, AMCP, and 802.11 MAC. LCV-MMAC avoids control channel contention and thus negotiation if control channel saturation is detected while still using data channels. The IEEE 802.11 MAC protocol has poor performance as it is not able to cope

Table 1. IEEE 802.11 system parameters

Parameter Name	Parameter Value
SIFS	$10\mu s$
DIFS	$50\mu s$
Long Retry Limit	7
Short Retry Limit	4
CW_{min}	31
CW_{max}	1023
EIFS	$364\mu s$
DATA packet	8000 bits
Bit rate for DATA packets	2 Mbps
Channel switching delay	$224\mu s$
RTS packet	160 bits + Phy header + MAC header
CCTS/RCTS,ACK packet	112 bits + Phy header
CRN packet	160 bits + Phy header + MAC header

with the contention and collisions in the network. MMAC has the worst performance when the number of nodes in the chain exceeds 4 due to the fact that each node intends to send packets to multiple outgoing destinations resulting in a Head of Line (HOL). After a successful contention, the node transmits on the reserved channel for one neighbour only. Fig 1 shows that compared to IEEE 802.11 MAC, the average aggregated throughput advantage of LCV-MMAC on a large chain network (more than 14 nodes) is 59.64%. The DCA protocol provides only 50.35% aggregated throughput advantage over IEEE 802.11 MAC for large chain networks.

Fig 2 shows that as the number of nodes in the chain increase, the average delay experienced by 802.11 MAC increases which is due to increased contention along the route. Multi-channel MAC protocols alleviate both contentions and collisions, which can significantly reduce the delay. LCV-MMAC and other multi-channel protocols have almost constant average delay with an increase in the number of nodes in the chain. MMAC requires nodes to align their handshake and channel negotiations with the ATIM phase and therefore induces additional delay. When the number of nodes exceeds 10, nodes in the network are not able to synchronize their schedule with each other therefore they schedule their transmissions without aligning ATIM phase which reduces its delay. In order to analyze how effectively channel capacity is shared among all the flows, we have used Jain's Fairness Index [11]. This is defined in Equation 4.

$$\text{Fairness index } (FI) = \frac{\left(\sum_{i=1}^{m} x_i\right)^2}{m \sum_{i=1}^{m} x_i^2} \tag{4}$$

Where m denotes the number of contending flows in the network, x_i is the throughput achieved by flow i. Absolute fairness is achieved when $FI = 1$. The worst case unfairness occurs when $FI = \frac{1}{m}$.

As shown in Fig 3, both DCA and MMAC do not have good FI with value ranging from 0.45 to 0.85. 802.11 MAC also suffers in terms of FI. The reason is the capture behaviour of these protocols which is contention based causing control channel saturation. LCV-MMAC and AMCP solve these problems and therefore have better fairness with $FI = 1$.

6.2 Grid Topology

We have placed 100 nodes in rows and columns in a 10×10 grid and the distance between these nodes is $250m$. We established a varying number of TCP NewReno connections between randomly chosen source and destination from 2 to 16 in steps of 2. TCP connections were randomly started between 0 and $1000ms$ to mitigate periodic congestion effects. Each experiment is simulated for a period of 300 seconds, and each experiment is repeated 10 times with different seed for MAC backoff timer and with different randomly generated connections. We have used a different seed for the random source/destination generator for each repetition. A total of 10 different scenarios with different randomly generated connections were tested for each repetition.

Fig 4 shows that multi-channel MAC protocols have significantly better aggregated throughput compared to 802.11 MAC. This is due to the use of multiple channels giving a significant bandwidth increase. With an increase in the number of flows, contention in the network increases and therefore degrades the aggregated throughput. LCV-MMAC provides higher aggregated throughput compared to both 802.11 MAC and AMCP under low as well as high loads. It achieves comparable performance with DCA and MMAC under moderate traffic load. DCA alleviates the hidden terminal problem by using two transceivers, and has better performance in the grid topology compared to other multi-channel MAC protocols. The average improvement of the AMCP protocol over 802.11 MAC is about 41% for 10 connections. On the other hand, the LCV-MMAC protocol improves the aggregated throughput for 10 connections which is comparable with both DCA and MMAC protocols.

As shown in Fig 5 with an increase in the number of flows, the average delay experienced by 802.11 MAC increases due to increased contention. The average packet delay of multi-channel protocols is significantly smaller as there are fewer packet collisions and hence fewer retransmissions, in particular under high traffic loads. However, multi-channel MAC protocols show no significant advantage under lower traffic loads. LCV-MMAC shows comparable average delay with DCA and AMCP by avoiding control channel contention, and therefore channel negotiation procedure when it detects the control channel saturation.

Fig 6 shows that with an increase in the number of connections, the FI of all MAC protocols degrades. When the number of flows in the network is small, different flows are not likely to compete with each other. When the number of flows in the network increases, contention among the flows increases resulting in continuous backoffs and retransmission. The increase in the number of backoffs and retransmissions increases unfairness. LCV-MMAC achieves fairness comparable to AMCP. Under high contention LCV-MMAC avoids channel contention

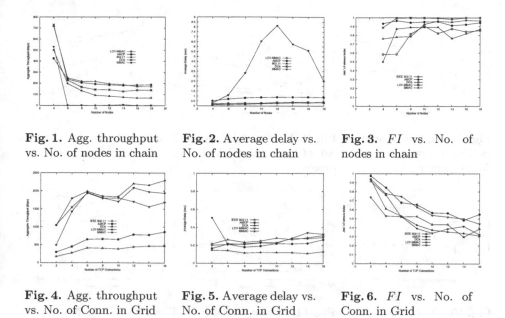

Fig. 1. Agg. throughput vs. No. of nodes in chain

Fig. 2. Average delay vs. No. of nodes in chain

Fig. 3. FI vs. No. of nodes in chain

Fig. 4. Agg. throughput vs. No. of Conn. in Grid

Fig. 5. Average delay vs. No. of Conn. in Grid

Fig. 6. FI vs. No. of Conn. in Grid

and therefore avoids frequent backoffs and retransmission which provides fair competition for data channels.

6.3   Random Topology

We simulated two kinds of random geometric graphs. In the first random topology, in a flat area of $500m \times 500m$, 100 nodes are placed randomly. The second topology is the same as the first topology; however, the area is reduced to $250m \times 250m$ to simulate a denser scenario. We varied the seed for both random topology generators and the random source/destination. For 10 random TCP connections, we tested 10 random geometric graphs yielding a total of $10 \times 10 = 100$ topologies/scenarios.

Low Density. LCV-MMAC performs approximately 6 times better than 802.11 MAC. It can be seen in Fig 7 that LCV-MMAC achieves 50% improvement over DCA by mitigating the control channel saturation. With an increase in the number of flows, local contention increases which results in an increased number of retransmissions and backoffs, which cause an increase in average delay. It can be seen from Fig 8 that by avoiding channel switching during the control channel saturation, LCV-MMAC achieves lower average delay than 802.11 and AMCP.

It is apparent from Fig 9 that the FI of all MAC protocols drops with an increase in the number of flows in the network. In the low density random network scenario, the FI of LCV-MMAC is comparable with AMCP and IEEE 802.11.

High Density. Since a $250m \times 250m$ topology is almost a single cell topology, all 100 nodes share the same control channel. When the number of flows increases

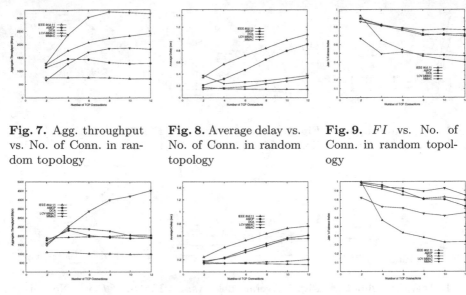

Fig. 7. Agg. throughput vs. No. of Conn. in random topology

Fig. 8. Average delay vs. No. of Conn. in random topology

Fig. 9. FI vs. No. of Conn. in random topology

Fig. 10. Agg. throughput vs. No. of Conn. in random topology

Fig. 11. Average delay vs. No. of Conn. in random topology

Fig. 12. FI vs. No. of Conn. in random topology

in the network, the control channel becomes saturated. As 802.11 MAC cannot support simultaneous transmissions, it has the worst aggregated throughput in denser network scenarios as shown in Fig 10. Control channel saturation in DCA does not exploit the effective use of available data channels. It is apparent from Fig 10 that by avoiding channel switching and therefore channel contention if the control channel is saturated, LCV-MMAC achieves more than 100% aggregated throughput improvement over all MAC protocols.

The average improvement of LCV-MMAC over the 802.11 MAC protocol for the $500m \times 500m$ topology, and the $250m \times 250m$ topology are 76%, and 75% respectively as shown in Fig 7, and Fig 10. LCV-MMAC improves the throughput over MMAC by 47% and 60% for both topologies. Fig 10 and Fig 7 show that for both topologies all the protocols achieve almost constant aggregated throughput, however LCV-MMAC in high density topology shows an increase for aggregated throughput with an increase in number of flows. The average delay achieved by LCV-MMAC is significantly lower than 802.11, AMCP, and MMAC as shown in Fig 11. It is apparent from Fig 12 that the control channel saturation degrades the FI of DCA significantly. For more than 4 flows, its FI drops below 0.5. LCV-MMAC tries to reduce the control channel saturation and channel contention, and has better FI than DCA. LCV-MMAC can suffer into multi-channel hidden terminal problem because there is no channel co-ordination during the control channel saturation.

7   Conclusion

The main objective of most of the research in multi-channel MAC protocols is to find out ways which can use multiple channels in an efficient way, thus further improving the aggregated throughput and fairness. LCV-MMAC improves the aggregated throughput in different network scenarios including random geometric graphs compared to single channel and some other multi-channel MAC protocols. LCV-MMAC uses an efficient channel selection technique according to the control channel saturation and avoids unnecessary channel contention. LCV-MMAC clearly demonstrates significantly better aggregated throughput performance and fairness compared to DC, MMAC, and AMCP in chain, grid and random network scenarios under high traffic load. LCV-MMAC, however, can suffer into multi-channel hidden terminal problem due to lack of co-ordination during control channel saturation.

References

1. Kyasanur, P., Vaidya, N.H.: Capacity of multi-channel wireless networks: impact of number of channels and interfaces. In: Proc. of the Int'l Conf. on Mobile Computing and Networking, pp. 43–57 (2005)
2. Garcia-Luna-Aceves, J., Sadjadpour, H.R., Wang, Z.: Challenges: Towards truly scalable ad hoc networks. In: Proc. of the Int'l Conf. on Mobile Computing and Networking, pp. 207–214 (2007)
3. So, J., Vaidya, N.: Multi-channel MAC for ad hoc networks: Handling multi-channel hidden terminals using a single transceiver. In: ACM Symposium on Mobile Ad Hoc Networking and Computing, pp. 222–233 (2004)
4. Wu, S., Lin, C.Y., Tseng, Y.C., Sheu, J.P.: Multi-channel MAC for ad hoc networks: Handling multi-channel hidden terminals using a single transceiver. In: Proc. of the Int'l Conf. on Symposium on Parallel Architectures, Algorithms and Networks, pp. 232–237 (2000)
5. Wu, P.J., Lee, C.N.: Connection-oriented multi-channel MAC protocol for ad-hoc networks. Computer Communications 32(1), 169–178 (2009)
6. Jingpu, S., Salonidis, T., Knightly, E.: Starvation mitigation through multi-channel coordination in CSMA multi-hop wireless networks. In: ACM Symposium on Mobile Ad Hoc Networking and Computing, pp. 214–225 (2006)
7. Kuang, T., Williamson, C.: A bidirectional multi-channel MAC protocol for improving TCP performance on multihop wireless ad hoc networks. In: Proc. of the Int'l Conf. on Symposium on Modeling and Simulation of Wireless and Mobile Systems, pp. 301–310 (2004)
8. Vaidya, N., Kyasanur, P.: Routing and interface assignment in multi-channel multi-interface wireless networks. In: Proc. of the IEEE Conf. International Wireless Communications and Networking Conference, pp. 2051–2056 (2005)
9. Zhai, S., Chen, X., Fang, Y.: Call admission and rate control scheme for multimedia support over IEEE 802.11 wireless LANs. Wireless Networks 12(4), 451–463 (2006)
10. Zhai, S., Chen, X., Fang, Y.: Improving transport layer performance in multihop ad hoc networks by exploiting MAC layer information. IEEE Transanctions on Communications 6(5), 1692–1701 (2007)
11. Jhon, R.: The Art of computer systems analysis. Jhon Wiley and Sons (1991)

Modelling Wireless Sensor Networks for Performability Evaluation

Fredrick A. Omondi, Enver Ever, Purav Shah, and Orhan Gemikonakli

School of Science and Technology, Middlesex University
The Burroughs
London
NW4 4BT
{f.adero,e.ever,p.shah,o.gemikonakli}@mdx.ac.uk

Abstract. The higher demand for use of Wireless Sensor Technology, in the presence of complexities of various deployment environments, and application areas such as wireless multimedia sensor networks, call for the need to improve performance and availability of WSNs. This paper seeks to justify the need to model both performance and availability of WSNs together, parting from the current independent approaches and provides a systematic modelling approach for performability of WSNs. This has further been necessitated by positive research findings facilitating repair and replacement of faulty and dead sensor nodes and communication links. Two different analytical solution approaches are employed for performability modelling of a WSN cluster, and simulation results presented are in agreement with the analytical approximations.

Keywords: Wireless Sensor Networks, Modelling, Performance, Availability, Reliability, Performability.

1 Introduction

Wireless Sensor Networks (WSNs) have recently found applications in numerous areas including seismic, acoustic, chemical, physiological sensing as well as interconnected devices that are able to ubiquitously retrieve multimedia content. Most common applications have included battlefield surveillance, home security, habitat monitoring, forecast systems, health monitoring, industrial systems, traffic control and animal tracking among many others. Over the years, the performance of WSNs has been mainly hindered by either/or node, link and network failures resulting from hardware and software failures. In [1], failures have been further categorised into network, cluster and node failures, depending upon where the failures occur. Complete Sensor network failure can occur if the base station or a number of member cluster heads fail.

In order to improve the lifetime of sensor nodes, two main approaches are used; one to develop mechanism to prolong the battery life and another to replace dead or failed nodes. Recent research by [2], [3], [4] show that in order to extend the lifespan of the nodes, the limited battery energy is conserved by

J. Cichoń, M. Gębala, and M. Klonowski (Eds.): ADHOC-NOW 2013, LNCS 7960, pp. 172–184, 2013.
© Springer-Verlag Berlin Heidelberg 2013

only utilising it to cover the period of operations. The use of sleep and active modes to help prolong node energy usage has also been proposed in [2], [5]. Other proposed methods include the use of Back-Up Cluster Heads (BCH) for node failures [6], reduced transmission and data aggregation power [7], and the use of energy aware protocols [4]. Recent research indicate the use of mobile nodes in repairing failing nodes [8], [9], [10], [11], thus emphasizing the need for performance and availability modelling in WSNs. Recent research shown in [2], [6], [12], [13] consider performance and availability separately and thus, there exists a need to consider performance and availability together.

This paper presents a modelling approach for performance and availability studies together for WSNs in the presence of failures, repairs and restoration. The rest of this paper is organised as follows: Sections 2 and 3 reviews current research trends in the areas of Performance and Availability in WSNs respectively, Section 4 discuses the necessity of performability in WSNs, and the applicability of existing performability modelling methods, Section 5 presents the approaches used for performability modelling of a WSN cluster using two different analytical solutions as well as simulation, Section 6 discusses the results obtained using three different methods of evaluation and finally, Section 7 concludes the current work and provides future directions.

2   Related Work for Performance Evaluation of WSNs

Performance modeling and analysis has been and continues to be of great practical and theoretical importance in supporting research as well as in the design, development and optimization of computer and communication systems and applications. The current trend towards the use of WSNs in various application areas also brings with it the need for more performance and availability modelling for optimization of deployment of WSNs.

In [14], a simulation technique is used to evaluate the performance of known hierarchical routing protocols like LEACH, PEGASIS and VGA. Results indicate that there exists a trade-off for the right choice of routing protocols in order to achieve the required performance. However, recently, there has been a growing interest in unequal clustering techniques to improve the overall network lifetime significantly and combat the hotspot problem that exists in multihop WSNs [15]. It is worth noting that network coverage which is dependent upon good connectivity is a key factor for better performance in WSNs.

In [2], a Markov model for WSNs whose nodes may enter sleep mode was presented, which was used to investigate the system performance in terms of energy consumption, network capacity, and data delivery delay. It also investigated the trade-off which exists between the performance metrics and the sensor dynamics in sleep/active mode of WSNs.

In [16], it was noted that due to limited hardware consumption, optimizing node packet buffer and maximizing performance is necessary to improve transmission Quality of Service (QoS) in WSNs. In addition, recently there are some studies on development of middleware that separates the interaction behaviour of sense-and-react WSN applications, in an attempt to improve the performance as well as reliability for mission critical applications [17]. A packet buffer evaluation method using Queuing Network Models was proposed where, blocking probabilities and system performance indicators of each node were then calculated using approximate iterative algorithm for blocking probabilities. Given that this study concentrated only on single server model in WSNs and methods for calculating packet buffer capacity for nodes, they also indicate that Sink nodes require higher performance. However the effects of node failures are not considered.

In order to address convergence related issues, a new structure was proposed which converges WSNs and Passive Optical Networks(PON) in [18] . The performance of the structure is modelled and analysed using two M/M/1 queues in tandem. The results indicate how WSN and PON dimensions affect the average queue length, hence may be used as a guideline for resource allocation.

3   Current Research in Reliability/Availability of WSNs

The main drawback to the provision of high availability demanded by WSN applications has been limited lifetime, service attacks by intruders, software and hardware failures just to mention a few. There has been recent research in this area [12], [1], [19]. Research in [6], [20] which presents use of Back up and Secondary Cluster Heads (BCH) respectively as a form of redundancy when a cluster head fails do not take into account the performance degradation due to replacement and transfer delays between failing CH and BCH in the event of failure. In [12], a Markov model characterizing fault-tolerant sensor node for applications with high reliability requirements is proposed which is based on the novel concept of determining the coverage factor using sensor fault detection algorithm accuracy, but fails to presents performance and availability simultaneously. In [13], reliability and producibility of WSN were investigated and it was concluded that star topology showed better reliability and producibility but at the cost of limited network size. This limitation was solved using cluster topology for multi-hop communication which is also limited by a central point of failure at the CH which automatically disconnects child nodes from the sink node. Further studies on the effect of unreliable WSN links on dependability parameters and the adoption of non linear battery discharges were also proposed. Almasaeid et al. [9] proposed to minimize the number of additional nodes needed to repair the connectivity by achieving a certain level of fault tolerance using "on the minimum K-connectivity algorithm".

In order to improve coverage and connectivity when nodes begin to fail, it was proposed in [8] to use a mobile robot to replace the failed sensor nodes with new ones, where the robot is able to strategically place the new sensor nodes in a central location that would enable maximum habitat coverage. In another study, Song et al. [21] presented the design and implementation of a reconfigurable robot that can serve as a mobile node for wireless sensor networks making it adaptable to changing terrain conditions in real-world applications. Jun. et.al

[11] developed a jumping WSN robot node for use in repair of broken network connections which can provide powerful maintenance support for applications in unfriendly environments.

It is evident that failing nodes can be repaired through software configurations and replacement using mobile nodes in cases of complete failure. These developments further promotes the need for performability evaluation of WSNs.

4   The Necessity of Performability

Previous research indicate limited work in the area of WSN performance and availability modelling. Apart from modelling, other studies have attempted to tackle known issues of energy, routing, topology, reliability and dependability in an attempt to optimize the performance of WSNs.

In [11] and [22], mechanisms for repairing and replacing failing nodes to sustain longer network life time have been developed. Together with recent work in the areas of performance and availability, it is now possible to combine performance and availability together in order to model the system's behaviour and analyse performance with existence of failures and repairs. The modelling and evaluation of performability of WSNs has not been considered before and such models can be used to achieve efficient and reliable configurations to optimizevarious aspects of WSNs.

Because of the wireless nature, and complicated configuration, WSNs may need to deal with failures. For such scenarios, pure performance models that ignore failure and repair recovery generally overestimate the system's ability to perform [23]. On the other hand, pure availability analysis tends to be conservative since performance considerations are not taken into account [23]. In order to obtain realistic composite performance and availability measures, consideration of performance changes associated with failure and recovery behaviour is proposed. In order to analyse the performance degradation caused by failures, performability modelling is the way forward [23], [24], [25].

Performance and availability evaluation has successfully been used to model and analyse communication and computing systems over the years [23], [24], [25], [26], [27]. There is however no record of previous research on WSNs performability. This is mostly attributed to several deployment challenges, top among them battery power depletion which normally reduce the lifespan of WSN networks. With successful research being done to improve the life span of the nodes, WSNs tend to inhibit similar characteristics of communication networks hence the available modelling and solution techniques may be successfully used to model these networks.

To obtain realistic solutions, state space models have been employed successfully to model complicated systems exhibiting transitions between various independent states [28] [20]. These models may be broadly categorised as; Markovian, Non-Markovian and Non-Homogeneous Markov models [28]. In the literature Quasi Birth and Death processes(QBDs) have been used extensively to model performance and reliability of various systems.

5   Modelling Approaches for Performability

In this section use of Open Queuing Networks is discussed to model the behaviour of a WSN cluster network based on previous works [2], [16], [18], [29]. In Chiassereni et al. [2] open queuing networks are successfully used to model a Markov sensor network, whose nodes may enter sleep mode. The system performance is analysed in terms of energy consumption, network capacity and data delivery delay. In [29] analysis of overall packet arrival rate approximations is considered at the servers using various methods and Poisson approximation is specified as an accurate approach in large scale networks in which the nodes receive arrival streams from a number of other nodes. This is based on the fact that the superposition of many independent and relatively sparse processes converge to Poisson distriburion as the number of component processes tend to go infinity. In their Study Tie et al. [16] successfully modelled WSN cluster node behaviour using M/M/1/N with holding nodes and the model was found to be consistent with real data. In another study, Zhenfei et al. successfully modelled the convergence of WSNs with PON using two M/M/1 queues in tandem. The model is in turn used to derive various performance metrics.

In order to improve reliability of WSNs in hash environments, Munir et al. [12] propose a fault-tolerant sensor node model for applications with high reliability requirements. In [12], sensor failure probabilities are assumed to follow exponential distribution with rate λ_s over a period of time t. Note that in this study, time between failures and repair times are assumed to be exponentially distributed. To allow a Markovian chain analysis, it is possible to assume that the time to failure of all components have an exponential distribution. This signifies that the distribution of the next failure time of a component does not depend on how long the component has been operating. The next break-down is the result of some suddenly appearing failure (software, signal, configuration related failures), not of gradual deterioration.

5.1   System Model and Solution Techniques

A reference scenario is shown in Figure 1 in which one cluster head hops through another to get to the sink. Reduced function nodes (end sensor nodes) are also able to connect to the cluster head directly or through full function nodes.

In this structure, FF and RF devices(here after referred to as sensor nodes) forward sensed information directly to the cluster head (CH) for further processing and upward transmission to the sink directly or through other CHs. CHs may also generate data packets based on observations of their habitat. In this study, we assume that the cluster heads are aware of the routes to their neighbours and that best routes are chosen to save on energy while ensuring minimal transmission delays.

The resulting job arrivals at the cluster heads is a collection of jobs from the cluster nodes, the sensed information by the cluster itself and the forwarded data from other cluster heads. The jobs are assumed to be independent and identically distributed random variables with rate λ. The operation is assumed similar at

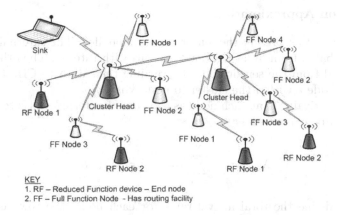

Fig. 1. Network topology of the reference scenario

all other CHs. For this study the behaviour of a single CH (k) is modelled as an open queue network using M/M/1 queues. We assume that all sensor nodes are connected directly to the CH. Since the number of cluster nodes are taken to be more than 30, it is possible to assume that the resulting superposition of all the job arrivals at node k ($k = 1, 2, \ldots N$ where, N is the total number of cluster nodes and $N \geq 30$) follow Poisson distribution with mean arrival rate λ_k [29]. In this study CH is assumed to be the node number zero($k = 0$).

Figure 2 shows the proposed queuing model for analysing a single CH behaviour. Once jobs are processed at CH k, they are transmitted to the sink directly or through an intermediary CH r. where CH r represents next available route to the sink. At CH r the process is similar to that at CH k. Considering that the nodes are prone to failures, it is assumed that when a node fails due to any reason, it is taken into repair process immediately [6], [8], [10]. This could be through use of back up nodes, software reconfigurations or replacement of failing nodes. Service times, failure times, and repair times are all assumed to be exponentially distributed. μ_k , ξ_k and η_k are the service, failure and repair rates respectively of node k, and the queues are assumed to have no blocking for incoming packets. Packets are handled on First Come First Served basis(FCFS) and the interruption policy is such that service is resumed from the point of interruption or repeat with re-sample.

Jobs leaving node k are rerouted to node r with the probability $q_{k,r}$ for service at node r. If jobs are not routed to node r then $q_{k,r} = 0$. It is assumed without loss of generality that as far as the queue length distribution is concerned $q_{k,k} = 0$. Also $q_{k,K+1} = 1 - \sum_{r=1}^{k} q_{k,r}$ is the probability of a job to leave the system after being serviced at node k. It is also assumed that the exit probability $q_{k,K+1}$, is non zero for at least one value of k. Q is the routing probability matrix of size $K \times K$, such that, $Q = q_{k,r}; (1 \leq k, r \leq k)$. To analyse the performability of this system, steady state conditions are considered.

5.2   Poisson Approximation

Figure 2 is used to represent the sensor cluster modelled. The external arrival to the CH (probably from other CHs) are arrivals with rate σ_k. The other arrivals are originated from the sensor nodes forwarding their data to CH. The rate of traffic from node r within the cluster to node k is $\lambda_r q_{r,k}$.

The total arrival rate at each node as the sum of external and internal traffic rates λ_k, can be expressed as:

$$\lambda_k = \sigma_k + \sum_{r=1}^{N} \lambda_r q_{r,k}; k = 1, 2, , K \tag{1}$$

In order to define the total arrival rates for each node, the row vectors $\lambda = (\lambda_1, \lambda_2, , \lambda_N)$ and $\sigma = (\sigma_1, \sigma_2, , \sigma_N)$ can be employed. Let also E_k be the unit matrix of size $K \times K$ then;

$$\lambda(E_k - Q) = \sigma \tag{2}$$

Letting the effective average service rate at the CH be $\hat{\mu}_k$, and taking into account the losses resulting from failures and repairs it can be shown that $\hat{\mu}_k$ is given by equation 3, [26], [30], [28].

$$\hat{\mu}_k = \mu_k . \eta_k / (\eta_k + \xi_k) \tag{3}$$

For steady state, the effective service rate must be greater than the effective arrival rate at the CH. Thus $\hat{\mu}_k > \lambda_k; k = 1, 2, \ldots K$ is the condition for steady state analysis. In earlier studies, [26], [27], [30] the mean queue length (MQL) for such a system is expressed as:

$$MQL = \frac{\lambda_k[(\xi_k + \eta_k)^2 + \xi_k \hat{\mu}_k]}{(\xi_k + \eta_k)[\eta_k . \hat{\mu}_k - \lambda_k(\xi_k + \eta_k)]} \tag{4}$$

With the values of MQL and λ_k known, the response time (R) for the cluster head can be calculated as:

$$R = \sum_{N=1}^{K} MQL / \sum_{N=1}^{K} \lambda_k \tag{5}$$

5.3   Spectral Expansion Solution Technique

Please note that, since all the sensor nodes forward the information to the CH, in the system considered, the matrix Q has a special form and the total amount of arrivals to CH can be calculated as $\lambda_0 = K \times \lambda$, where K is the number of sensor nodes in the WSN cluster and λ is the average packet generation rate of the sensor nodes [18]. There has been lots of similar studies on M/M/1 with breakdown and repairs though not in WSN area [26], [30], [28]. The state transition diagram for the cluster head is given in figure 3.

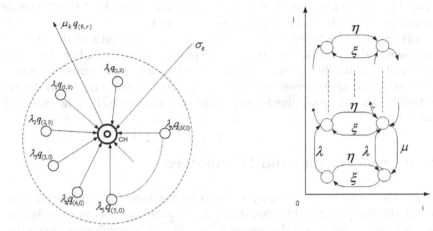

Fig. 2. Queuing Model for a single CH

Fig. 3. State transition diagram for CH performability model

The model treats sleep and breakdown states as short and long breakdown periods respectively since data will continue to arrive during both states. However service is only possible when server is operational. The system state at time t may be described using a pair of integer valued random variable $I(t)$ and $J(t)$ specifying the cluster head failure and repair configurations, and the number of jobs in the system respectively. The operative states $I(t)$ in this case represents the assumed failed and working periods of the CH. $Z = [I(t), J(t)]; t \geq 0$ is an irreducible Markov process on a lattice strip (a QBD process), that models the system. Its state space is $(0, 1)$x$(0, 1, \ldots)$. Similar models [24], [25], [26], [27], [29] are analysed for exact performability evaluation of various Multi sever systems with single repairman and for both finite and infinite L for some repair strategies. It is possible to extend the exact solution methodology for performability evaluation for of WSNs.

Since the possible operative states of the CH and the number of data arrivals are represented in the horizontal and vertical directions of the lattice respectively, the transition matrices can be derived as:

i A is the matrix of instantaneous transition rates from (i, j) to state $(l, j),(i = 0, 1; l = 0, 1; i \neq l; j = 0, 1, \ldots)$, with zeros in the leading diagonal, caused by a change in the state [25], [24]. These are the purely lateral transitions of the model Z. A clearly depends on parameters ξ and η. The state transition matrices A and A_j are of size $(2) \times (2)$ and can be given as shown below.

ii Matrices B and C are transition matrices for one step upward and one step downward transitions respectively [25], [24]. The transition rate matrices do not depend on j for $j \geq M$, where M is a threshold having an integer value [24]. The respective transition matrices are shown below:

$$A = A_j = \begin{bmatrix} 0 & \eta \\ \xi & 0 \end{bmatrix}, \ B = B_j = \begin{bmatrix} \lambda & 0 \\ 0 & \lambda \end{bmatrix} \text{ and } C = C_j = \begin{bmatrix} 0 & 0 \\ 0 & \mu \end{bmatrix}$$

Elements of matrix B are dependent on the data arrival rate (λ) at the CH while elements of matrix C depend on the CH service rate (μ).

Spectral expansion solution can be employed and the details of the method used can be found in [26] [30]. From the state probabilities, a number of steady-state availability, reliability, performability measures can be computed quite easily. For illustration, we have concentrated on the mean queue length and the response time which can be obtained using equation 5 where MQL is the expected value of $J(t)$.

6   Numerical Results and Discussions

In this section numerical results are presented for the model considered. A comparative analysis is performed for two different solution approaches, namely Poisson approximation and spectral expansion [29], [24]. The results are very close and further verification with simulation results are also in good agreement.

A dedicated software written in C++ language was used to simulate the actual system. Simulation results were then compared with the analytical results obtained by applying spectral expansion and Poisson approximation solution approaches to the Markov model of the system. All results obtained reveal good agreement with both spectral and Poisson approximation techniques. VC++ 10.0, and the NAG library (for spectral expansion only) were used to achieve all the results presented in this study.

The following parameters are used throughout this section unless otherwise stated. The values of failure and repair rates are chosen to ensure that repair rate $\eta_k = 0.5$, is much higher than failure rate $\xi_k = 0.001$. Service rate is taken as $\mu_k = 300$ and arrival rates are chosen carefully to ensure the system remains stable. The results are shown for Poisson approximation, spectral expansion method and simulation in all the figures.

Figure 4 shows MQL as a function of λ. In order to analyse the effects of arrival rate on the cluster size, the experiment is achieved by varying number of nodes in each run. The results indicate that fewer nodes are able to accommodate higher arrival rates as opposed to the system getting saturated at low arrival rates when many nodes are used to cover the habitat. A trade off therefore exists in order to determine the best performance with optimum coverage. In other words, the model can be used to specify the size of a cluster when a specific flow is expected from the sensing nodes.

Figure 5 shows response time as a function of number of nodes for various λ values. The three solution techniques show that the results are in good agreement with best response times realised with fewer nodes in the cluster. The model may therefore be used to select appropriate response time operation region for WSNs. Please note that, the unit of response time is dependent to the units of arrival and service rates. Since these values can be dependent on the type of application, in this study a generic approach is adopted.

In most cases, a WSN cluster is initially populated with the maximum required nodes for best coverage. In figure 6, the arrival rate is maintained at specific

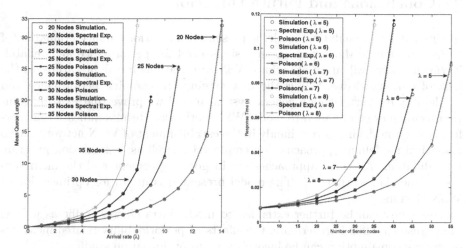

Fig. 4. MQL changes with variations in arrival rate

Fig. 5. Variations of response time with changing number of network nodes

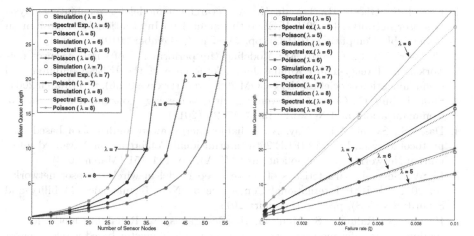

Fig. 6. Variations of MQL with changing number of network nodes

Fig. 7. Variations of MQL with Failure rate changes

values of $\lambda = 5, 6, 7, 8$, indicating the maximum limit for required sensor nodes per cluster to maintain appropriate traffic that the CH can handle.

In figure 7, MQL is presented as a function of the failure rate, where $\lambda = 5, 6, 7, 8$ and $K = 30$. Results show the effects of failures clearly demonstrating the importance of repair facilities and high reliability for accepted levels of system availability, without compromising the system performance significantly.

7   Conclusions and Future Directions

In this paper, an analytical model is presented for performability evaluation of WSNs. Although Markov chain analysis is used for performance and reliability/availability evaluation of various WSN applications in the literature, to the best of our knowledge, this is the first attempt to combine performance and availability metrics. Using a generic system model we prove that existing solution techniques can be used to model WSN networks. Results indicate that performability modelling is very handy in the establishment of WSN networks. Two analytical modelling approaches are employed as well as a simulation program for validation. All three approaches are in good agreement and the maximum discrepancy is less than 2%. The model presented is useful for optimization of WSN clusters.

This study can be further extended to model intra cluster traffic as well as inter cluster traffic. Furthermore, the effects of performability related measures on energy consumption can be incorporated in optimisation studies.

References

1. Kim, D.S., Ghosh, R., Trivedi, K.: A hierarchical model for reliability analysis of sensor networks. In: 2010 IEEE 16th Pacific Rim International Symposium on Dependable Computing (PRDC), pp. 247–248 (December 2010)
2. Chiasserini, C.F., Garetto, M.: Modeling the performance of wireless sensor networks. In: Twenty-Third Annual Joint Conference of the IEEE Computer and Communications Societies, INFOCOM 2004, vol. 1(xxxv+2866), p. 4 (March 2004)
3. Shin, J., Sun, C.: Creec: Chain routing with even energy consumption. Journal of Communications and Networks 13(1), 17–25 (2011)
4. Dash, S., Swain, A.R., Ajay, A.: Reliable energy aware multi-token based mac protocol for wsn. In: 2012 IEEE 26th International Conference on Advanced Information Networking and Applications (AINA), pp. 144–151 (March 2012)
5. Li, W.: Several characteristics of active/sleep model in wireless sensor networks. In: 2011 4th IFIP International Conference on New Technologies, Mobility and Security (NTMS), pp. 1–5 (Febraury 2011)
6. Hashmi, S., Rahman, S., Mouftah, H., Georganas, N.: Reliability model for extending cluster lifetime using backup cluster heads in cluster-based wireless sensor networks. In: 2010 IEEE 6th International Conference on Wireless and Mobile Computing, Networking and Communications (WiMob), pp. 479–485 (October 2010)
7. Gao, Q., Zuo, Y., Zhang, J., Peng, X.H.: Improving energy efficiency in a wireless sensor network by combining cooperative mimo with data aggregation. IEEE Transactions on Vehicular Technology 59(8), 3956–3965 (2010)
8. Dini, G., Pelagatti, M., Savino, I.: Repairing network partitions in wireless sensor networks. In: IEEE Internatonal Conference on Mobile Adhoc and Sensor Systems, MASS 2007, pp. 1–3 (October 2007)
9. Almasaeid, H., Kamal, A.: On the minimum k-connectivity repair in wireless sensor networks. In: IEEE International Conference on Communications, ICC 2009, pp. 1–5 (June 2009)

10. Liu, X., Feng, Y., Lv, Q., Zhao, T.: Cascaded movement strategy for repairing coverage holes in wireless sensor networks. In: 2011 International Conference on Information Technology, Computer Engineering and Management Sciences (ICM), vol. 2, pp. 108–111 (September 2011)
11. Zhang, J., Song, G., Qiao, G., Li, Z., Wang, A.: A wireless sensor network system with a jumping node for unfriendly environments. International Journal of Distributed Sensor Networks 2012, Article ID 568240, 8 pages
12. Munir, A., Gordon-Ross, A.: Markov modeling of fault-tolerant wireless sensor networks. In: 2011 Proceedings of 20th International Conference on Computer Communications and Networks (ICCCN), July 31- August 4, pp. 1–6 (2011)
13. Bruneo, D., Puliafito, A., Scarpa, M.: Dependability evaluation of wireless sensor networks: Redundancy and topological aspects. In: 2010 IEEE Sensors, pp. 1827–1831 (November 2010)
14. Almazydeh, L., Abdelfattah, E., AL-Bzoor, M., Al-Rahayfeh, A.: Performance evaluation of routing protocols in wireless sensor networks. International Journal of Computer Science and Information Technology 2(2) (2010)
15. Ever, E., Luchmun, R., Mostarda, L., Navarra, A., Shah, P.: Uheed - an unequal clustering algorithm for wireless sensor networks. In: SENSORNETS, pp. 185–193 (2012)
16. Qiu, T., Feng, L., Xia, F., Wu, G., Zhou, Y.: A packet buffer evaluation method exploiting queueing theory for wireless sensor networks. Comput. Sci. Inf. Syst., 1028–1049 (2011)
17. Russello, G., Mostarda, L., Dulay, N.: A policy-based publish/subscribe middleware for sense-and-react applications. Journal of Systems and Software 84(4), 638–654 (2011)
18. Wang, Z., Yang, K., Hunter, D.: Modelling and analysis of convergence of wireless sensor network and passive optical network using queueing theory. In: 2011 IEEE 7th International Conference on Wireless and Mobile Computing, Networking and Communications (WiMob), pp. 37–42 (October 2011)
19. Masoum, A., Jahangir, A.H., Taghikhaki, Z.: Survivability analysis of wireless sensor network with transient faults. In: 2008 International Conference on Computational Intelligence for Modelling Control Automation, pp. 975–980 (December 2008)
20. Thein, T., Chi, S.D., Park, J.S.: Increasing availability and survivability of cluster head in wsn. In: The 3rd International Conference on Grid and Pervasive Computing Workshops, GPC Workshops 2008, pp. 281–285 (May 2008)
21. Song, G., Ye, X., Niu, Y., Meng, T.: A reconfigurable mobile node for wireless sensor networks in unfriendly environments. In: 2010 International Conference on Environmental Science and Information Application Technology (ESIAT), vol. 1, pp. 618–621 (July 2010)
22. Houaidia, C., Idoudi, H., Saidane, L.: Improving connectivity and coverage of wireless sensor networks using mobile robots. In: 2011 IEEE Symposium on Computers Informatics (ISCI), pp. 454–459 (March 2011)
23. Trivedi, K.S., Ma, X., Dharmaraja, S.: Performability modelling of wireless communication systems. International Journal of Communication Systems 16(6), 561–577 (2003)
24. Ever, E., Kirsal, Y., Gemikonakli, O.: Performability modelling of handoff in wireless cellular networks and the exact solution of system models with service rates dependent on numbers of originating and handoff calls. In: International Conference on Computational Intelligence, Modelling and Simulation, CSSim 2009, pp. 282–287 (September 2009)

25. Kirsal, Y., Gemikonakli, O. (eds.): Approaches to Modeling and Analysis for Performability Evaluation of Handoff Schemes in Wireless Cellular Networks, Middlesex University (September 2009),
www.cms.livjm.ac.uk/pgnet2009/Proceedings/Papers/2009021.pdf
26. Chakka, R., Mitrani, I.: Approximate solutions to open networks with breakdowns and repairs. Technical report, University of Newcastle, Department of Computer Science
27. Mitrany, I.L., Avi-Itzhak, B.: A many-server queue with service interruptions. Operations Research 16(3), 628–638 (1968)
28. Sheng-li, L., Jing-bo, L., De-quan, Y.: The m/m/1 repairable queueing system with variable breakdown rates. In: Chinese Control and Decision Conference, CCDC 2009, pp. 2635–2637 (June 2009)
29. Chakka, R.: Performance and Reliablility Modeling of Computer systems Using Specral Expansion, PhD thesis, University of Newcastle Upon Tyne (1995)
30. Thomas, N., Mitrani, I.: Routing among different nodes where servers break down without losing jobs. In: Proceedings of the International Computer Performance and Dependability Symposium, pp. 246–255 (April 1995)

Integration of DANUM-Based Carrier-Grade Mesh Networks and IMS Infrastructure

Przemyslaw Walkowiak, Maciej Urbański, Andrzej Figaj, and Pawel Misiorek

Institute of Control and Information Engineering
Poznan University of Technology
M. Sklodowskiej-Curie 5, 60-965 Poznan, Poland
{Przemyslaw.Walkowiak,Maciej.Urbanski,Andrzej.Figaj,
Pawel.Misiorek}@put.poznan.pl

Abstract. We propose a system, referred to as the CARMNET system, that combines the utility-aware flow control and resource allocation for wireless networks with the IMS architecture in order to optimize the Internet access for mobile users. The system utilizes the Delay-Aware Network Utility Maximization System (DANUMS), which is an application of the NUM framework for wireless mesh networks. The proposed integration of DANUMS with telecom operator infrastructure enables a centralized management of user profiles and requirements across many wireless mesh networks. The utility functions predefined for each type of traffic allow DANUMS to manage the network resources in accordance to the user-perceived utility. The solution is based on the IP Multimedia Subsystem (IMS) architecture and the application of the SIP protocol, which are widely used by telecom operators. As a result of compliance with widely used protocols, the solution can be easily deployed in existing infrastructures.

Keywords: Wireless mesh/multi-hop networks, IP Multimedia Subsystem, Session Initiation Protocol, Network Utility Maximization.

1 Introduction

The broadband access to the Internet tends to the vision of ubiquitous Internet access, where the primary transmission medium will be Wireless Mesh Networks (so called WMNs). However, even in the case, in which the network access is easy accessible, there is still the need for additional control and management of network resources realized by means of Authentication, Authorization and Accounting (AAA) functions [18] and Quality of Service (QoS). In order to address this problem, we present a system, referred to as the CARMNET system, that integrates the functionalities of the utility-oriented wireless network resource management and user management performed by means of IMS infrastructure. The vision of the CARMNET system as a carrier-grade wireless network co-operated by users is presented in [3]. The system utilized the Delay-Aware Network Utility Maximization System, as a the wireless resource management system, which

J. Cichoń, M. Gębala, and M. Klonowski (Eds.): ADHOC-NOW 2013, LNCS 7960, pp. 185–196, 2013.

is able to fairly serve heterogeneous traffic by following resource virtualization approach [10]. The goal of the proposed integrated system is to enable carrier-grade wireless networking. As a result of combining DANUM-based resource virtualization with the IMS functionalities, the system enables future introduction of the efficient charging system for the mobile users of wireless networks. The CARMNET system, is developed as the main outcome of the CARMNET project [19].

1.1   The CARMNET System Vision

The CARMNET system enables its end-users to share their network resources, in particular to share their Internet access. Figure 1 presents the CARMNET scenario in which some network nodes (referred to as the CARMNET Internet sharing nodes) offer the Internet connection to users of other nodes. The Internet sharing is optimized as a result of the application of the utility-aware resource allocation subsystem which allows to compare the utility of flows with different requirements with regard to end-to-end delay and throughput.

On the other hand, the CARMNET system uses an enhanced IMS architecture to provide the session and user management. The CARMNET idea is to combine the unique features of utility-aware flow control and resource allocation (provided by DANUMS) and the AAA functionalities (provided by the IMS subsystem) in order to provide the utility-based charging, in a way that encourages mobile users to share their Internet connection and which is, at the same time, suitable for telecom operators.

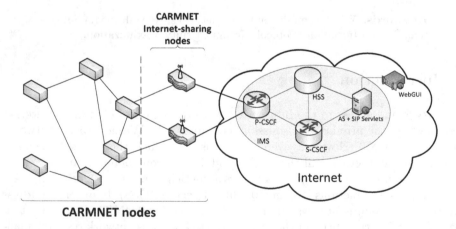

Fig. 1. Scenario of the CARMNET system usage

1.2 DANUMS as a Resource Managament System

The Delay-Aware Network Utility Maximization System [10] is a solution based on NUM model [6] that is aimed at serving heterogeneous traffic with different utility requirements.

DANUM System is interoperable with widely used protocols of the most typical wireless networking stack, such as TCP, UDP, and 802.11 MAC [10]. That feature allows to make the wide-scale application of DANUMS technically possible, especially because it may be easily installed on Linux-based network nodes. The DANUM model is based on optimization of states of virtual queues which accumulate virtual units depended on the flows' utility. Such a mechanism may serve as the basis for charing an user for the network utility rather than for the traffic volume [3].

1.3 Advantages of IMS Infrastructure Reuse

SIP is a standard in telecom business. IMS systems are already a crucial part of telecom infrastructure, thus it is easier to introduce new features based on SIP than to implement new protocols. IMS subsystem may be enhanced with new features by implementing new servlets running on an application server. The IMS/SIP infrastructure provides out-of-the-box functionalities such as user authentication/authorization that may be reused in the CARMNET system. The SIP User Agent (SIP UA) applied in our solution is used only for transferring additional data needed by CARMNET wireless nodes by means of a simple SIP message schema. Therefore, a very thin client implementation is needed. The biggest benefit of using SIP in DANUM systems is the fact that in such a case the impact on the operator infrastructure is minimal. The migration form non-DANUM to DANUM systems does not require large investments in the infrastructure.

1.4 Paper Outline

The paper is structured as follows. In Section 2, the related work in the relevant research area is presented. Section 3 describes the overview of the DANUM model. In Sections 4 and 5, the integrated CARMNET system model and implementation are presented, respectively. The paper is completed by the parts concerning validation tests (Section 6) and conclusions (Section 7).

2 Related Work

The problem of integrating the carrier-grade IMS platform with wireless networks aimed at widening the area of Internet access provided by telecom operator is still intensively investigated. Most of already proposed solutions focus on enabling the mobility of user devices and Quality of Service (QoS) for IMS services like voice and video calling. The authors of [7] propose the integration of the Mobile IP standard [13] (that enables the seamless handover between networks) and the IMS infrastructure allowing to automatically trigger the registration procedure in IMS services in a case of user location change. The more complex solution is proposed in [1], which, in addition to providing seamless and transparent mobility, allows maintaining Quality of Service of the ongoing session in WiMAX and UMTS networks. However, the session QoS is verified only against the throughput which can be achieved in the preferred network.

The authors of [2] propose introduction of QoS and AAA into carrier-grade mesh networks as a result of using the additional IMS infrastructure components like Resource and Admission Control Subsystem (RACS) and Network Attachment Sybsystem (NASS). The RACS supports the resource reservation methods, admission control, and policy decisions, whereas the NASS supports automatically device configuration and network access control. Both the proposed subsystems and IMS infrastructure enable AAA and QoS functions for network clients, which have to successfully perform the SIP registration procedure to get access to the network and other provided services.

3   Overview of the DANUM Model

The section provides a shortened description of the DANUM model. A more detailed presentation may be found in [10] and [11].

3.1   The Main Assumptions of the DANUM Model

The DANUM model formulation is based on modeling rate and delay as functions of a new optimization variable [10]. The model assumes network utility management modeled as the explicit transfer of 'virtual money' called *denarii* [10], associated with the new optimization variable. Based on the new utility function formulation (see [10] for details), we define the primal DANUM problem aimed at finding *denarius* rates y_f that solve:

$$\max_{y_f} \sum_f U_f(y_f), \tag{1}$$

subject to the constraints associated with the system of virtual queues [10]. As shown in [10] the DANUM optimization problem may be decomposed into the following two components:

- Flow control scheme – with the aim of finding the optimal *denarius* rate as a solution to the problem presented below:

$$\arg\max_{y_f} (U_f(y_f) - y_f q_{a_f}^f), \tag{2}$$

- Routing and scheduling scheme – with the aim of finding the optimal resource allocation:

$$\arg\max_{\mathbf{p} \in \Theta} \sum_{(i,j) \in L} \max_f (p_{i,j}^f [q_i^f - q_j^f]), \tag{3}$$

where q_i^f is a Lagrange multiplier associated with the *denarius* queue of flow f stored at node i, a_f is a source node of flow f, $\mathbf{p} = (p_{i,j}^f)$ is a vector of per-flow utility-dependent virtual transmission rates, L is a set of available links, and Θ is a convex set representing possible allocations of virtual transmission rates [10].

The DANUM indirect flow control complements (rather than substitutes) the functions of regular transport protocols. Instead of applying direct changes of flow rates, the mechanism of virtual units (*denarii*) rate control is used. The detailed description of the DANUM system implementation, including the specification of additional signaling and monitoring protocols may be found in [11] and [10].

3.2   Examples of Utility Functions for Heterogeneous Traffic

The approach of solving the DANUM problem imposes the application of formulas for utility functions [10] for file transfer and multimedia streams. In the case of TCP flows, the utility is defined as follows (see Figure 2(a)):

$$U_T(x,d) = -\frac{w_T}{xd^2},\tag{4}$$

where x is the sending rate, $d = RTT$ is a delay approximation, and w_T is a TCP 'aggressiveness' parameter [10]. Parameter w_T is used to tune the 'aggressiveness' of TCP flows when they compete with UDP flows configured by setting the analogous parameter w_U (see Equation (5)).

We model the utility of a UDP-based streaming media flow as a non-concave function of delay d and rate x (see Figure 2(b)):

$$U_U(x,d) = \frac{w_U}{\left(1 + e^{a(x_t - x)}\right)\left(1 + e^{b(d - d_t)}\right)},\tag{5}$$

where w_U corresponds to the 'aggressiveness' of the streams when they compete with TCP flows; a, b are parameters controlling the slope of utility; and x_t, d_t are threshold values for rate and delay, respectively [10].

4   Integration of WNM and IMS Approach

Wireless Mesh Networks may provide a way to increase operator network coverage without the need to invest in the new infrastructure. On the other hand, telecom operators may provide core network management functionalities. In the presented solutions each node is acting as a SIP User Agent, however, instead using it to manage call session, it is applied to control the mesh network.

The IMS infrastructure is used to authenticate, authorize and store user profile information as well as for accounting/charging. The registration is performed by SIP UA using the authentication/authorization functions of IMS AAA by means of the standard SIP REGISTER message. The novel approach is to apply

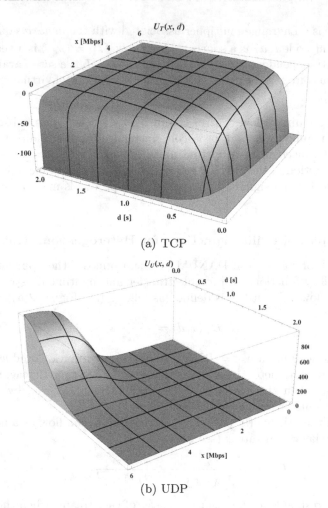

(a) TCP

(b) UDP

Fig. 2. The shapes of the utility functions for TCP and UDP flows

SIP SUBSCRIBE/NOTIFY messages to transport the User Profile XML. Among other information it contains metrics of flows generated/transmitted by given node, which may be used for charging.

4.1   The Architecture of the System

The architecture of the CARMNET system consists of multiple components located both on the client- and server-side in the Internet (see Figure 3). ow.

DANUMS Loadable Kernel Module (LKM) is an implementation of the DANUM model under Linux environment. It is located at the kernel level which allows a deep integration in the networking stack necessary to introduce custom queueing and scheduling subsystems. For the network path resolution, the OL-SRd [20] - the popular implementation of the OLSR protocol is used. As OLSR

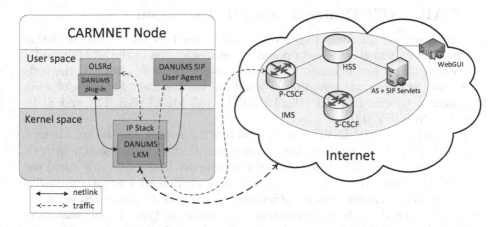

Fig. 3. Interactions between CARMNET components

messages are used to distribute data needed by the packet scheduler implemented as LKM, the special way of communication between user-space (OLSRd) and kernel-space (DANUM LKM) had to be applied. The Netlink [17] protocol serves that role.

The direct communication of DANUM LKM located in the client's node with the IMS infrastructure located in a network is a difficult task because of the need of implementing a high-level SIP protocol in the low-level kernel module in a way preserving the high efficiency of the Linux kernel. Therefore, we have implemented a new SIP User Agent (SIP UA) in the user-space running on the client side, which is responsible for asynchronous communication between LKM and IMS. The communication between LKM and SIP UA is realized using the Netlink protocol, whereas the communication between SIP UA and IMS is realized using the CARMNET XML protocol encapsulated in the standard SIP protocol.

The user interface (WebUI) is a WWW application dedicated to users of the CARMNET network. Users are allowed to configure their own profiles and to bind the utility function (and its arguments) to the type of traffic (e.g., WWW, e-mail, Skype). Moreover, WebUI allows to show the utility unit account balance and reports about transmitted traffic and its 'price'.

Two types of servlets are located on the application server. The first one is responsible for managing AAA functions and user profiles in the CARMNET network. After connecting to the network, each client must be authenticated and authorized in the SIP servlet before starting the Internet session. During session each node in the network reports to the SIP servlet about what type and how many traffic it served. All information about users are stored in the HSS. The Diameter protocol is used for communication purposes. Because of the fact that WebUI is located outside the main IMS infrastructure, it should have direct access to the HSS component. Therefore the second type of servlets on application server is implemented, which provides SOAP services allowing to access the user profiles through WebUI.

5 CARMNET Protocol as a SIP Extension

We have designed a custom high-level protocol based on the Session Initiation Protocol (SIP), which is the most popular protocol related to the IP Multimedia Subsystem (IMS) architecture. We have identified the following SIP methods as a basis for the implementation of a session management for users of wireless mesh networks: INVITE [16], UPDATE [15], MESSAGE [12], and SUBSCRIBE/NOTIFY [14].

The purpose of SIP INVITE is to establish the session and its parameters, whereas the UPDATE method allows further modification of that session without altering state of the SIP dialog. In order to support DANUMS, one could use SIP INVITE method to initialize session for each flow and negotiate quality properties. The dynamic changes of parameters could be implemented using the UPDATE method. Such an approach is most similar to typical 'call' sessions in which SIP is used. However, this solution suffers from a high message complexity. The per-flow control information is already transmitted by DANUMS in more suitable and optimized matter [9,11].

The MESSAGE type request was created for purpose of the Instant Messaging communication. This request can be issued with or without the already established SIP dialog. The specification of SIP protocol indirectly allows use of the MESSAGE method as a transport protocol. Such a scenario enables the encapsulation of an independent protocol, which would only inherit AAA functions provided by IMS platform. This solution was used as the first approach to the CARMNET system development, but it was proved to be too much inconsistent from the standard protocol [12].

The SIP Methods for Event Notification [14] introduce a framework for asynchronous data exchange between end-points. The SUBSCRIBE method initializes dialog in which a subscription is established for the specified duration. The subscriber is notified by means of the NOTIFY method of the every change of state in the subscribed package. In case in which some participant stops responding, the subscription is gracefully ended by reaching its timeout. This feature makes such a model suitable for wireless networks.

Figures 4, 5, and 6 illustrate the exchange of messages among the components located both on a wireless node (i.e., DANUMS LKM, OLSRd, SIP User Agent) and on the IMS platform (i.e., P-CSCF, S-CSCF, and HSS servers and SIP servlet). In Figure 4 the phases of user registration and resource consumption reporting are shown. Figure 5 describes the process of CARMNET profile synchronization. Finally, the exchange of messages performed in order to inform the node about its neighbors is presented in Figure 6.

In order to enable monitoring and accounting of virtual currency, the scheme presented in Figure 4 has been implemented. Right after connecting to the CARMNET network, the node 'uses' its *SIP User Agent (SIP UA)* to send the REGISTER message to the CARMNET IMS platform. When *CARMNET SIP Servlet* is informed about *SIP UA* registration, it initializes the subscription

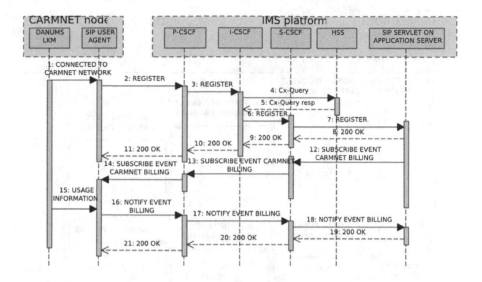

Fig. 4. Registration and usage reporting

with the same expiration time as the one used in the REGISTER method. The subscribed event allows *CARMNET SIP Servlet* to receive incremental information about the traffic being forwarded by *SIP UA* hosting node.

DANUM utility functions and specific features of DANUM subsystem operations (e.g., per-flow packet management) provide some level of customization. The CARMNET user profile accumulates these user-defined properties. The profile can be edited by means of *WebGUI* (see Figure 3) interface. To provide synchronization of the CARMNET profile, the SIP UA, once registered, requests the subscription of event connected to change of the CARMNET profile state (Figure 5). After the first subscription request, the current state is communicated by CARMNET SIP Servlet immediately.

As soon as the new neighbor is detected by OLSRd, the subscription of this event is provided (Figure 6). This information allows to determine whether the traffic of newly detected node should be served by other nodes (e.g., it allows to check whether the new node is reliable according to the current state of its account). Since the information supplied by OLSR protocol is just the node IP address, the associated SIP UA has to be determined by means of the HSS service.

6   Validation Tests

The validation tests confirming the technical deployability of the proposed solution were conducted in the wnPUT2 testbed. The testbed consists of 15 nodes based on x86 PC computers managed by the framework [8] compatible with the DES-Cript experiment scenario description language [4,5]. The proposed protocol was implemented using Fraunhofer Fokus OpenIMS Core [21] and deployed

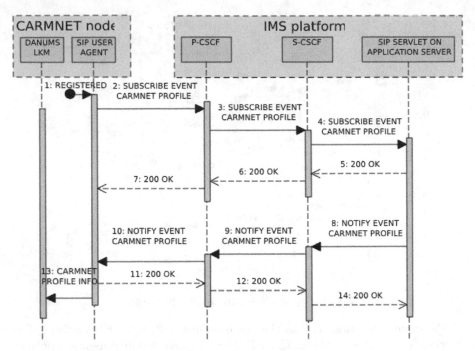

Fig. 5. CARMNET profile synchronization

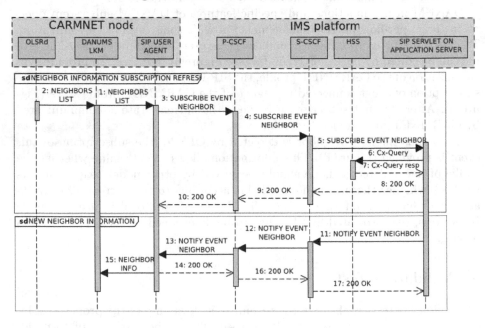

Fig. 6. Neighbor information dispersion

on the wnPUT2 wireless mesh network testbed located on the campus of Poznan University of Technology. As the environment of SIP servlets, the Sailfin application server was chosen. The DANUMS Linux kernel module with CARMNET SIP User Agent were run on each node and connected to the OpenIMS-based infrastructure deployed in a separate network.

7 Conclusions

The paper main contribution is the presentation of the wireless network control system integrating the delay-aware wireless network resource management system with the IMS-based AAA functionalities. The presented system combines utility-aware wireless network flow control and resource management with IMS-based session and user management. One of the most important advantages of the CARMNET system is its compliance with key relevant standards, including protocols of typical TCP/IP stack, and the core IMS standards. The system optimizes the wireless Internet access by taking into account both the perspective of telecom operators and the perspective of mobile Internet users.

Acknowledgement. This work was partly supported by a grant CARMNET financed under the Polish-Swiss Research Programme by Switzerland through the Swiss Contribution to the enlarged European Union (PSPB-146/2010, CARMNET), and by Poznan University of Technology under grant 45-102/12 DS-MK.

References

1. Alamri, N.M., Adra, N.: Integrated MIP-SIP for IMS-based WiMAX-UMTS vertical handover. In: 2012 19th International Conference on Telecommunications (ICT), Jounieh, Lebanon, pp. 1–6 (2012)
2. Bayer, N., Roos, A., Karrer, R.: Towards carrier grade wireless mesh networks for broadband access. In: 2006 1st Workshop on Operator-Assisted (Wireless Mesh) Community Networks, Berlin, Germany, pp. 1–10 (2006)
3. Glabowski, M., Szwabe, A.: Carrier-Grade Internet Access Sharing in Wireless Mesh Networks: the Vision of the CARMNET Project. In: The Ninth Advanced International Conference on Telecommunications, Roma, Italy (in press 2013)
4. Güneş, M., Juraschek, F., Blywis, B.: An experiment description language for wireless network research. Journal of Internet Technology (JIT), Special Issue for Mobile Internet 11(4), 465–471 (2010)
5. Güneş, M., Juraschek, F., Blywis, B., Watteroth, O.: DES-Cript - a domain specific language for network experiment descriptions. In: International Conference on Next Generation Wireless Systems, NGWS 2009, Melbourne, Australia (2009)
6. Kelly, F.: Charging and rate control for elastic traffic. European Transactions on Telecommunications 8(1), 33–37 (1997)
7. Le, L., Li, G.: Cross-Layer Mobility Management based on Mobile IP and SIP in IMS. In: 2007 International Conference on Wireless Communications, Networking and Mobile Computing (WiCOM), Shanghai, China, pp. 803–806 (2007)

8. Nowak, A., Walkowiak, P., Szwabe, A., Misiorek, P.: wnPUT Testbed Experimentation Framework. In: Bononi, L., Datta, A.K., Devismes, S., Misra, A. (eds.) ICDCN 2012. LNCS, vol. 7129, pp. 367–381. Springer, Heidelberg (2012)
9. Szwabe, A., Misiorek, P.: Integration of Multi-path Optimized Link State Protocol with Max-Weight Scheduling. In: IEEE International Conference on Information and Multimedia Technology (ICIMT 2009), Jeju Island, South Korea, pp. 458–462 (2009)
10. Szwabe, A., Misiorek, P., Walkowiak, P.: Delay-Aware NUM system for wireless multi-hop networks. In: European Wireless 2011 (EW 2011), Vienna, Austria, pp. 530–537 (2011)
11. Szwabe, A., Misiorek, P., Walkowiak, P.: Protocol Architecture for DANUM Systems. Technical Report IAII-595, Poznan University of Technology, Institute of Control and Information Engineering (2010)
12. Campbell, B., Rosenberg, J., Schulzrinne, H., Huitema, C., Gurle, D.: Session Initiation Protocol (SIP) Extension for Instant Messaging. RFC 3428 (Proposed Standard) (2002)
13. Perkins, C.: IP Mobility Support for IPv4, Revised. RFC 5944 (Proposed Standard) (2010)
14. Roach, A.: SIP-Specific Event Notification, RFC 6665 (Proposed Standard) (2012)
15. Rosenberg, J.: The Session Initiation Protocol (SIP) UPDATE Method, RFC 3311 (Proposed Standard) (2002)
16. Rosenberg, J., Schulzrinne, H., Camarillo, G., Johnston, A., Peterson, J., Sparks, R., Handley, M., Schooler, E.: SIP: Session Initiation Protocol. RFC 3261 (Proposed Standard) (2002)
17. Salim, J., Khosravi, H., Kleen, A., Kuznetsov, A.: Linux Netlink as an IP Services Protocol. RFC 3549 (Informational) (2003)
18. Vollbrecht, J., Calhoun, P., Farrell, S., Gommans, L., Gross, G., de Bruijn, B., de Laat, C., Holdrege, M., Spence, D.: AAA Authorization Framework. RFC 2904 (Informational) (2000)
19. Carrier-grade delay-aware resource management for wireless multi-hop/mesh networks (CARMNET), http://carmnet.eu
20. OLSR-d: An ad-hoc wireless mesh routing daemon (2009), http://www.olsr.org
21. Fokus Open IMS Core, http://www.openimscore.org/

Group-Based Anonymous On-Demand Routing Protocol for Resource-Restricted Mobile Ad Hoc Networks

George Moldovan, Anda Ignat, and Martin Gergeleit

University of Applied Sciences Wiesbaden Rüsselsheim, Wiesbaden, Germany
{george.moldovan,anda.ignat,martin.gergeleit}@hs-rm.de

Abstract. As the number of applications and the support for ad hoc networks increases, so does the role and manufacturer heterogeneity of the nodes. This creates a valid concern for the network security, privacy and efficiency. We propose as solution a group-based anonymous routing protocol targeting low resource networks, where groups could contain nodes with similar hardware or trust levels. Distinctive features are the support for overlapping groups, the possibility of a sender restricting the route to a destination to one or more groups, and the existence of a node revocation mechanism. To achieve this, two main primitives are utilized: secret handshakes with key-agreement from pairing-based cryptography and private set intersection from pre-distributed keys. We analyze the effects of varying the group size on the network connectivity, on the average route length and on the network privacy. To validate the protocol's efficiency, we measure its performance on resource-restricted TelosB motes.

Keywords: Group Partitioning, Anonymous Routing, Mobile Ad Hoc Networks, Resource Efficiency.

1 Introduction and Contribution

Wireless sensor networks (WSNs) are set to play an important role in the future. Their application scenarios vary from personal area networks, assisted living facilities designed to facilitate and prolong the autonomy of its users, to monitoring a wide range of events, such as medical or environmental occurrences. In these applications a WSN represents a heterogeneous mix of nodes with various roles, software and manufacturers. For example, in an assisted living residence, part of the sensors would be provided by the medical service, specialized in reading the user's medical data. Another group would be part of the lighting and temperature control, profiling and determining the comfort settings for each specific resident. Sensors provided by the electrical company would aggregate the power usage of various devices and help monitoring and minimizing the overall energy consumption. If the nodes' identities and their roles are not anonymized, any observer would be able to create a detailed profile of the network and its

J. Cichoń, M. Gębala, and M. Klonowski (Eds.): ADHOC-NOW 2013, LNCS 7960, pp. 197–208, 2013.

users. Furthermore, the platform heterogeneity could affect other network parameters, like its overall lifetime or responsiveness: nodes designed and equipped for frequent environment sampling and large bandwidth communication might inadvertently lead their neighbors, designed for less resource demanding tasks, to quickly deplete their energy reserves or to prematurely change their operation modes and rates by switching to an energy saving method. This heterogeneity therefore justifies a reasonable concern for the security and privacy inside the network, as well as for the overall network efficiency. A possible solution to this challenge is the ability to define groups in the network, and the ability of routing messages through selected nodes, determined by trust levels, hardware or energy resources.

In this paper, we present a group-based on-demand routing protocol for mobile ad hoc networks (GAR), whose contribution is the novel support for partitioning the network into multiple, overlapping private groups of nodes, and the possibility of restricting the creation of a route from a sender to a destination through a single or multiple available in-network groups. In addition, the protocol defines how untrusted nodes can be isolated from the rest of the network by the network administrator, and maintains the anonymity of nodes that send, forward or receive messages, the unlinkability between nodes, messages and communication routes, as well as the privacy of all the groups which are not shared by two nodes. Since motes like the TelosB [17] are representative for low-cost, low computing power WSNs, we target resource efficiency and thus minimize in GAR the amount of expensive operations required to establish the network and forward messages. Thus, in the previous example, GAR enables, for high amount of private data, the creation of a restriction to nodes which are part of two groups - one representing high trust levels, one representing the required hardware resources. For public information with high priority, like fire alarms, it allows nodes to forward messages indiscriminately. GAR also allows the decoupling of low-trust metering nodes from the consuming nodes, by defining two groups which share only the nodes that aggregate consumption values, thus preventing the profiling of users.

The rest of the paper is organized as follows: Sect. 2 briefly refers to related work. Section 3 describes the cryptographic blocks used by the protocol and assumptions about the network. Section 4 contains the detailed description of GAR. In Sect. 5 we analyze the protocol's susceptibility to anonymity and linkability attacks, the effect of groups on the network parameters, and the performance on our target nodes. Section 6 concludes and outlines the future work.

2   Related Work

We present a brief overview of the history of secure and anonymous routing in static and dynamic networks, by referring some of the significant progresses made during its development. For a more elaborate survey we refer to [16].

Chaum [3] proposed *mix* networks, which consist of chains of proxy servers forming random paths to a destination, through which a layered encrypted message is forwarded. This makes tracking a particular message difficult. An early implementation of the a mix network is proposed by Goldschlag et al. [6], the *onion routing*. The protocol provides anonymous connections between nodes in a network and is resistant to eavesdropping and traffic analysis. The sender of a message builds an onion which sets the path through the network to the destination; each router peels a layer from the onion, which privately reveals the next route hop. Kong and Hong introduced ANODR in [7] a non-source, identity-free protocol. In ANODR, the sender broadcasts a route request message containing an expanding onion and a trapdoor information that can be only opened by the destination. The expanding onion creates the route from the source to the destination, which is used for returning the route reply. The destination embeds a proof for the originator, which guarantees that the trapdoor was indeed correctly opened. To prevent an attacker from following onions during the route creation phase, ANODR uses an expensive public-key encryption. This leads to a significantly high latency route creation method when running on resource-restricted nodes. Zhang et al. proposed MASK [10], which performs a proactive neighbor detection during which it establishes a shared key using a secret handshake. The secret handshake allows two nodes to determine if they are part of the same group; if not, the communication ends. The reactive route creation phase, which uses the already established links and keys, has a significantly smaller cryptographic overhead than ANODR but it also has as disadvantage the fact that the destination's true identity is revealed as part of the public route request message, and thus it does not provide source anonymity.

Both GAR and MASK use Balfanz's secret handshake protocol [1], but with different purposes: in MASK, every node is assigned to a single group. Since members of the network cannot be part of multiple groups, the network is split into disjoint parts. In our approach we use our previous work [8] as a way of defining groups, which allows us to admit multiple groups per node. This eliminates the danger of disjoint sub-networks in favor of selective group-restricted routing or communication, while always allowing these restrictions to be lifted. The purpose of the secret handshake in our protocol is that of a node revocation mechanism only (Sect. 4.2), a feature provided by neither ANODR nor MASK.

3 Preliminaries

3.1 Secret Handshake from Bilinear Pairings

An admissible bilinear pairing [2] $\hat{e} : \mathbb{G}_1 \times \mathbb{G}_1 \to \mathbb{G}_2$, with $(\mathbb{G}_1, +)$ and $(\mathbb{G}_2, \cdot)$ cyclic groups, is a map which is bilinear: $\forall g_1, g_2, g_3 \in \mathbb{G}_1, \hat{e}(g_1 + g_2, g_3) = \hat{e}(g_1, g_3) \cdot \hat{e}(g_2, g_3)$; non-degenerate: $\hat{e}(g_1, g_1) \neq 1$ and is a generator; and efficiently computable. In practice, the Weil and the Tate pairings are known implementations of these bilinear pairings [5].

Balfanz et al. [1] introduced a secret pairing-based handshake: given an admissible bilinear map $\hat{e}$, two hash functions $H_1 : \{0,1\}^* \to \mathbb{G}_1$ and $H : \mathbb{G}_2 \to \{0,1\}^n$

(a collision-resistant hash function such as SHA-1), a network administrator generates a secret master key $t \in \mathbb{Z}_q$ and generates for all users pairs of the form $\langle P, S \rangle$, where P is a random generated pseudonym, and S is a secret point generated by computing $t\, H_1(P)$. Two users A (the initiator) and B running the protocol start by exchanging unused pseudonyms P_A and P_B, and the randomly generated numbers n_A and n_B. B then computes the key $K_{BA} = \hat{e}(H_1(P_A), S_B)$, which forms the value $V_{BA} = H(n_A \parallel n_B \parallel 0 \parallel K_{BA})$ and sends is to A. A generates its own key $K_{AB} = \hat{e}(S_A, H_1(P_B))$ and hash V_{AB}, then checks if the two hashes match. This holds only if both A and B had their secret point generated using the same master key t.

3.2   Private Set Intersection from Pre-distributed Keys

The protocol we introduced in [8] allows two entities, each containing a set of private elements, to compute the intersection of their sets, while not leaking any information about elements which are not included in the resulting subset. The set elements, provided by a global administrator, are modeled by a pair $\langle g, k \rangle$, where g represents an ID associated to a real element, and k is a unique corresponding key, each chosen from a large domain.

If A wants to compute the intersection of its set with B's set, they first have to establish a shared key K_{AB}. Afterwards, B computes $E_{CK_B}(g_B)$ for every element of its set, where E is a symmetric encryption scheme, and $CK_B = k_B \oplus K_{AB}$. The resulting set $set(B)_{K_{AB}} = \{E_{CK_B}(g_B), \forall g_B \in set(B)\}$ is sent to A ordered lexicographically. A creates its corresponding set $set(A)_{K_{AB}}$, with $CK_A = k_A \oplus K_{AB}$. The elements in the intersection of $set(A)_{K_{AB}}$ and $set(B)_{K_{AB}}$ can thus be identified by A, while determining the non-shared pairs from their encryption is computationally unfeasible as long as E is cryptographically secure.

3.3   Network Administrator

The existence of a trusted network administrator (NA) is required. The NA needs to define for each node, prior to its introduction into the network, its pseudonym–secret point pairs required by the secret handshake (Sect. 4.1), as well as the groups to which it belongs. The groups are stored as a set of pairs $\langle g, k \rangle$ (Sect. 3.2). In order to prevent an attacker from distinguishing between different nodes solely based on the group-set size, the NA is required to make all group-sets of equal size G. If a node is part of less than G groups, the NA must randomly generate the appropriate number of unique fill-in groups. A second global constant set by the NA is the current network wide *epoch* duration. An epoch is a time interval at the end of which the nodes clear any information regarding message routes and neighbor lists, and perform the neighbor detection, handshake and route discovery again. Section 4.2 elaborates on the role of the epochs.

Physically, the NA can be composed of multiple entities with identical functionalities. To guarantee that they will function securely and reliably even if some of them suffer a failure or are compromised, a threshold secret sharing scheme like [18] should be employed, where the shared secret is represented by the master key used to generate the secret points.

4 Protocol Design

4.1 Overview

Secret Handshake and Group Intersection. The first stage of the protocol consists of a node detecting all its neighbors, and performing the secret handshake from Sect. 4.1 with each of them. For two neighboring nodes A and B, a secure common key is computed.

After nodes A and B establish a shared key K_{AB} and perform the group-set intersection described in Sect. 3.2, the resulting subset is known only to the initiator of the process. If A is the initiator, it needs to share with B the result - the indexes of the common groups. To prevent abuse, A also needs to provide a proof which confirms its claim. The indices are extracted from $set(B)_{K_{AB}}$ as $I_B = \{in_B^0, \ldots, in_B^w\}$, where w is the size of the intersection. The proof is computed as $H(E_{PR}(g_B^0)\| \ldots \|E_{PR}(g_B^w))\}$ with $PR = H(P_A\|P_B) \oplus k$, which proves that the ID and corresponding key of all shared groups are known to A. If the resulting subset is empty, the communication ends.

Route Pseudonym and Masking Keys. In order to prevent message coding attacks, A and B can each now proceed to separately generate, similar to [10], a common set containing pairs $\langle l_{AB}, s_{AB} \rangle$, where $l_{AB}^i = H(K_{AB} \| 2 \cdot i)$ represents the ID of a link established between A and B, while $s_{AB}^i = H(K_{AB} \| 2 \cdot i + 1)$ represents an associated key. Every message m exchanged between A and B will be constructed as $\langle l_{AB}^i, \{m\}_{s_{AB}^i} \rangle$. The difference between the two approaches is that in our version i must be automatically incremented on both sides after each data packet. A second distinction is that s does not have the role of encrypting the message (which is the responsibility of the sender and that of the receiver), but that of masking its form on every hop. $\{m\}_s$ is therefore defined as $m \oplus s$ (similar to the one-time pad). Messages with length greater than s are encrypted with consecutive masking keys.

Route Detection. Depending on the group restriction that might be selected by the source, there are two different route request messages (RREQ) formats that can be used. Both are sent by public broadcast under the current active pseudonym of the sender.

If the sender does not wish to restrict the groups through which a message can be forwarded, the format of *group agnostic* RREQ is

$$\langle RREQ, P_{src}, seq, SR_{(src, dest)} \rangle,$$

where *seq* is a globally unique random generated number which identifies the RREQ, and $SR_{(src, dest)}$ is a cryptographic function constructed by the

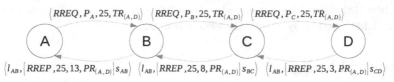

Forwarding Table of B during the RREQ

Incoming Pseudonym	Inc. Route Pseudonym	Outgoing Pseudonym	Out. Route Pseudonym	Seq. Nr.
P_A		P_C		25
...	...	...	...	...

Forwarding Table of B after the RREP

Incoming Pseudonym	Inc. Route Pseudonym	Outgoing Pseudonym	Out. Route Pseudonym	Seq. Nr.
P_A	13	P_C	8	
...	...	...	...	...

Fig. 1. GAR route request and reply message exchange

message source for the destination using a common shared key[1]. Depending on the resources available on the node, it can be a one-way keyed hash function, a symmetric encryption function or a trapdoor function [4]. By using SR, the targeted receiver of the RREQ is anonymously defined. Every node receiving the RREQ will also save the pseudonym of the previous sender and seq, and will try to open the function. If it does not succeed, it will replace the pseudonym from the previous forwarder of the RREQ with its own, and will then broadcast the message.

Once the RREQ reaches its destination, the target node will be able to open $SR_{(src,\ dest)}$. It then generates a proof $PR_{(dest,\ src)}$ that only the source of the RREQ will be able to decrypt, and constructs a route reply message (RREP):

$$\langle l, \{RREP,\ seq,\ R,\ PR_{(dest,\ src)}\}_s\rangle.$$

This is then sent by unicast using the masked channel to the node from which it previously received the RREQ. The seq, which is used to link RREPs to a route, gets discarded and replaced by the one-hop route pseudonym R. Figure 1 presents the content of the routing tables locally stored during and after the route creation. One row in the routing table corresponds to single route. It contains four entries: the two neighbors, and for each one, the corresponding route pseudonym. Once the route is created, a message m will be sent as $\langle l,\ \{R,\ m\}_s\rangle$.

If the sender wants to restrict the route through one or more groups, it can force it by using a *group restricted* RREQ of the form:

$$\langle RREQ,\ P,\ seq,\ LG,\ SR_{(src,dest)}(H(LG))\rangle,$$

with $LG = \{E_{seq\oplus k^{in_i}}(g^{in_i}), \ldots, E_{seq\oplus k^{in_k}}(g^{in_k})\}$ being the encrypted list of allowed groups, where $in_i, \ldots,\ in_k$ identify the groups to which the route search is restricted. The function SR_{Key} represents a derivative of $H(LG)$: $SR_{Key\oplus H(LG)}$. This prevents malicious nodes from replacing values in LG, since it would close the function and prevent the RREQ from finding any source in the network. The nodes which are unsuccessful in opening SR need to check if they have at least

---

[1] We assume that the sender and destination have securely exchanged proper cryptographic keys before being introduced into the network.

one of the necessary groups before replacing P with their own pseudonym and forwarding the message. The checking is done by computing the function E with the key derived from seq for each of their groups, and verifying that there exists at least one match when compared to LG. This is the same method as in Sect. 3.2 and therefore allows nodes to identify only common groups.

Route Repair. When a hop that is part of a route through the network becomes unavailable, a route error message (RERR) is generated and sent backwards on the route. Each node receiving a message of the form $\langle RERR, R\rangle$ will purge the corresponding route from its mapping table. Once the RRER reaches the initiator, it has to perform a new route detection.

4.2 Node Revocation

Nodes are required to choose a new pseudonym–secret point pair at the start of every epoch. Once a node runs out of pairs, it needs to request a new set from the NA. However, a malfunctioning node might simply always use the same pair, with the outcome being the temporary loss of its pseudonym unlinkability or even anonymity. If the node is malicious, it might willingly decide to use the same valid pair(s) in order to be accepted by the surrounding nodes, instead of risking to receive new pairs which might be incompatible with those of its neighbors. In both examples, the nodes need to be excluded from the network. This can be achieved by changing the way in which the secret points are generated and the way in which handshakes are performed: let $revDate$ be the ending date of the current epoch. By having the NA generate the secret point for a pseudonym by computing $\langle P, t\, H_1(P \parallel revDate)\rangle$ and having both A and B compute $K_{AB} = \hat{e}(H_1(P_B \parallel revDate), S_A)$, and respectively $K_{BA} = \hat{e}(H_1(P_A \parallel revDate), S_B)$, during the handshake stage, pseudonyms are bound to expiring secret points. Using such a pseudonym beyond $revDate$ will result in every handshake being refused. This not only forces nodes to replenish their pseudonym–secret point pairs regularly, but it also serves as a mechanism for the NA to exclude a certain node by refusing to issue it new such pairs.

4.3 Binding Groups to a Node

To prevent malicious nodes from using groups which were not assigned to them by the NA, a method of binding the groups to specific node is needed. This is achieved by binding each pseudonym to the group-set of the node through a signature that is generated by the NA. Besides generating and loading a pair $\langle P, S\rangle$ onto a node, the NA will also compute a third element called a transfer signature $T = Sign_{PK_{NA}}(E_{P \oplus k^{in^0}}(g^{in^0}), \ldots, E_{PS \oplus CS^{in^G}}(g^{in^G}))_{lex}$. During the intersection routine between A and B, the group-sets $set(A)_{K_{AB}}$ and $set(B)_{K_{AB}}$ are generated by deriving $CK_A = k_A \oplus P_A$ and $CK_B = k_B \oplus P_B$. By having the signatures exchanged and checked, both nodes can be assured that the list presented to them is authentic.

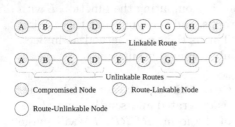

Fig. 2. Node unlinkability for an *active* route A-I (top) and for compromised nodes on an *inactive* route (bottom)

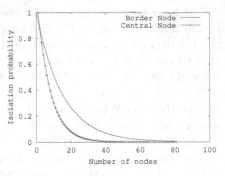

Fig. 3. Probability that a node is isolated

5   Analysis

5.1   Route Unlinkability and Node Anonymity

Route Unlinkability. In this section, we evaluate the route unlinkability degree relative to the number of compromised nodes of a given route. Anonymous routing requires the unlinkability of messages from routes. For external attackers, this is achieved by applying the masking key on a message on its way towards the destination. On compromised nodes, the local routing table offers information about route pseudonyms and the corresponding neighboring nodes.

A compromised node, part of an *active* route, can therefore link the active pseudonym of each of its direct neighbors to any of its routes. For multiple compromised nodes, shared routes be can identified through internal message coding attacks[2]. The implication is represented in Fig. 2: nodes B, E and H are linkable to the A-I route by their compromised neighbors A and C, D and respectively I. We can quantify this linkability of nodes to a specific active route length d (sometimes referred to as *traceability*) as $R_{Active} = \frac{\sum C_i}{d}$, where C_i represents the length of a compromised segment and the left- and right-most bordering nodes. In the example, $R_{Active} = \frac{7}{9} = 0.77$.

For *inactive* routes, it is only possible to link consecutive compromised nodes to a route. This scenario is depicted in Fig. 2, bottom: while both A-B and D have at least a route in common with C, it can't be certainly established whether they share at least one common route between them, since the mapping between the B-C and C-D route pseudonyms is stored only on Cs internal message forwarding table. Therefore the traceability ratio for inactive routes is $R_{Inactive} = \frac{\max(C_i)}{d}$. In the scenario described in Fig. 2, $R_{Inactive} = \frac{4}{9} = 0.44$.

Node Anonymity. A node exposes to its neighbors a tuple with a randomly generated pseudonym and a set of groups. If the group or group-tuple stored

---

[2] The masking keys are applied on-hop only. Every forwarding node handles m in its original form, which allows attackers to identify common routes.

on a node is unique and identifiable through a combined attack [3], it becomes unambiguously linked to the node's real identities and pseudonyms. To prevent that, the NA needs to guarantee that each group or tuple is shared by at least k nodes, thus preventing an attacker from linking pseudonyms, a group or group-sets to specific nodes with a probability higher than $\frac{1}{k}$ - the k-anonymity model [13]. To maintain this model, the groups or tuples shared by less than k nodes need to be merged with other groups that share similar grouping criteria in the network.

5.2   Groups and Node Connectivity

Node Connectivity. In most cases, distinct groups are usually shared by two or more existing group-tuples. In particular, there must exist one global group containing every entity in the network, which allows each node to always reach the NA regardless of what other groups it might share with its neighbors. However, we analyze the connectivity degree in a mobile network by assuming a scenario where a certain unique group is contained by only one k-anonymous tuple, an therefore is of minimum size. We model the mobility of the network according to the waypoint mobility model [14] and assume that the distribution of the nodes has reached a stationary distribution [19].

Bettstter's approximation [14] for the probability distribution of a mobile node's position in a two dimensional area $a \times a$, $f(x,y) \approx \frac{36}{a^6}((x - \frac{a}{2})^2 - \frac{a^2}{4})((y - \frac{a}{2})^2 - \frac{a^2}{4})$, shows how this converges, even though initially uniform distributed, towards an asymptotically distribution in the middle of the movement area. For a node A at position (u,v), the probability that a second node lies in communication range r_o is $P_{\langle A,ro \rangle} = \int_{v-r_0}^{v+r_0} \int_{u-\sqrt{r_0^2-(y-v)^2}}^{u+\sqrt{r_0^2-(y-v)^2}} f(x,y)\, dx\, dy$. In a network with k nodes, the probability that A has d neighbors can be approximated by the Poisson distribution $P(degree = d|A) \approx \frac{(\mu(A))^d}{d!} e^{-\mu(A)}$, where $\mu(A) = kP_{\langle A,ro \rangle}$, and the probability of having no neighbors is given by $P(degree = 0|A) \approx e^{-\mu(A)}$. The plot in Fig. 3 shows $P(degree = 0|A)$ with A located near the edges of the specified area, with $a = 500$ m and $r_0 = 70$ m (expected outdoor range of the TelosB TI CC2420 radio). The probability of having isolated nodes exponentially approaches 0 when k increases linearly. Thus, for our scenario, the probability of having isolated nodes due to the group size is negligible as long as k is not smaller than 40 nodes. For networks of different sizes and movement models, k needs to be similarly determined by NA. We note that, even though the probability of a node being isolated is higher for mobile nodes, the state is not definitive, and that the overall connectivity degree of an arbitrary node is significantly higher than that of a node part of a static uniformly distributed network [15].

Route Length. The length of a route created between two nodes varies depending on whether it was created with or without group restrictions. For most

---

[3] Multiple attackers, part of various groups, performing handshakes with the target and exchanging identified groups.

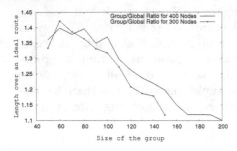
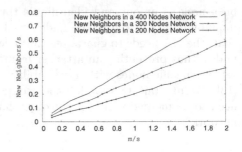

Fig. 4. Groups overhead over unrestricted routing

Fig. 5. Number of new neighbors in a mobile network

of their lifetime in a mobile network, nodes are more likely to be found near the center of the area than near the borders. For routes restricted to certain groups this translates into a higher density and therefore into shorter and direct paths between sender and destination. However, if we assume that the sender is positioned near the left-most border of the movement area and the destination near the right-most border, the overall density of the area containing forwarding nodes significantly declines. We modeled this scenario for various sizes of k and we compared the average length of the created route with that of an unrestricted route in Fig. 4. As expected, when k increases, the ratio between the group-restricted and unrestricted routes approaches one. For small values of k the route length increases in average with 25% to 45%.

5.3 Computational and Network Timings

We implemented[4] GAR on a TelosB IEEE 802.15.4 wireless sensor node [17], using a MSP430F1611 microcontroller running at a frequency of 8 MHz, 48 KB Flash, 10 kB SRAM and a 16x16 multiplier. The effective rate of the CC2420 radio in an active, medium size network lies under 90 kbps [12]. The resulting timings for the cryptographic blocks used by GAR, depicted in Table 1, allow us to model and determine the computational overhead and latency induced by it in each stage of the routing protocol as summarized in Table 2.

On the TelosB platform, the bilinear pairing computation takes 3.2 s per node, while the time delay induced by exchanging the pseudonyms and the random generated number takes in average 25 ms. When determining the group intersection, the signature validation is measured at 1.057 s. The rest of the operations consist of hashing G groups, each one represented as a 160 bit hash value. Lastly, a 320 bit ECDSA signature needs also to be exchanged, with a total expected time of $10(G+2)$ ms. Figure 5 depicts the average number of new neighbors discovered every second by a node traveling from one border to another in a simulated mobile network. For medium sized networks, mobility values of up

---

[4] For the bilinear pairing functions we used the optimized RELIC library [11].

Table 1. TelosB MSP430F1611 timing results of cryptographic functions

160 bit Tate Pairing of degree 4	3.2 s
ECDSA Sig. Check	1.057 s
128 bit RC6 20 Rounds	17 ms
SHA-1 160 bit hashing	5 ms
160 bit key HMAC-SHA1 160	7 ms

Table 2. Network timings results based on the cryptographic timings

Secret Handshake	3.23 s
Group Intersection	$10(G+108)$ ms
Unrestricted RREQ	$(d+1)25$ ms
Restricted RREQ	$15(u + 4)d$ ms
Message Forwarding	$5(d+7)$ ms

to 1.6 m/s allow a TelosB node to perform the detection, secret handshake and group intersection steps of the protocol in a reasonable time. For larger networks, the speed at which it can still detect all new neighbors in useful time decreases linearly with its size.

The size of a RREQ not restricted to any specific group is 376 bit, consisting of a 80 bit random generated RREQ-ID, the current 160 bit pseudonym and a 128 bit destination encryption. Each node receiving the route discovery tries to symmetrically decrypt the encrypted SR. In this case, a RREQ needs $(d+1)25$ ms to reach a destination that is d nodes away from the initiator. If the RREQ is restricted to u groups, the size grows with $160u$ bits. Including the hashing costs needed for deriving the decryption key, this leads to approximately $15(u + 4)d$ ms. After the route creation, the computational overhead needed to forward a message through the sensor network is marginal. The masking is performed twice (for incoming and outgoing messages) by each forwarding node as a XOR operation on the fixed size message 128 bit message, with negligible overhead. On the sender and receiver side, each one has to perform an encryption and decryption of m respectively. The time needed to forward a 128 bit message thus accounts to $5(d+7)$ ms.

For average values of d and u ($d = 15$, $u \leq 5$), the overall costs of GAR translates into latencies of approximatively 1 s for the creation of unrestricted routes, and respectively 2.5 s for restricted ones, and message forwarding latencies of under 150 ms. This guarantees a prompt reporting of events and message exchange between our targeted resource-restricted mobile nodes in the network.

6   Conclusion

This paper presented GAR, an anonymous on-demand routing protocol for ad hoc networks whose focus is on allowing the network administrator to establish and impose group restrictions on all network nodes. By using a model to compute possible group sizes, we were able to analyze their influence on the network connectivity and on the average length of group-restricted routes. With regard to the resource restrictions of WSNs, the evaluation shows that for medium sized networks with average mobility, the group intersection and group restricted route requests of GAR still allow the in-time detection of new neighbors, the quick creation of routes and exchange of information. As future work we plan to evaluate the protocol's suitability and performance in a real network composed of TelosB nodes.

References

1. Balfanz, D., Durfee, G., Shankar, N., Smetters, D., Staddon, J., Wong, H.: Secret Handshakes from Pairing-based Key Agreements. In: IEEE Symposium on Security and Privacy, pp. 180–196 (2003)
2. Boneh, D., Franklin, M.: Identity-based Encryption from the Weil Pairing. SIAM J. of Computing 32(3), 586–615 (2003)
3. Chaum, D.: Untraceable Electronic Mail, Return Addresses, and Digital Pseudonyms. Communications of the ACM 24(2), 84–88 (1981)
4. Menezes, A.J., Vanstone, S.A., van Oorschot, P.C.: Handbook of Applied Cryptography. CRC Press (1996)
5. Galbraith, S.D., Harrison, K., Soldera, D.: Implementing the Tate Pairing. In: Fieker, C., Kohel, D.R. (eds.) ANTS 2002. LNCS, vol. 2369, pp. 324–337. Springer, Heidelberg (2002)
6. Goldschlag, D.M., Reed, M.G., Syverson, P.F.: Onion Routing. Communications of the ACM 42(2), 39–41 (1999)
7. Kong, J., Hong, X.: ANODR: ANonymous On Demand Routing with Untraceable Routes for Mobile Ad-hoc Networks. In: Proceedings of the 4th ACM International Symposium on Mobile Ad Hoc Networking & Computing, pp. 291–302 (2003)
8. Moldovan, G., Ignat, A.: An Anonymous Efficient Private Set Intersection Protocol for Wireless Sensor Networks. In: Proceedings of the 25th International Conference on Architecture of Computing Systems - ARCS Workshops, pp. 39–49 (2012)
9. Perkins, C.E., Royer, E.M., Das, S.: Ad hoc On-Demand Distance Vector Routing. RFC3561 (2003)
10. Zhang, Y., Liu, W., Lou, W., Fang, Y.: MASK: Anonymous On-Demand Routing in Mobile Ad Hoc Networks. IEEE Transactions on Wireless Communications 5(9), 2376–2385 (2006)
11. Aranha, D.: F., Gouvêa, C.,P.,L.: RELIC is an Efficient LIbrary for Cryptography, http://code.google.com/p/relic-toolkit/
12. de Meulenaer, G., Gosset, F., Standaert, F., Pereira, O.: On the Energy Cost of Communication and Cryptography in Wireless Sensor Networks. In: Proceedings of the 2008 IEEE International Conference on Wireless & Mobile Computing, Networking & Communication, pp. 580–585 (2008)
13. Sweeney, L.: k-Anonymity: A Model for protecting privacy. Int. J. Uncertain. Fuzziness Knowl.-Based Syst., 557–570 (2002)
14. Bettstetter, C., Wagner, C.: The Spatial Node Distribution of the Random Waypoint Mobility Model. In: Mobile Ad-Hoc Netzwerke, 1. deutscher Workshop über Mobile Ad-Hoc Netzwerke WMAN 2002, pp. 41–58 (2002)
15. Bettstetter, C.: On the Connectivity of Ad Hoc Networks. The Computer Journal, 557–570 (2004)
16. Liu, J., Kong., J., Hong., X., Gerla, M.: Performance evaluation of anonymous routing protocols in MANETs. In: Wireless Communications and Networking Conference WCNC, pp. 646–651 (2006)
17. Polastre, J., Szewczyk, R., Culler, D.: Telos: enabling ultra-low power wireless research. In: Fourth International Symposium on Information Processing in Sensor Networks, pp. 364–369 (2005)
18. Shamir, A.: How to share a secret. Communications of the ACM 22(11), 612–613 (1979)
19. Navidi, W., Camp, T.: Stationary distributions for the random waypoint mobility model. IEEE Transactions on Mobile Computing 3(1), 99–108 (2004)

Greedy Routing Recovery Using Controlled Mobility in Wireless Sensor Networks[*]

Nicolas Gouvy[1,2], Nathalie Mitton[2], and Jun Zheng[3]

[1] Université Lille Nord de France, LIFL, France
[2] Inria Lille - Nord Europe, France
[3] National Mobile Communications Research Lab, Southeast University, China
{nicolas.gouvy,nathalie.mitton}@inria.fr, junzheng@seu.edu.cn

Abstract. One of the most current routing families in wireless sensor networks is geographic routing. Using nodes location, they generally apply a greedy routing that makes a sensor forward data to route to one of its neighbors in the forwarding direction of the destination. If this greedy step fails, the routing protocol triggers a recovery mechanism. Such recovery mechanisms are mainly based on graph planarization and face traversal or on a tree construction. Nevertheless real-world network planarization is very difficult due to the dynamic nature of wireless links and trees are not so robust in such dynamic environments. Recovery steps generally provoke huge energy overhead with possibly long inefficient paths. In this paper, we propose to take advantage of the introduction of controlled mobility to reduce the triggering of a recovery process. We propose Greedy Routing Recovery (GRR) routing protocol. GRR enhances greedy routing energy efficiency as it adapts network topology to the network activity. Furthermore GRR uses controlled mobility to relocate nodes in order to restore greedy and reduce energy consuming recovery step triggering. Simulations demonstrate that GRR successfully bypasses topology holes in more than 72% of network topologies avoiding calling to expensive recovery steps and reducing energy consumption while preserving network connectivity.

Keywords: greedy routing, hole bypassing, wireless sensor networks, controlled mobility.

1 Introduction

Miniaturization, costs decrease and advances in low-power electronic and radio communication technologies have made possible the emergence of new kinds of networks such as Wireless Sensor Networks (WSN). WSN are sets of a handful to thousands of sensors communicating through the radio medium in a multi-hop fashion. Each sensor embeds a low-power processor with limited computing and memory capabilities, a radio device and sometimes a localization device. Sensors

---

[*] This work was partially supported by a grant from CPER Nord-Pas-de-Calais FEDER Campus Intelligence Ambiante and Université Lille 1.

J. Cichoń, M. Gębala, and M. Klonowski (Eds.): ADHOC-NOW 2013, LNCS 7960, pp. 209–220, 2013.
© Springer-Verlag Berlin Heidelberg 2013

in WSN forward their data regularly, on event or on request. It is essential for the nodes to collaborate in order to route data in a reliable and energy-efficient way to a given destination. A popular routing family for WSN is geographic routing which requires nodes to be aware of their location. Geographic protocols generally include a greedy mechanism making each forwarding node to forward the packet to one of its neighbors closer to the destination than itself. Routing fails if their is no neighbor closer.Multiple mechanisms have been developed to address the greedy routing failure such as network planarization [1], trees [2], or hole detection mechanisms [3]. However they do not fit with WSN mobility at all as they strongly rely on nodes static position to overcome failures.

A new approach is to introduce controlled mobility enabled sensors. [4] shows that deploying resourceful mobile devices in a WSN provides better results than increasing network density. Still, only a few works use this controlled mobility in order to optimize route topology. Moreover none of them integrates a recovery mechanism as the classical approaches which have been developed in static WSN are unsuited to mobility at all. As a response, we propose the Greedy Recovery Routing routing protocol. GRR adapts the network topology in order to make energy savings with regard to the routed traffic. It enhances existing greedy routing protocol CoMNet and extends it with a light recovery mechanism. Both (enhanced CoMNet and light recovery steps) take advantage of node controlled mobility. GRR aims to bypass geographic routing failure by relocating nodes such as the greedy routing can be reused. GRR shows the following properties:

- *Localized*: Routing decisions rely only on local information: forwarding node geographical location, the ones of its neighbors and of the destination.
- *Scalable*: GRR is memoryless, no routing path information has to be stored.
- *Energy Efficient*: At each routing steps GRR chooses next hop and computes its relocation taking all costs into account.
- *Guaranteed Connectivity*: GRR guarantees network connectivity through the use of a Connected Dominated Set or the Relative neighborhood Graph.
- *Less Calls to Hard Recovery*: GRR implements a light recovery mechanism which aims to reduce the triggering of expensive hard recovery steps (based on face or tree traversals) by restoring greedy routing when possible. It will eventually fill routing holes restoring an end-to-end greedy routing.

The remaining of this paper is organized as follows. Section 2 reviews existing works on delivery guarantee and mobile routing in WSN. We detail the models used in the paper in Section 3, while the prerequisites are exposed in Section 4. Section 5 presents and details our approach. Simulation results are detailed in Section 6. Finally Section 7 concludes the paper and presents future work.

2   Related Work

Delivery Guarantee in Static Networks

To the best of our knowledge, the most popular recovery strategy for geographic routing is face routing [1]. It makes a packet traverse the faces of the planarized network graph until greedy routing is possible. Faces traversal is done using the right-hand-rule. However, face is energy-consuming since it may generate long detours and makes the packet follow a succession of short edges [5]. Moreover, it requires reliable network planarization which is nearly impossible to provide in real world due to the non-uniform wireless links.

In hull trees [2] each node has an associated convex hull that "contains the location of all its descendant in the tree". When greedy routing fails, forwarding node checks its hull tree downstream to find a path to destination. If the destination belongs to the hull tree of the forwarding node, packet is forwarded to the first corresponding child node. Otherwise packet is forwarded upstream. This approach can be very memory consuming depending on network density and size and number of hull trees employed. Moreover a moving node could easily destroy trees.

Authors in [3] propose an approach which does not rely on planarization nor trees. At network bootstrap, an algorithm is applied locally on each node in order to mark those where packets may possibly get stuck. And then, each marked node identifies both its upstream and downstream nodes on the boundary of the hole. Consequently, when the greedy routing fails, the packet can be routed along the hole until greedy becomes possible again. Yet this approach requires a lot of messages at network bootstrap and in order to adapt to topology changes. To the best of our knowledge, none of the existing approaches behaves well under the hypothesis of mobility unlike Greedy Routing Recovery which successfully takes advantage of mobility to both optimize greedy routing -enhancing a previously proposed approach- and reduce the number of calls to expensive recovery steps.

Routing with Controlled Mobility

Mobility has long been considered as a hazard in WSN, causing a degradation of performances or routing failures. A handful of proposals considers the use of controlled mobility in order to adapt the network topology with regard to the routing path. Existing routing protocols such as MobileCOP [6] find an initial route using a Cost-Over-Progress (COP) metric [7], and then iteratively move each forwarding node to the midpoint of its upstream and downstream nodes on the route. These routing protocols may not be efficient as they can cause energy consuming zig-zag movements and the network may be disconnected (a node may move out of range of its neighbors). In addition, none of these approaches considers the cost of moving in the routing decision. CoMNet [8] is the first fully localized COP-based geographic routing protocol that takes the moving cost into account while guaranteeing network connectivity. It has been extended in [9] in order to consider mobility consequences on a multiple hop point of view. Nevertheless, none of the controlled mobility enabled protocols can recover from a local minimum.

GRR bases on CoMNet: it enhances it in its greedy part and extends it with a light recovery mechanism to exploit controlled mobility in order to restore greedy routing when possible.

3 Models

The network is modeled as an undirected simple finite graph $G(V, E)$, with V the set of nodes and E the set of edges. $(A, B) \in E$ if A and B are in transmission range. The Euclidean distance between A and B is noted as $|AB|$. We denote by $N(A)$ the set of neighbors of A: $N(A) = \{V \in E \mid (A, V) \in V\}$ and $N_D(A)$ the subset of $N(A)$ which are closer to the destination D than A: $N_D(A) = \{B \in N(A) \wedge |BD| < |AD|\}$. Every node in V is aware of its geographical location and can be either a mobile or stationary sensor. Relocation of a node A is noted as A'. Although our approach is model-independent, we use the following widely employed cost models as a proof of concept:

Transmission Cost. We denote by $C_s(.)$ the energy consumption or cost for radio transmission between two nodesdistant of r [10]:

$$C_s(r) = r^\alpha + c \quad \text{if } r \neq 0 \tag{1}$$

where c represents the energy overhead due to radio device, α is a real constant (> 1) that represents the signal attenuation. The associated optimal radio transmission radius [11] for radio transmission is $r^* = \sqrt[\alpha]{\frac{c}{\alpha-1}}$.

Mobility Cost. We denote by $C_m(.)$ the cost to relocate a node B to B'. We use the model adopted in previous similar works [6], in which k is a constant :

$$C_m(|BB'|) = k * |BB'| \tag{2}$$

4 Preliminaries

4.1 Greedy Routing

The greedy step of GRR proposes an enhanced CoMNet (COnnectivity preservation Mobile routing protocols for actuator and sensor NETworks) [8]. CoMNet is a geographic routing protocol for WSN which takes advantage of nodes controlled mobility in order to adapt network topology to the routing traffic. Precisely, CoMNet uses a Cost-Over-Progress (COP) [7] approach: current node A chooses $B \in N_D(A)$ which minimizes the ratio of the global cost (packet transmission cost plus node relocation cost) to the progress made towards the destination. Indeed, B satisfies the following optimization problem:

$$B = arg_{min_{K \in N_D(A)}} \frac{C_s(|AK|) + C_m(|KK'|)}{|AD| - |K'D|} \tag{3}$$

CoMNet comes in three different variants to fit the best to various environments:

- $CoMNet - Move_{(DSr)}$ aligns nodes on the Source Destination (SD) line with all hop lengths to be equal to the optimal transmission distance r^*.
- $CoMNet - ORouting$ on the $Move$ aligns nodes on the (SD) line.
- $CoMNet - Move_r$ makes next hop node B is relocated on the intersection of the circle $C(A, r^*)$ of radius r^* centered at A and the (BD) line.

4.2   Connectivity

Relative Neighborhood Graph (RNG): RNG [12] is a graph reudction that can be computed locally. An edge (U, V) exists if distance $|UV|$ is less than or equal to the distances $|UW|$ and $|VW|$ for any other vertex W:

$$\forall W \neq U, V \in E : |UV| \geq max[|UW|, |VW|] \tag{4}$$

It reduces the average node degree to $\simeq 3$ while preserving networking connectivity. A moving node which stays connected to its RNG neighbors will keep network connectivity unchanged.

Connected Dominated Set (CDS): CDS is a connected subset of V that covers the same area. If nodes in the CDS (*i.e. dominant*) are static, we ensure that all other nodes which move stay in transmission range of the CDS. It guarantees that their is always a path between every pair of nodes of the network. In [13] authors have proposed a fully localized algorithm. Giving a node A, $N(A)$ and S the subset of $N(A)$ with higher priority (such as id, battery level, etc...) than A: $S \leftarrow N(A) - \{U \in E \mid p(U) < p(A)\}$, A is dominant if one of the following statement do not hold:

- S is not empty: $S \neq \{\emptyset\}$
- S is connected: $\forall A \in S, \exists B \in S$ s.a. $A \neq B \wedge |AB| < radio\_range$
- every node in $N(A)$ is in S or in range of S: $\forall B \in N(A), B \in S \vee N(B) \cap S \neq \emptyset$

5   Geographic Routing Recovery

5.1   Overview

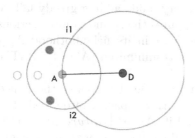

Fig. 1. $N_D(A)$ is empty, greedy routing to D fails on A. Previous routing nodes are in white.

Greedy routing failure is due to the lack of a neighbor closer to the destination than the current forwarding node. Figure 1 represents such a case: forwarding node A has no neighbor closer to destination D. A represents a local extremum of the routing path toward destination node D, it faces a hole. Greedy forwarding requires at least one node in the area defined by the intersection of $C_1(A, r*)$ the disk centered at A and of radius $r*$, and $C_2(D, |AD|)$ the disk centered at D of radius $|AD|$ to succeed.

Hence we propose Greedy Routing Recovery which combines greedy routing and a light recovery mechanism which aims to restore the greedy forwarding.

In greedy forwarding, GRR routes packet and relocates nodes in a CoMNet way. Forwarding nodes are aligned on the source-destination line in order to save energy. However, when greedy fails, GRR switches to light recovery step. In light recovery, forwarding nodes relocates next hop on the intersection locations i_1 (or i_2) in order to bypass the routing hole. When greedy routing become possible again, GRR switches back to it. GRR uses nodes controlled mobility to both optimize network topology to the traffic and create greedy routing paths in order to avoid expensive recovery steps.

Algorithm 1. GRR(A,D, p) - Node A has a packet p for node D

```
1:  if isInGreedyMode(p) then
2:     if N_D(A) ≠ ∅ then
3:        next, next' ← SelectGreedy(A, D);
4:     else
5:        addFailureLocation(p);
6:        next, next' ← SelectRecovery(A, D);
7:     end if
8:  else
9:     moveToLocation(p) {execute relocation order while it does not disconnect the network}
10:    if N_D(A) ≠ ∅ and |AD| ≤ |D, failureLocation(p)| then
11:       next, next' ← SelectGreedy(A, D);
12:    else
13:       next, next' ← SelectRecovery(A, D);
14:    end if
15: end if
16: forward(p,next,next') {forward packet p to node next with relocation order in next'}
```

More precisely GRR works as follows. Upon reception of a packet in greedy forwarding (Algo 1, l1), a node A checks its neighborhood toward destination $N_D(A)$. If it exists a node $next$ closer to the destination than itself, A computes its new location $next'$ and forwards the packet to it before or after relocating it (Algo 1, l3) depending on CoMNet variants. If there is no neighbor closer to the destination, A marks the packet into light recovery mode adding greedy failure location into its header (Algo 1, l5). A then computes the i_1 and i_2 intersections locations. It forwards the packet to the node $next$ in its neighborhood $N_{(A)}$ whose transmission and relocation cost to the i_x is minimized (Algo 1, l6). The packet includes a relocation order to position i_x.

Upon reception of a packet in light recovery, a node A moves to the joined location. When stopped, A checks wether it can turn the packet into greedy forwarding. This is possible only if $N_A(D) ≠ ∅$ and if $|AD| < |failure_{location}D|$) (Algo 1, l9-10). Otherwise the packet is forwarded to the node $next$ in its neighborhood $N_{(A')}$ whose transmission and relocation costs to the i_x is minimized. We have to mention that network connectivity is guaranteed despite nodes movement using either the CDS or the RNG neighbors as described in Section 4.2. Every moving node moves up to its new location while its relocation does preserve network connectivity.

Figure 2 illustrates a complete GRR routing from S to D. Routing is greedy from S to B and nodes A, B and C are relocated on the (SD) line. However,

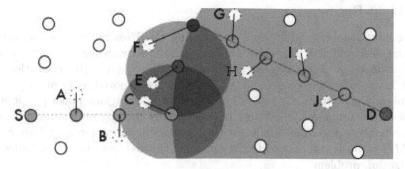

Fig. 2. GRR routing from node S to D. Original node location is dashed. Green nodes have been relocated during greedy. Blue ones during light recovery.

greedy fails on C as none of its neighbors is closer to D than itself. Consequently, C switches the packet into light recovery. Node E is selected: C forwards the packet to it with the order to relocate in i_x. E relocates. When E stops, it checks if greedy is still impossible. That is still the case: using greedy recovery E forwards the packet to F and makes it relocate. On F, greedy routing is possible since G exists, furthermore F is not further to D than C. On the next routing, the SD path will be completely greedy.

In the following Section, we detail the sub-algorithms used to select and relocate next forwarder while in greedy forwarding or in light recovery.

5.2   Greedy

Although GRR relies on CoMNet [8] relocation patterns in its greedy part, routing is different as GRR computes the COP associated to each pattern for every possible next hop at each step of the routing as described in Algorithm 2. Forwarding node A computes for each neighbor B in $N_D(A)$ (Algo 2, l4) its relocations according to the three different CoMnet variants (Algo 2, l6-8) and retains the associated lowest COP (Algo 2, l9). Finally A returns the node (and its relocation) in $N_D(A)$ which minimizes the COP (Algo 2, l10-13, l16).

Algorithm 2. SelectGreedy(A,D) - Run at node A toward destination D.

```
 1: next ← −1; {next hop elected}
 2: nextℓ ← {0,0,0}; {next hop computed relocation}
 3: minCOP ← +∞ {lowest COP over all N_D(A)}
 4: for all {B ∈ N_D(A)} do
 5:     Bℓ_or, Bℓ_mr, Bℓ_mdsr ← {0,0,0};
 6:     Bℓ_or ← ORouting(A, B, D); Bℓ_mr ← Move_R(A, B, D); Bℓ'_mdsr ← Move_DSR(A, B, D)
 7:     bCOP ← min[COP(A, B, Bℓ_or), COP(A, B, Bℓ_mr), COP(A, B, Bℓ_mdsr)];
 8:     if (bCOP < minCOP) then
 9:         next ← v, nextℓ ←{relocation which minimizes COP}
10:         minCOP ← bCOP
11:     end if
12: end for
13: return next, nextℓ {return next node next with its computed relocation in next'}
```

5.3   Light Recovery

During light recovery, GRR stops considering CoMNet relocations patterns as all of them would fail. However GRR makes the forwarding A node compute for every node B in its $N(A)$ –and not only in $N_D(A)$– the relocation cost on the intersection between $C_1(A, |r^*|)$ $C_2(D, |AD|)$ noted as i_1 or i_2 plus the transmission cost from A to B. Those specific i_x locations make possible the bypass of the routing hole as they restore greedy routing on the next routing. Moreover, those locations minimize the moving distance for next hop B and the $|AB'|$ radio transmission cost is optimal. Elected next hop B satisfies this optimization problem:

$$B = arg_{min_{K \in N(A)}} C_s(|AK|) + C_m(|Ki_x|) \qquad (5)$$

where i_x is replaced by respectively i_1 or i_2.

Algorithm 3 details light recovery. Forwarding node A first computes the intersection locations i_1 or i_2 (Algo 3, l4-5). Then, A computes for each of its neighbor B in $N_D(A)$ (Algo 3, l6) the total cost of its relocation both in i_1 and i_2 and the associated transmission cost from B to A (Algo 3, l7-8). A finally forwards packet and relocation order to the node which minimizes both costs the most (Algo 3, l9-18, l20).

Algorithm 3. SelectRecovery(A,D) - Run at node A when $N_D(A)$ is empty.

```
 1: next ← −1 {next hop elected}
 2: nextı ← {0,0,0} {next hop computed relocation}
 3: minCOST ← +∞ {lowest total cost computed over all N_D(A)}
 4: (i_1, i_2) ← intersections(C_1(A, |r*|), C_2(D, |AD|));
 5: for all {B ∈ N_D(a)} do
 6:    tCOST1 ← C_s(|AK|) + C_m(|Ki_1|); tCOST2 ← C_s(|AK|) + C_m(|Ki_2|)
 7:    if (tCOST1 < minCOST) then
 8:        next ← B; nextı ← i_1; minCOST ← tCOST1
 9:    end if
10:    if (tCOST2 < minCOST) then
11:        next ← B; nextı ← i_2; minCOST ← tCOST2
12:    end if
13: end for
14: return next, nextı {return next node next with its computed relocation in next'}
```

---

6   Experiments and Performance Analysis

6.1   Simulation Environment

We simulate GRR using the last release of WSNet [14] network simulator. Different node degrees δ (from 10 to 25) are considered and maximum node speed is set to $1m.s^{-1}$. Relocations engender delays in packet delivery which are not in the scope of this study. Maximal radio range is set to 50m. Simulations are performed on 25 randomly generated 500x500m maps with nodes uniformly deployed on which 3 holes of diameters 100m has been created. Greedy routing fails on those maps for the selected source and destination couple. Initial battery level of every node is 1J and every node is capable of moving. $Cs(.)$ is computed using Eq. 1 in which we use common values: *i.e.* c = 3.10^9J and α=2, which leads to an optimal transmission range of $r^* = 22.36$m. $Cm(.)$ is computed using Eq. 2, with mobility parameter k computed as follows:

1. if $C_s(.) = C_m(.)$, k is solution to $C_s(r^*) = C_m(r^*)$.
2. if $C_s(.) >> C_m(.)$, then k is solution to $C_s(r^*) = 10^2 C_m(r^*)$.
3. if $C_s(.) << C_m(.)$ then k is solution to $C_s(r^*) = 10^{-2} C_m(r^*)$.

6.2 Routing Success Rate

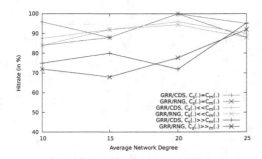

Fig. 3. Percentage of topologies where GRR succeed with regard to δ

Figure 3 shows the percentage of topologies on which routing succeeds using GRR with regards to various network densities and cost models. Note that each of the topologies used in those simulations is connected at network bootstrap. In other words, it exists at least one multi-hop path between every pair of nodes in the network at network start. Even so, a classical geographic greedy heuristic would fail because of network holes.

With regards to the connectivity guarantee mechanism and density, results are very similar. When the network density is low, most nodes belong to the CDS and so mobility is limited in GRR-CDS. RNG neighbors in GRR-RNG are also very sparse and so mobility is limited. With the increasing density, the percentage of topologies on which GRR (all variants) succeeds increases as more and more nodes are free to move with GRR-CDS. Similarly in GRR-RNG, neighbors are more numerous - and consequently closer - and so mobility freedom increases.

With regards to cost model, we see that the percentage of routing success is maximal when $C_s(.) == C_m(.)$. Considering that A is the forwarding node in recovery mode, the reason is that it makes A consider every node in $N(A)$ equally. However, success rate is minimal when $C_s(.) >> C_m(.)$ as it provokes the selection of a node which is close to A in order to reduce transmission costs. But this node has to travel a long distance to reach i_x and restore greedy. Yet this distance can be impossible to travel due to the connectivity guarantee mechanism. Middle case is when $C_s(.) << C_m(.)$: A selects as next forwarder a node close to i_x.

6.3 Number and Length of Recovery Occurrences

Figure 4 depicts the number of transitions between greedy and recovery forwarding steps with regards to time. Results show that at network bootstrap the

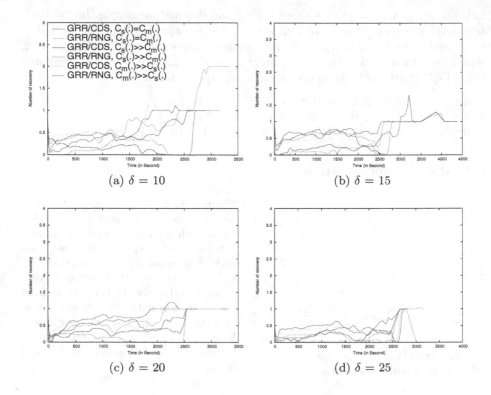

(a) $\delta = 10$ (b) $\delta = 15$

(c) $\delta = 20$ (d) $\delta = 25$

Fig. 4. Average number of recovery occurrences along time

number of recovery occurrences tends to decrease very quickly due to an initial greedy path restoration thanks to our recovery mechanism. Then, nodes that have been relocated die since they have spent energy to relocate and are now highly employed for greedy forwarding. Consequently the number of recoveries increases again after a certain time since the light recovery routing scheme has to restore greedy forwarding. However this might not be possible since movable nodes might have already moved: that is the reason why sometimes light recovery occurrences does not decrease any more.

This analysis is confirmed by Figure 5 which represents the average path length (in number of hops) in recovery step. We see that after an initial decrease, the number of hops in recovery stabilizes and then increases again after a certain time when relocated nodes start to die. The routing hole is bypassed but the GRR routing protocol has to circumvent an increasing hole along time. Please note that routing does not succeed at all without GRR light recovery.

6.4   Average Packet Cost

Figure 6 represents the total cost – including both transmission and relocation costs of nodes on routing path – of a delivered packet with regards to time.

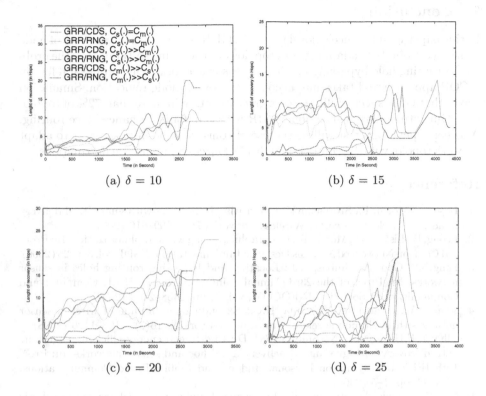

(a) $\delta = 10$ (b) $\delta = 15$

(c) $\delta = 20$ (d) $\delta = 25$

Fig. 5. Average length of recovery step in hops along time

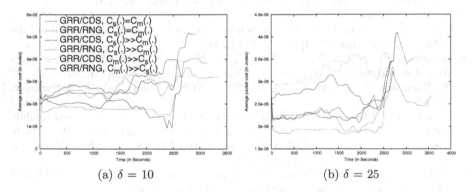

(a) $\delta = 10$ (b) $\delta = 25$

Fig. 6. Average packet total cost along time

Similarly to previous behavioral results, results show an initial energy overhead whatever the cost model, the recovery mechanism or the network density. This is due to the initial path relocation which makes node move a lot. In a second step, packet cost tends to stabilize up to a final stage where packet cost increases dramatically because GRR has to bypass major holes.

7   Conclusion

In this paper we introduce a novel protocol, GRR which takes advantages of node controlled mobility to adapt network topology to the traffic and addresses the problem of routing hole bypassing in wireless sensor networks. Our proposal relies on a COP approach and takes into account the cost of node relocation. Simulation shows that GRR successfully restores greedy routing in more than 72% of topologies. Our future work will combine GRR with a mobility enhanced face routing. Another topic of interest will be the use of controlled mobility in order to adapt network topology to the context of multiple sources and multiple destinations.

References

1. Bose, P., Morin, P., Stojmenović, I., Urrutia, J.: Routing with guaranteed delivery in ad hoc wireless networks. Wireless Networks 7(6), 609–616 (2001)
2. Leong, B., Liskov, B., Morris, R.: Geographic routing without planarization. In: Proc. 3rd Conf. on Networked Systems Design & Implementation, NSDI, vol. 3, p. 25 (2006)
3. Fang, Q., Gao, J., Guibas, L.J.: Locating and bypassing routing holes in sensor networks. In: Proc. of the 23rd Annual Joint Conf. of the IEEE Computer and Communications Societies, INFOCOM, vol. 4, pp. 2458–2468 (2004)
4. Wang, W., Srinivasan, V., Chua, K.C.: Extending the lifetime of wireless sensor networks through mobile relays. Trans. on Networking 16(5), 1108–1120 (2008)
5. Elhafsi, E.H., Mitton, N., Simplot-Ryl, D.: End-to-end energy efficient geographic path discovery with guaranteed delivery in ad hoc and sensor networks. In: Proc. 19th IEEE Int. Symp. on Personal, Indoor and Mobile Radio Communications, PIMRC, pp. 1–5 (2008)
6. Liu, H., Nayak, A., Stojmenović, I.: Localized mobility control routing in robotic sensor wireless networks. In: Zhang, H., Olariu, S., Cao, J., Johnson, D.B. (eds.) MSN 2007. LNCS, vol. 4864, pp. 19–31. Springer, Heidelberg (2007)
7. Kuruvila, J., Nayak, A., Stojmenović, I.: Progress and location based localized power aware routing for ad hoc and sensor wireless networks. Int. Journal of Distributed Sensor Networks 2(2), 147–159 (2006)
8. Hamouda, E., Mitton, N., Simplot-Ryl, D.: Energy Efficient Mobile Routing in Actuator and Sensor Networks with Connectivity preservation. In: Frey, H., Li, X., Ruehrup, S. (eds.) ADHOC-NOW 2011. LNCS, vol. 6811, pp. 15–28. Springer, Heidelberg (2011)
9. Gouvy, N., Mitton, N.: MobileR: Multi-hop energy efficient localised mobile georouting in wireless sensor and actuator networks. In: Simplot-Ryl, D., Dias de Amorim, M., Giordano, S., Helmy, A. (eds.) ADHOCNETS 2011. LNICST, vol. 89, pp. 147–161. Springer, Heidelberg (2012)
10. Rodoplu, V., Meng, T.: Minimum energy mobile wireless networks. IEEE Journal on Selected Areas in Communications 17, 1333–1344 (1999)
11. Stojmenović, I., Lin, X.: Power-aware localized routing in wireless networks. IEEE Trans. on Parallel and Distributed Systems 12(11), 1122–1133 (2001)
12. Toussaint, G.: The relative neighbourhood graph of a finite planar set. Pattern Recognition 12(4), 261–268 (1980)
13. Carle, J., Simplot-Ryl, D.: Energy-efficient area monitoring for sensor networks. Computer 37(2), 40–46 (2004)
14. Fraboulet, A., Chelius, G., Fleury, E.: Worldsens: development and prototyping tools for application specific wireless sensors networks. In: Proc. 6th ACM/IEEE Int. Conf. on Information Processing in Sensor Networks, IPSN, pp. 176–185 (2007)

A Test-Bed Analysis of Simultaneous PMIPv6 Handover in 802.11 WLANs Environment

Michal Hoeft and Jozef Wozniak

Department of Computer Communications
Gdańsk University of Technology
{michal.hoeft,jowoz}@eti.pg.gda.pl

Abstract. Providing mobility in access networks is a challenge that we have to deal with. Due to networks' convergence and migration to all-IP networks, mobility management at the network layer is required. However, there is a need for cooperation mechanisms between the network layer and lower layers to support multimedia services and make handover more efficient. This paper presents experimental research on simultaneous handover performance in an integrated PMIPv6 and 802.11 WLAN system.

Keywords: Mobility, Simultaneous handover, Proxy Mobile IPv6, IEEE 802.11, Test-bed.

1 Introduction

In view of increasing numbers of mobile terminals and, at the same time, rapidly increasing popularity of multimedia and real-time services offered to users, mobility management has become a real challenge for network designers and operators. A substantial amount of research is focused on either layer 2 handovers or layer 3 handovers. However, it is quite natural that integration of these two layers and separately performed procedures could bring a significant acceleration of the handover and switching procedures.

In some circumstances, especially in large networks with a huge number of mobile terminals, layer 2 handover might be inefficient. Moreover, if the access points use L2 links, it is hard to deploy a mobile system in networks with IP connections. To integrate mobility system entities connected within an IP network, layer 3 solutions have to be used. What is more, a great number of IP mobility protocols are based on lower layer triggers that start a handover procedure at network layer. However, L2 triggers are not always well defined, e.g., for Proxy Mobile IPv6. A closer integration of L2 and L3 solutions might be the only solution for the new demanding applications.

Vehicle-to-infrastructure communication is one of example scenarios where simultaneous handover of layer 2 and layer 3 is required [1],[2],[3]. In this case, due to a large area of a mobility domain and a number of mobile terminals, seamless handover should be provided in layer 3, where frequent changes of attachment point benefit from network based mobility, like Proxy Mobile IPv6.

J. Cichoń, M. Gębala, and M. Klonowski (Eds.): ADHOC-NOW 2013, LNCS 7960, pp. 221–232, 2013.

There are more papers that focus on similar topics. The 802.21 framework in the Proxy Mobile IPv6 environment is commonly used in different scenarios – for providing mobility in heterogeneous network [4] or advanced load balancing based on network layer flows [5]. A new proposition of simultaneous Proxy Mobile IPv6 and 802.11 network is described in [6]. Although the authors have proposed a new schema and the use of Inter-Access Point Protocol (IAPP) to transfer handover information, they do not provide mathematical analyses, simulations or results of the experiments. The paper [7] presents an analysis of the operations affecting the handoff delay in PMIPv6 environment and introduces MIH services to optimize it. The proposed schema is validated in the OPNET network simulator.

This paper focuses on a study of simultaneous 802.11 WLANs and Proxy Mobile IPv6 handovers that takes place in a test-bed environment. The body of the paper is organized as follows: Section 2 contains an analysis of the handover in 802.11 WLANs. Additionally, new trends and standards are also described . Section 3 presents the idea and main components of Proxy Mobile IPv6. Section 4 introduces the proposed schema for simultaneous handover for 802.11 WLAN and PMIPv6. Section 5 gives the details of the tested system. Section 6 presents the results of the experiment. Conclusions can be found in Section 7.

2   IEEE 802.11 WLANs Overview

The IEEE 802.11 is one of the commonly accepted wireless local area network technologies. IEEE 802.11 standard's handover (L2 handover) refers to wireless network attachment point change during a Mobile Host (MH) movement. The handover results in de-association from the old (current) AP and re-association to a new one, adequate to a new MH's location . The interval between the time of the optimal start of the handover and the time of the actual start of the handover procedure is called the detection phase. It is hard to both define and measure it. Handover procedure begins with a scanning and its aim is to detect a new AP. The MH has to employ authentication and association processes. Subsequently, 802.1x authentication can be conducted on top of 802.11 association. It introduces additional 4-way handshake for key exchange and derivation. In order to provide a better QoS, some additional procedures introduced by the IEEE 802.11e standard can be used.

The most time-consuming operations are detection, scanning and 802.1x authentication. The literature suggests that there are several solutions for reduction of their impact on handover's time [8],[9],[10],[11],[12],[13].

Some aspects of influence minimization of detection and scanning phases upon handover time are introduced in IEEE 802.11k [10] and 802.11v [8] standards. The first one was released in December 2008 and refers to radio resources measurements. Its mechanisms are responsible for providing information about radio environment and available resources for upper layer.

The second one was published in July 2007, and completed in March 2010. The IEEE 802.11v standard introduces BSS Transition Management that enables APs to trigger handover to the nearest, less loaded or more appropriate AP.

Additionally, non-standardized algorithms and mechanism of minimization of the scanning time can be found in the literature e.g., [11][12][13].

The 802.1x Authentication introduces a significant temporal overhead caused by authentication messages exchange with an AAA server.

The IEEE 802.11r standard [9] addresses handover efficiency and 802.1x authentication reduction. It is discussed in more details in Subsection 2.1.

Other proprietary mechanism proposed by hardware vendors could also be used (most often based on dedicated wireless network controllers). Since they are vendor specific and dedicated for particular scenarios, they lie outside of the scope of this paper.

2.1   Fast BSS Transition

The Fast BSS Transition mechanism was proposed in the IEEE 802.11r amendment [9] to improve handover efficiency. It introduces two protocols, FT Protocol for standard re-association and FT Resource Protocol for re-association with QoS resources reservation. They are used for simpler and shorter authentication procedures of handover inside one mobility domain. The complete, time-consuming authentication procedure is conducted only for the first entrance into the mobility domain (*FT initial mobility domain*). Subsequently, security association and key hierarchy are created for all other APs in the mobility domain. The first AP that MH is associated with is called R0KH (R0 Key Holder) and it is responsible for managing level 0 Pairwise Master Key (PMK-R0). During re-associations with the nAP, a PMK-R1 is generated and transported by R0KH to the new AP (named R1KH – level 1 Key Holder). Time-consuming communication between the nAP and the AAA server is replaced with shorter R0KH – R1KH communication.

In the mobility domain, APs advertise both capabilities and policies for supporting the FT protocols and methods.

Two algorithms of communication between the old and new AP are proposed:

- Over-the-Air – the MH uses wireless link to communicate directly with the new AP (Fig. 1a),
- Over-the-DS – the MH uses existing association with the old AP, and FT Action Frames over wired link to reach the new AP (Fig. 1b).

In the first case (Fig. 1a) the Mobile Host, after deciding to change the access point, sends 802.11 Authentication Request message containing Information Elements required by FT Protocol [14]. Robust Security Network IE (RSNIE) carries PMK-R0 obtained from the first AP. Mobility Domain IE (MDIE) contains domain details from Beacon and Probe Response sent by nAP including a domain identifier, capabilities and policies. Fast BSS Transition IE (FTIE) contains R0KH identifier that is obtained during the initial association and SNonce created by the Mobile Host.

The response from the nAP contains the same types of Information Elements as the request described in the previous paragraph, however, their role might be changed. In this case, RSNIE contains confirmation of the PMK-R0 identifier. MDIE contains the domain identifier, capabilities and policies. FTIE is made up of an R1KH identifier supplied by nAP, ANonce produced by nAP and SNonce sent by MH.

In Association Request messages exchange, besides RSNIE, MDIE and FTIE, another type of Information Elements is used – Resource Information Container Request (RIC-Request). RSNIE contains PMK-R1, MDIE is made up of the same information as Authentication Request. FTIE carries ANonce, SNonce, R0KH identifier, R1KH identifier and MIC calculated on MH side. If QoS methods from 802.11e standard are implemented in the domain, RIC-Request carries Information Elements – Traffic Specification and Traffic Classification IEs.

The Association Response message includes RSNIE, MDIE, FTIE and RIC-Response. RSNIE and MDIE are similar to appropriate IEs in Association Request. FTIE is made up of ANonce, SNonce, R0KH identifier, R1KH identifier and MIC calculated on AP side. RIC-Response contains Traffic Specification and Schedule IEs.

The content and type of IEs is the same using both Over-the-air and Over-the-DS content.

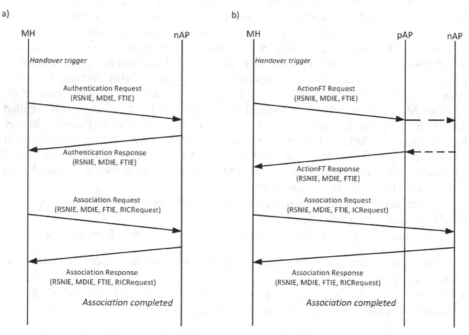

Fig. 1. 802.11r handover scenarios: a) FT over the Air, b) FT over the DS

3   Proxy Mobile IPv6 Handover

In large IP networks, there is a need for network layer mobility management. One of the popular mobility protocols is the Proxy Mobile IPv6 [15]. It reuses well-known, mature concepts of Mobile IPv6 [16] and implements a Network-Based Mobility approach (NBM). As opposed to the Host-Based Mobility approach (HBM), where mobile terminal is used in order to provide mobility, NBM performs mobility without any cooperation with the Mobile Host. The L3 handover should be transparent for mobile terminals and should not affect any routing or addressing configuration. One of the advantages of the NBM concept is the mobility management integration into a network operator center. For that reason, PMIPv6 is considered of use in the LTE architecture [17].

Another advantage is that PMIPv6 eliminates requirements for changes in MH's IPv6 protocol suits.

Proxy Mobile IPv6 defines two additional network elements – Local Mobility Anchor and Mobile Access Gateway. The first one is similar to the Home Agent from Mobile IPv6. It is a gateway between a group of mobile terminals and the rest of the network. It is responsible for maintaining configurations and routing information. The LMA terminates bi-directional IP tunnels where data packets from/to MHs are transported. The Mobile Access Gateway is an element that maintains connectivity between MHs and LMA. All MAGs act as a default access router, thus after the handover the MH does not notice that there has been a change of the attachment point.

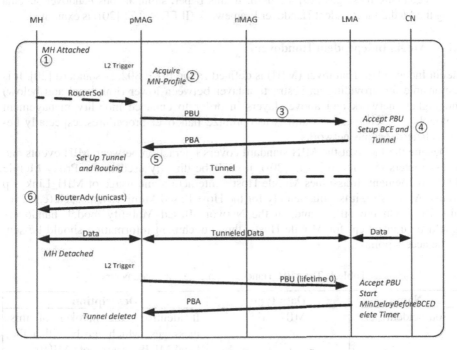

Fig. 2. Proxy Mobile IPv6 message flow

LMAs and MAGs communicate by means of Proxy Binding Update (PBU) and Proxy Binding Acknowledgment (PBA) – messages of Proxy Mobile IPv6 protocol. An example message flow for registration and handover is presented in Fig. 2.

Mobile Host attachment should be detected and signaled by L2 Trigger (1). The ICMPv6 Router Solicitation message may arrive at any time and it does not cause any ordering relation in the signaling flow. After authentication and authorization of the MH with its profile (2), the pMAG sends the Proxy Biding Update message (3). The PBU message contains MH-Identifier and lifetime of the Biding Cache Entry (BCE) for the corresponding mobility session. The LMA accepts this PBU message, creates the BCE and the endpoint of the bidirectional tunnel (4). Next, it responds (5) with the Proxy Binding Acknowledgment message that contains a Home Network Prefix(es) (HMP). On receiving the PBA message, the MAG finishes setting up its endpoint of the bidirectional tunnel. Once the tunnel is established, the MAG sends ICMPv6 Router Advertisement (6) with HNP and address configuration mode.

4   IEEE 802.11 WLANs and PMIPv6 Simultaneous Handover Schema

Although communication between L2 and PMIPv6 elements is an important and time-consuming factor, standardization documents (e.g. RFC) do not propose any way of informing MAG entities about MH attachments. In practice, syslog messages [18] or Radius Accounting messages [19] are used. In this paper, simultaneous handover schema using the Media Independent Handover Framework (IEEE 802.21 [20]) is examined.

4.1   Media Independent Handover

Media Independent Handover (MIH) is defined in the IEEE 802.21 standard [20]. It is responsible for providing an abstraction layer between lower (link-local and below) and higher (network and above) layers in order to enhance mobility management protocols and schemes. Its aim is to optimize handover procedures, especially between heterogeneous networks.

Despite the fact that the MIH standard covers interaction between MIH events and access routers (Section 6.3.6 in [20]), it can't be directly used in the Proxy Mobile IPv6 environment. It assumes Mobile Host's interaction and usage of MIH_Link_Up events. As a result, it is suitable only for the Host-Based Mobility approach (e.g. Mobile IPv6). On the other hand, in the Network-Based Mobility model, handovers should be transparent for Mobile Hosts, thus attachment information should be sent by the access point.

Table 1. The MIH_Handover_Init message parameters

Name	Data type	Description
SourceIdentifier	MIHF_ID	It identifies the invoker of this messages, which can be either the local MIHF or a remote MIHF.
LinkIdentifier	LINK_TUPLE_ID	Identifier of the link that is associated with an AP.

In the presented implementation, an MIH user is integrated with AP software. The MIH user, after a successful attachment of MH, informs the appropriate MAG and generates a trigger for the upper layer. In order to increase handover efficiency, a new message, following the pattern of MIH_Link_Handover_Complete, was used. The original MIH_Link_Handover_Complete message was designed for a heterogeneous environment. Because conducted experiments include homogenous handovers, the format of this message is modified (Table 1.) to eliminate unnecessary fields. The SourceIdentifier field contains MAC address of a new access point, LinkIdentifier is an identifier of MH's interface connected to the AP. This MIH message is called MIH_Handover_Init.

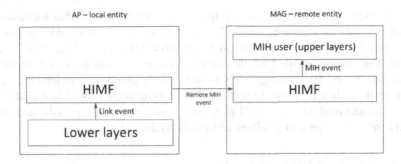

Fig. 3. MIH Remote Events

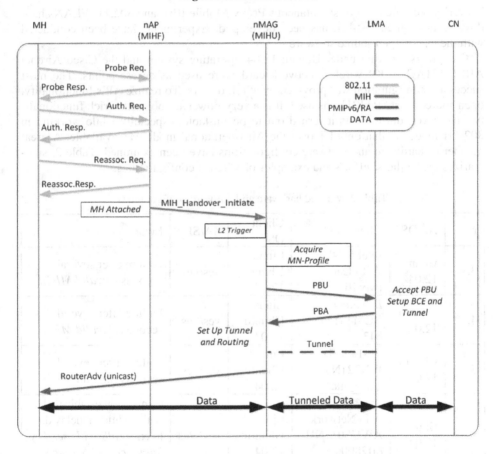

Fig. 4. Simultaneous handover for PMIPv6 and 802.11 WLAN

The notification is generated by the Media Independent Handover Function (MIHF) when an attachment to the new access point is completed (Fig. 3). Upper layer entities take different actions on this notification. In the described PMIPv6 environment, the MIH user makes use of the event to start PBU/PBA procedures.

Fig. 4 presents the message flow of implemented mechanisms. After successful authorization and association (802.11 messages), the new AP sends the MIH_Handover_Init message with its MAC (SourceIdentifier) and the client's LinkIdentifier. The value from the LinkIdentifier field is used to obtain Mobile Node Profile and create the PBU message with Mobile Node Identifier Header Option. The subsequent procedure is similar to standard PMIPv6 message flow. The nMAG establishes a bidirectional tunnel with LMA, and after that sends a dedicated Router Advertisement message to the unicast address of Mobile Host.

5   Test-Bed and Experiments

To verify the efficiency of simultaneous Proxy Mobile IPv6 and 802.11 WLANs handovers, an original test-bed has been developed. Experiments have been conducted with the use of open source software.

Computers working under Ubuntu 12.04 operating system and the Cisco Aironet AIR-CB21AG-W-K9 wireless network card were used as access points. The most successful configuration was provided by ath5k drivers. To realize AP's functionality, open source HostAPd [21] was used. It is a very powerful tool with a rich functionality . However, during tests it turned out to be unstable, especially while working in 802.11r mode, so that only FT over the Air operational mode was available. A great number of hardware and software configurations have been examined. Table 2. summarizes the authors' efforts and examples of verified configurations.

Table 2. Verified hardware and software configurations

No	AP OS	Client NIC	Client OS	SA/SH	Error
3.	Ubuntu 12.04	Intel 5100 WiFi Link / iwlwifi	Linux Ubuntu 12.04	yes/yes	Failure after several seconds (*Invalid MIC*).
4.	Ubuntu 12.04	*Broadcom* BCM4312 / b43	Linux Ubuntu 12.04	yes/yes	Failure after several seconds (*Invalid MIC*).
5.	Ubuntu 12.04	TP-Link TL-WN721N / ath9k_htc	Linux Ubuntu 12.04	yes/yes	Failure after several seconds (*Invalid MIC*).
6.	Ubuntu 12.04	Alfa Network AWUS051NH / rt2800usb	Linux Ubuntu 12.04	yes/ yes	Connection established with additional delay due to error (*nl80211: set_key failed; err=-2 No such file or directory*)

SA/SH – successful association, successful handover

For the purposes of this paper, the authors used the Proxy Mobile IPv6 system which is the authors' own implementation prepared at the Department of Computer

Communications of Gdańsk University of Technology (see [19]). It is an expandable Python application based on Scapy – Python networking module.

MAGs and APs have been extended so that they are able to communicate by means of MIH messages in accordance with the scheme described in Section 4. All system elements have been synchronized using NTP protocol.

The topology of the experimental network is presented in Figure. 5.

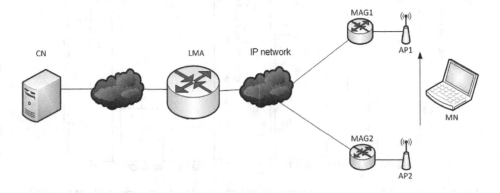

Fig. 5. Test-bed architecture

6   Results

The performed experiments aimed to verify the handover influence on the connectivity gap. To achieve higher reliability, each test scenario was repeated 30 times. All confidence intervals are stated at the 95% confidence level. Different scenarios of L2 and simultaneous L2+L3 handovers have been examined.

Fig. 6 illustrates the results of the handover time for different authentication models (WPA2-PSK, WPA2-EAP, 802.1x). The parameter L2 Handover time comprises only the layer 2 handovers. Fig 6a presents the scenario including the scanning phase (see Sec. 2). During the experiments, this phase takes an extremely long time – more than 4.5s. Due to unacceptably long scanning time, L2 handover times were very long. This drawback might be caused by specific software/hardware configuration [22]. In contrast, Fig. 6b) presents the results of the L2 handovers scenario where the scanning phase is omitted. This was achieved by means of the *roam* command, with a new AP MAC address as a parameter, in the *wpa_supplicant* console. As can be seen, the L2 handover time is significantly reduced in comparison to previous scenarios.

Fig. 7. plots the final results and the cumulative distribution function of simultaneous handover. The parameter L2+L3 Handover time comprises both the layer 2 and the layer 3 handovers. The average handover time for 802.11r scenario is 0.15±0.04s, whereas for 802.1x it is 0.48±0.26s. As can be seen, the average handover time is significantly reduced (by more than two-thirds) when using 802.11r mechanisms in comparison to the traditional 802.1x authentication.

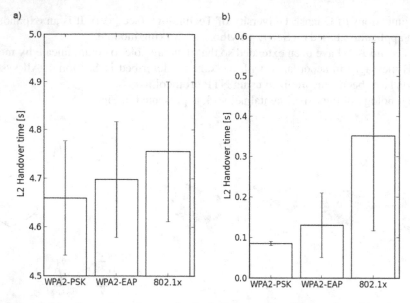

Fig. 6. Layer 2 handover time, a) scenarios including scanning; b) scenarios without scanning

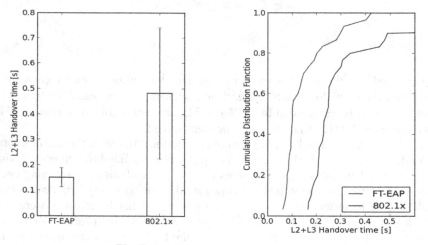

Fig. 7. Results of simultaneous handovers

7   Conclusions

We have evaluated the effectiveness of the simultaneous IEEE 802.11 and Proxy Mobile IPv6 handovers in the test-bed architecture with two MAG components. Simultaneous handover has been implemented with the use of the IEEE 802.21 standard, for which a dedicated message was used. Basing on the described architecture, several experiments that verified handover times have been conducted.

It was shown that 802.11r and Proxy Mobile IPv6 integration has a great impact on handover efficiency. The average handover time for layer 2 and layer 3 was reduced to 150ms, which is an acceptable value for majority of multimedia and real-time applications. Thus, the implemented solution gives mobile users the opportunity to use multimedia application.

The proposed and described layer 2 and layer 3 integration by means of the Media Independent Handover Framework is easily expandable and in future studies might be enhanced to cooperate with other wireless systems, like LTE .

References

1. Lu, H., Zhang, S., Lin, X.: Mobility-assisted fast handover for Proxy Mobile IPv6 in vehicle-to-infrastructure communications. In: Intelligent Transportation Systems 2012, September 16-19, pp. 921–926 (2012)
2. Zhu, K., Niyato, D., Wang, P., Hossain, E., Kim, D.I.: Mobility and handoff management in vehicular networks: A survey. Wirel. Commun. Mob. Comput. 11(4) (April 2011)
3. Cespedes, S., Lu, N., Shen, X.: VIP-WAVE: On the Feasibility of IP Communications in 802.11p Vehicular Networks. IEEE Transactions on Intelligent Transportation Systems 14(1), 82–97 (2013)
4. Jose, J., Prithiviraj, A.: PMIPV6-HC-MIH: An approach for improving handover performance in NGWN. In: 2012 International Conference on Computing, Electronics and Electrical Technologies (ICCEET), March 21-22, pp. 910–914 (2012)
5. Kim, M.-S., Lee, S.: Load balancing based on layer 3 and IEEE 802.21 frameworks in PMIPv6 networks. In: 2009 IEEE 20th International Symposium on Personal, Indoor and Mobile Radio Communications, pp. 788–792 (September 2009)
6. Lee, J.-C., Park, J.-S.: Fast Handover for Proxy Mobile IPv6 based on 802.11 Networks. In: 10th International Conference on Advanced Communication Technology, ICACT 2008, vol. 2, pp. 1051–1054 (February 2008)
7. Kim, I., Jung, Y.C., Kim, Y.-T.: Low Latency Proactive Handover Scheme for Proxy MIPv6 with MIH. In: Ma, Y., Choi, D., Ata, S. (eds.) APNOMS 2008. LNCS, vol. 5297, pp. 344–353. Springer, Heidelberg (2008)
8. Meschke, R., Krohn, M., Daher, R., Gladisch, A., Tavangarian, D.: Novel handoff concepts for roadside networks using mechanisms of IEEE 8002.11k & IEEE 802.11v., ICUMT 2010, 1232–1238 (October 2010)
9. IEEE 802.11r: Amendment 2: Fast Basic Service Set (BSS) Transition, IEEE Std 802.11r-2008 (July 2008)
10. IEEE Standard for Information technology– Local and metropolitan area networks– Specific requirements– Part 11: Wireless LAN Medium Access Control (MAC)and Physical Layer (PHY) Specifications Amendment 1: Radio Resource Measurement of Wireless LANs. IEEE Std 802.11k-2008 (Amendment to IEEE Std 802.11-2007), pp.1–244 (June 2008)
11. Garcia, E., Ferrer, J.L., Lopez-Aguilera, E., Vidal, R., Paradells, J.: Client-driven load balancing through association control in IEEE 802.11 WLANs. Eur. Trans. Telecomms. 20, 494–507 (2009)
12. Ramani, I., Savage, S.: SyncScan: Practical Fast Handoff for 802.11 Infrastructure Networks. In: Proceedings of IEEE Infocom (2005)

13. Machań, P., Wozniak, J.: Proactive handover for IEEE 802.11r networks. In: 2011 4th Joint IFIP Wireless and Mobile Networking Conference (WMNC), October 26-28, pp. 1–7 (2011)

14. Machan, P., Wozniak, J.: Performance evaluation of IEEE 802.11 fast BSS transition algorithms. In: 2010 Third Joint IFIP Wireless and Mobile Networking Conference (WMNC), October 13-15, pp. 1–5 (2010)

15. Gundavelli, S., Leung, K., Devarapalli, V., Chowdhury, K., Patil, B.: Proxy Mobile IPv6, IETF, RFC5213 (August 2008)

16. Perkins, C., Tellabs Ed., Johnson D., Arkko J.: Mobility Support in IPv6, IETF, RFC 6275 (July 2011)

17. ETSI TS 123 402, Architecture enhancements for non-3GPP accesses, version 11.5.1 (release February 11, 2013)

18. OAI PMIPv6, http://www.openairinterface.org/openairinterface-proxy-mobile-ipv6-oai-pmipv6 (available February 21, 2013)

19. Hoeft, M., Gierlowski, K., Gierszewski, T., Konorski, J., Nowicki, K., Wozniak, J.: Measurements of OF QoS/QoE parameters for media streaming in a PMIPv6 testbed with 802.11 b/g/n WLANs. Metrology and Measuring Systems 02/2012; XIX(02/2012), 283–294 (2012)

20. IEEE Standard for Local and Metropolitan Area Networks- Part 21: Media Independent Handover. IEEE Std 802.21-2008, pp.c1–301 (January 21, 2009)

21. hostapd, http://hostap.epitest.fi/hostapd/ (available February 21, 2013)

22. Martinovic, I., Zdarsky F.A., Bachorek A., Schmitt J.B.: Measurement and Analysis of Handover Latencies in IEEE 802.11i Secured Networks. In: Proceedings of the 13th European Wireless Conference (EW 2007), Paris, France (2007)

A Column Generation Approach to Maximize Capacity of Multi-rate Power Controlled TDMA Wireless Sensor Networks

Mejdi Kaddour

Laboratoire d'Informatique et des Technologies de l'Information
University of Oran
BP. 1524 El MNaouer, Oran 31000, Algeria
kaddour.mejdi@univ-oran.dz

Abstract. This paper describes a mixed-integer linear programming model to maximize wireless sensor networks (WSN) capacity by determining traffic routes and target coverage, adjusting data rates, and dynamically controlling transmission power. Radio transmissions are studied under signal-to-interference-plus-noise-ratio (SINR) model. In the considered WSN, a set of targets needs to be continuously monitored by a given number of already deployed sensors. The objective is to gather as much sensing data as possible during a unit time period. Since this type of problems is known to be NP-hard, we propose a computationally feasible column generation based approach to find near-optimal solutions when dealing with integer flow and coverage variables. Extensive computational experiments even show that optimal solutions are reached in most cases for reasonable network sizes. Numerical results illustrate the benefits from extending the number of available transmission power levels, the major impact of data rate adaptation on throughput, and demonstrate that achieving higher network capacity comes at the expense of deploying more sensors than the minimum required to connectivity and target monitoring.

Keywords: Wireless sensor networks, TDMA, SINR, power control, Q_coverage, MILP, column generation.

1 Introduction

A wireless sensor network (WSN) consists of spatially distributed autonomous sensors to monitor physical or environmental conditions, such as temperature, sound, pressure, etc., and to cooperatively relay data through a wireless network to a sink. The predominant networking technology in this area in based on IEEE 802.15.4 for physical and MAC layers, and Zigbee for the higher layers of the protocol stack [1]. In particular, IEEE 802.15.4 is a derivative of CSMA/CA contention-based medium access scheme. But many other MAC layer approaches in WSNs propose also collision-free access schemes. TDMA (Time Division Multiple Access) is such an example as the time is divided into frames each containing

J. Cichoń, M. Gębala, and M. Klonowski (Eds.): ADHOC-NOW 2013, LNCS 7960, pp. 233–244, 2013.
© Springer-Verlag Berlin Heidelberg 2013

a certain number of fixed size slots. TDMA is typically managed by a central entity which defines a schedule of the frame assigning each sensor a fixed number of slots for transmitting and receiving data. Transmission links could be scheduled at the same slot if they don't interfere mutually. In general, TDMA is well suited to WSNs where traffic patterns are regular and predictable over time.

One of the main challenges in designing TDMA for WSNs is to define a link schedule optimizing a given performance metric such as capacity, end to end delay, energy consumption or fairness index. Typically, one common goal of optimization models is to minimize the number of slots per frame as it implies maximizing capacity. The key issue here is how to mitigate interference by adapting transmit powers, traffic routes or channel bandwidth to the network topology and the traffic loads. We find two interference models in the literature i.e., i) the protocol interference model and ii) the physical model. Under the protocol model, a successful transmission occurs when the intended receiving node falls inside the transmission range of its transmitting node and falls outside the interference ranges of other non-intended transmitters. On the other hand, under the physical model, a transmission is successful if and only if signal-to-interference-and-noise-ratio (SINR) at the intended receiver exceeds a certain threshold so that the transmitted signal can be decoded with an acceptable bit error rate (BER). Physical model is widely considered as an accurate representation of the behaviour of the physical layer in real systems [2].

Optimization models based on the protocol interference model are usually based on conflict graphs. The idea is to find a maximum independent set representing links that can transmit simultaneously without causing collisions. This problem is also equivalent to the well-known graph coloring problem which is NP-hard [3]. Many works are based on linear programming formulations [4, 5]. Otherwise, the optimal scheduling problem under the physical model is generally formulated as a mixed-integer linear program (MILP) [6]. The general problem of determining a minimum-length schedule that satisfies given traffic demands and subject to SINR constraints in NP-hard [7]. Many polynomial-time approximation algorithms were proposed on this issue. The authors in [8] studied joint link scheduling and power control with the objective of throughput improvement while considering fairness. In [9], the authors considered the problem of assigning time slots to different users to minimize channel usage subject to constraints on data rate, delay bound, and delay bound violation probability under an SINR-based interference model. In [10], the authors modeled the combinatorially complex problem of joint routing, link scheduling, and variable-width channel allocation in both single and multi-rate multi-hop wireless networks as an MILP, and presented a solution framework using a column generation based approach.

Different from existing works, our contribution encompasses in a single unified optimization model both connectivity requirements (multi-path routing, power control, rate adaptation) and Q-coverage requirements in TDMA wireless sensor networks. In particular, we formulate an MILP referred to as *Joint Routing, Scheduling and Power control* (JRSPC), which can be optimally solved in most cases by a column generation based approach. The objective of JRSPC is to find the links which can transmit concurrently, and the association between targets and sensors that minimize the length of the TDMA frame[1]. Furthermore, it has the following properties: (1) the considered interference model is SINR-based; (2) sensors can adjust dynamically their transmission power to reduce interference or hop-distance from the sink; (3) transmission links can admit different data rates depending on the SINR at the receiver; (4) multi-path routing; (5) a target can be associated to multiple sensors providing redundancy and reliability to sensing data.

The rest of this paper is organized as follows. Section 2 introduces the definitions and the assumptions used throughout this paper. Section 3 gives formal details of our MILP model considering continuous, discrete and uniform power cases. Section 4 summarizes the results of various experiments showing the effects of sensor density, number of targets, and coverage on network capacity by applying different transmission strategies. Finally, section 5 concludes this article outlining possible future extensions.

2   Definitions and Assumptions

We consider a set of n sensors, denoted as $S = \{s_1, s_2, ..., s_n\}$ and a set of m targets, denoted as $T = \{t_1, t_2, ..., t_m\}$, arbitrary deployed on a given area. Each target must be covered by at least q sensors, known also as Q_coverage requirement. A target can be covered by a given sensor if the euclidean distance between them doesn't exceed a certain distance d_c. Each sensor generates a data packet of size θ bits each time it monitors an associated target. All the gathered data must be routed through multiple hops to a sink node, denoted as s_0. Also, we assume that each sensor is equipped with a single radio that can adjust dynamically its transmission power up to a certain limit P_{max}. We define the network connectivity graph as an directed graph $G = \{S \cup \{s_0\}, E\}$ with $E = \{(s_i, s_j) : s_i, s_j \in S \cup \{s_0\}\}$. There exists an edge between two nodes s_i and s_j if this latter lies within the transmission range of s_i resulting from of P_{max}.

According to the physical model, a transmission link (s_i, s_j) would be successful if $SINR_{i,j}$ measured at the level of node s_j is greater or equal to a threshold β. It is given by:

$$SINR_{i,j} = \frac{P_i d_{i,j}^{-\alpha}}{\eta W + \sum\limits_{i', i' \neq i} P_{i'} d_{i',j}^{-\alpha}} \tag{1}$$

where P_i is the transmission power of s_i, $d_{i,j}$ is the distance between nodes s_i and s_j, α is the path loss exponent (varies usually between 2 and 6), η is the power spectral density of the thermal noise, and W is the channel bandwidth.

---

[1] The main purpose of this work is to gain some insights about the maximal achievable capacity of sensor/target wireless networks with rate adaptation and power control, hence there is no compromise with respect to energy consumption or network lifetime.

Furthermore, we assume that each sensor node can transmit according to K modulation and coding schemes (MCS). Each MCS_k $(0 \le k < K)$ produces a certain data rate b_k. A transmission with rate b_k could be decoded successfully if the SINR measured at the receiver is above a corresponding threshold β_k. Note that the higher data rate is used, the higher $SINR$ is required. In addition, we assume an TDMA access scheme, where a central entity divides the radio channel into equal-length time frames and each frame into a number of slots. In each slot, one or more transmission links are scheduled, providing that the SINR requirement (1) is met at each receiver.

In the following section, we introduce an optimization model that minimizes the length of the TDMA frame by jointly determining the routes of the traffic flows, the active concurrent links at each slot, and the transmission power used by each sensor node at each assigned slot. Note that a minimal TDMA frame implies a maximal network capacity.

3   Joint Routing, Scheduling and Power Control Problem (JRSPC)

A straightforward or naive approach to the JRSPC problem would be to rely on a binary decision variable x_{ijt} which indicates if the link (s_i, s_j) is active during the slot t. The major drawback with this formulation is the huge number of binary variables that would arise: $n(n+1)T_{max}$. Indeed, since the number of slots in an TDMA frame is also a decision variable, the parameter T_{max} denoting the maximal number of slots in a frame should be chosen large enough to hold the resulting transmission schedule, otherwise, the problem will be infeasible.

Alternatively, in the same vein as [10], our model is based on the notion of *Configuration* which represents a set of transmission links that can be scheduled concurrently without violating the SINR requirement at each receiver. Despite the exponential number of possible configurations, which scales up following the cardinality of the power set of $S \cup \{s_0\}$ (2^{n+1}), we need in fact only a small subset of these configurations to resolve the problem as we will see in the following.

Let C be the set of all feasible configurations, and let $c = \{r_{ij}^c \mid \forall (s_i, s_j) \in E\} \in C$, where r_{ij}^c indicates the data rate of the link (s_i, s_j) during the time where the configuration c is scheduled. r_{ij}^c is equal to 0 when the corresponding link is inactive. The integer decision variable λ_c is defined as the number of slots in which the configuration c is active. In addition, let y_{ij} be a binary variable indicating if the target t_j is covered by the sensor s_i $(i \ne 0)$. Also, let f_{ij} be an integer variable counting the number of data packets forwarded from node s_i to node s_j during the whole TDMA frame. Our optimization problem can be stated now as follows:

$$\text{Minimize} \sum_{c \in C} \lambda_c \tag{2}$$

$$\text{s.t.} \qquad \sum_{s_i \in S} y_{ij} = q \quad \forall t_j \in T \tag{3}$$

$$\sum_{s_j \in S} f_{ji} + \sum_{t_k \in T} y_{ik} = \sum_{s_j \in S \cup \{s_0\}} f_{ij} \quad \forall s_i \in S \tag{4}$$

$$\sum_{s_i \in S} f_{i0} = mq \tag{5}$$

$$\sum_{c \in C} \lambda_c r_{i,j}^c - f_{ij}\theta \geq 0 \quad \forall s_i \in S, \forall s_j \in S \cup \{s_0\} \tag{6}$$

$$\lambda_c \ integer, \lambda_c \geq 0, \quad \forall c \in C \tag{7}$$

$$f_{ij} \ integer, f_{ij} \geq 0, \quad \forall s_i \in S, \forall s_j \in S \cup \{s_0\} \tag{8}$$

$$y_{ij} \in \{0,1\}, \quad \forall s_i \in S, \forall t_j \in T \tag{9}$$

Constraint (3) ensures that every target is covered exactly by q sensors. Constraint (4) represents flow conservation rule by stating that ingoing traffic into a sensor node, which is the sum of traffic forwarded by other sensors and the traffic generated locally by monitoring associated targets, is equal to outgoing traffic. Constraint (5) guarantees that all data packets are gathered by the sink. Lastly, the channel capacity constraint (6) ensures that the number of times each link (s_i, s_j) is scheduled over all the configurations is sufficient to forward traffic from node s_i to node s_j. Here, $r_{i,j}^c$ is the data rate of the link (s_i, s_j) in configuration c.

Note that this configuration-based ILP model doesn't provide an ordering of the transmission links as the straightforward model does. Nevertheless, this difference has no impact on overall capacity since the configurations can be scheduled at any order inside the frame without changing its length. Yet this configuration-based model is solvable if we can determine by some mean the set C. But enumerating all feasible configurations could be computationally very hard. We present in the next subsection a technique to alleviate this issue.

3.1 Column Generation Approach

Column generation (CG) is an exact and efficient algorithm for solving large-scale linear programs. That main idea relies on the fact that most of the variables will be non-basic (or equal to zero) in the optimal solution. Hence, only a subset of variables (or columns) needs to be considered in theory when solving the problem. The problem being solved is split into two problems: the master problem and the pricing problem. The master problem is the original problem with only a subset of variables being considered. The pricing problem is another problem created to generate only the variables which have the potential to improve the objective function, i.e., to find variables with negative reduced costs in case of a minimization problem. The algorithm alternates then between these two problems until no variable which would enhance the objective function is found.

In our case, the master problem is formulated as in (2)-(6) but considers only a subset C_0 of feasible configurations C. We describe later in this section how we obtain a subset C_0 for the first iteration, called also initial feasible solution.

As CG is an LP technique, we need also to make a linear relaxation on integer variables λ_c, f_{ij} and y_{ij}.

On another hand, the pricing problem relies on finding a new configuration which has the minimal reduced cost. Let u_{ij} be the dual variables associated with (6). The reduced cost of any configuration c not currently in C_0 is given by $\bar{e}_c = 1 - \sum_{(s_i, s_j) \in E} u_{ij} \times r_{ij}$, where r_{ij} are the data rates of the active links in c. If the minimal reduced cost has a strictly negative value, the corresponding configuration (column) is added to C_0 and the master problem is solved again, otherwise the optimal solution has been reached.

Now, we can write the pricing problem as follows:

$$\text{Minimize } 1 - \sum_{(s_i, s_j) \in E} u_{ij} \times r_{ij} \tag{10}$$

$$\text{s.t.} \qquad \sum_{s_j \in S} \sum_{k=1}^{K} x_{ijk} + x_{jik} \leq 1 \quad \forall s_i \in S \tag{11}$$

$$\sum_{k=1}^{K} x_{0i} = 0 \quad \forall s_i \in S \tag{12}$$

$$p_i d_{ij}^{-\alpha} - \beta_k \sum_{(s_u, s_v) \in E, u \neq i} p_u d_{uj}^{-\alpha} - L_1 x_{ijk} \geq \beta_k W \eta - L_1$$

$$\forall s_i \in S, \forall s_j \in S \cup \{s_0\}, 1 \leq k \leq K \tag{13}$$

$$p_i \leq P_{max} \sum_{s_j \in S} \sum_{k=1}^{K} x_{ijk} \quad \forall s_i \in S \tag{14}$$

$$x_{ijk} \leq L_2 p_i \quad \forall s_i \in S, \forall s_j \in S, 1 \leq k \leq K \tag{15}$$

$$r_{ij} = \sum_{k=1}^{K} b_k x_{ijk} \quad \forall s_i \in S, \forall s_j \tag{16}$$

$$x_{ijk} \in \{0, 1\} \quad \forall s_i \in S, \forall s_j \in S, 1 \leq k \leq K \tag{17}$$

$$p_i \geq 0 \quad \forall s_i \in S \tag{18}$$

where x_{ijk} indicates a transmission link between nodes s_i and s_j using rate b_k, p_i is the transmission power of node s_i, L_1 and L_2 are large positive constants. Constraint (11) states that a node cannot transmit and receive at the same time and restricts it to a single rate, whereas constraint (12) prevents the sink from transmitting. Constraint (13) is an enforcement of (1) by ensuring that when a link x_{ijk} is active, the SINR should be above the threshold β_k. Otherwise, this constraint becomes redundant. Note that K denotes the number of available MCSc and hence the number of available data rates. Constraints (14) and (15) bind between every transmission link and its power. If a transmission link is not active, power would be set to zero, and vice-versa. Finally, constraint (16) computes the data rates r_{ij} from the variables x_{ijk}.

Algorithm 1. Initial Basic Feasible Solution

$E = \{s_0\}$ ▷ set of connected sensors
$F = S$ ▷ set of non connected sensors
$c = 0$
while $F \neq \emptyset$ **do**
    find $s_i \in E$ and $s_j \in F$ where $d_{ij} \leq R_{max}$
          ▷ R_{max}: transmission range achieved with power P_{max} in absence of
            interference
    $E = E \cup \{s_j\}$
    $F = F - \{s_j\}$
    $r_{ij}^c \leftarrow b_{ij}$ ▷ b_{ij}: the maximal data rate supported on link (s_i, s_j) in absence of
            interference.
    $r_{i'j'}^c \leftarrow 0$ for $(i,j) \neq (i',j')$
    $c \leftarrow c + 1$
end while
return r

As stated above, the CG algorithm requires an initial basic feasible solution to be provided to the master problem, i.e., a set of configurations C_0 that leads to an initial solution verifying flow, capacity and coverage constraints. Algorithm 1 provides a simple solution by building a tree rooted at the sink spanning all the sensor nodes, and populates each configuration with a single transmission link having the highest achievable rate.

3.2 Near-Optimal Solution of the JRSPC Problem

The CG technique solves a linear relaxed version of the original MILP problem, but a more concrete solution to the problem would provide integer values for coverage and flow variables. Indeed, the coverage variable y_{ij} indicates whether or not a target is covered by a given sensor (0 or 1), and the flow variables f_{ij} specify the number of data packets traveling on the link (s_i, s_j) during the TDMA frame. A data packet should reach the sink as a whole entity and it's not a good idea to scatter it into several pieces as this may cause significant overhead.

Hence, we propose here a heuristic to figure out a near-optimal solution of the MILP. Let Z^{LP} and Z^{ILP} denote the optimal solutions of the problem, considering continuous and integer flow and coverage variables, respectively. Let Z^* be the solution obtained by resolving the ILP version of the master problem in section 3.1 after all the columns with negative reduced costs have been added and consequently Z^{LP} has been obtained. It's clear that $Z^{LP} \leq Z^{ILP} \leq Z^*$. So the desirable optimal solution lies somewhere between Z^{LP} and Z^*. As showed by our numerous experimentations, the gap between these two values is very small, and in most cases cannot hold an integer solution strictly less that Z^*. That means that we can reach the optimal solution of the original problem in those cases.

3.3   Discrete and Uniform Transmission Powers

The aforementioned pricing problem assumes continuous power levels, but in most real-world settings, only a finite number of power levels are available (e.g., CC2420 RF transceiver provides 8 programmable output power levels [11]). By assuming L available power levels, the corresponding pricing problem could be stated by substituting (13), (14) and (15) of the continuous case by the following:

$$\sum_{l=1}^{L} p_{il} \leq 1 \quad \forall s_i \in S \quad (19)$$

$$d_{ij}^{-\alpha} \sum_{l=1}^{L} p_{il} P_l - \beta_k \sum_{(s_u,s_v)\in E, u\neq i} \sum_{l=1}^{L} p_{ul} P_l d_{uj}^{-\alpha} - L_1 x_{ijk} \geq \beta_k W\eta - L_1$$

$$\forall s_i \in S, \forall s_j \in S \cup \{s_0\}, 1 \leq k \leq K \quad (20)$$

$$x_{ijk} \leq L_2 \sum_{l=1}^{L} p_{il} \quad \forall s_i \in S, \forall s_j \in S \cup \{s_0\}, 1 \leq k \leq K \quad (21)$$

$$p_{il} \in \{0,1\} \quad \forall s_i \in S, 1 \leq l \leq L . \quad (22)$$

where p_{il} is a binary indicating if the node s_i transmits with power level l, and P_l is a parameter denoting the power level l. Constraint (19) restricts each sensor to use only one power level within one configuration.

Now, assume that the sensor's radio interface enables only a single power level, or alternatively assume that changing the transmission power is a time-consuming operation in comparison with the duration of a single slot of the TDMA frame. In this case, the SINR constraint in the pricing problem can be adapted as follows:

$$d_{ij}^{-\alpha} P - \beta_k \sum_{(s_u,s_v)\in E, u\neq i} P x_{uvk} d_{uj}^{-\alpha} - L_1 x_{ijk} \geq \beta_k W\eta - L_1$$

$$\forall s_i \in S, \forall s_j \in S \cup \{s_0\}, 1 \leq k \leq K \quad (23)$$

where P is a fixed transmission power level.

4   Numerical Results

We evaluate in this section the numerical experiments conducted on the JR-SPC problem in various settings. The model was implemented in Java using the CPLEX Concert Technology (version 12.4) [12]. Sensors and targets are uniformly and independently deployed over a square area of 600 meters side length. The maximum transmission power is set to 20 mW, the path loss exponent is $\alpha = 2$, the power spectral density of the thermal noise and the channel bandwidth are set to 10^{-6} Watt/MHz and 1 MHz, respectively. Besides, the sensing range is fixed to 150 meters. On the basis on the modulation and coding schemes proposed in [13] as an extension to the 802.15.4 standard, the possible

Table 1. CPU times with CPLEX (sec)

Sensors	Targets	Coverage	Single-Rate			Multi-Rate		
-	-	-	Continuous	Discrete	Uniform	Continuous	Discrete	Uniform
20	40	1	3,01	8,34	2,53	15,02	42,38	29,54
20	100	1	3,05	9,75	3,31	11,12	37,06	26,07
30	100	1	17,26	70,62	19,10	103,48	407,82	222,76
30	100	2	16,15	61,23	16,38	132,27	493,15	338,13
40	160	2	116,44	418,44	89,56	763,71	4036,78	1962,56
40	160	3	145,95	430,49	100,68	767,10	4091,82	1872,65
50	200	1	409,61	4260,78	401,78	3308,52	43931,29	9807,48
50	200	3	735,08	3388,32	382,96	3500,82	45276,12	10829,87

transmission rates are $b_k = \{250$ kb/s, 500 kb/s, 1 Mb/s, 2 Mb/s$\}$ and require the following SINR threshold $\beta_k = \{1.3, 2.0, 4.0, 10.0\}$. Note that we consider also the standardized case of a single available data rate which then corresponds to 250 kb/s. We assume that a sensor generates a 125-byte data packet each time it monitors an associated target. Note that all the results presented in the following represent the obtained integer solution values Z^* (see section 3.2).

In each experiment, we consider 6 different power control and data rate strategies (S/C, S/D, S/U, M/C, M/D, M/U) which are the combination between (S)ingle/(M)ultiple rate(s) and (C)ontinuous/(D)iscrete/(U)niform power levels. In case of discrete power, we use the following values: 5 mW, 10 mW, 15 mW, 20 mW, whereas we use 20 mW in case of uniform power.

All experiments were run on a machine having two Intel Xeon X5670 processors (2.93GHz). Each row of Table 1 reports the average computation times over 10 problem instances using each of the above 6 strategies for a given number of sensors and targets and a given coverage level q. First, note that solving times depend heavily on the number of sensors and show exponential growth when adding more sensors. This is understandable since the more combinatorial part of the problem is the pricing which has to look over much more possible configurations. In general, there is a significant gap between single-rate and multi-rate and between continuous and discrete strategies which could be explained with the more combinatorial nature of the problem. Besides, we observe that the number of targets and the number of coverage levels, which both determine the traffic, exhibit little impact on CPU times.

The second set of experiments were designed to evaluate the throughput as a function of the number of targets. In order to get accurate and fair results, we compare the same problem instances by deploying first 20 targets and adding 20 more targets until reaching 100. The results are presented in terms of achievable throughput per target rather than by the length of the TDMA frame, as the former can be easily compared to reference data rates. For example, it is clear that the maximal throughput per target considering a single-rate strategy and 10 targets is 25 kb/s, because all the traffic flows have the sink as a destination which cannot receive faster than 250 kb/s. From Fig. 1a and 1b, we observe that throughput decays at a decreasing rate when the number of targets increases.

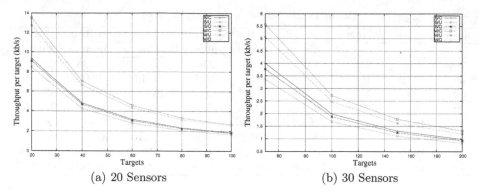

(a) 20 Sensors   (b) 30 Sensors

Fig. 1. Throughput per target vs. number of targets

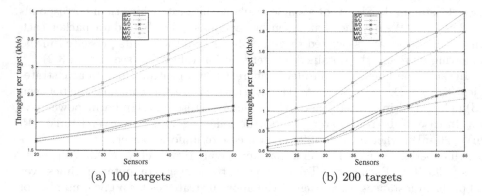

(a) 100 targets   (b) 200 targets

Fig. 2. Throughput per target vs. number of sensors

It appears also that multi-rate strategies outperform clearly single-rate ones (by about 150%) and that continuous power consistently enhances the throughput compared to discrete and uniform power. This is a noticeable result as it would suggest that the higher number of discrete power levels are available, the higher achievable throughput would be. A similar conclusion can be found in [14].

On another hand, the experiments reported in Fig. 2a and 2b illustrate the effect of sensor density on throughput. Assuming a fixed number of targets (respectively, 100 and 200), each network scenario was solved by first deploying 20 sensors covering all the targets and then gradually increasing the number of sensors. One would remark the quasilinear growth of throughput per target as a function of the number of sensors using each of the 6 strategies. The benefit from using multi-rate strategies is even more remarkable when the number of sensor increases. These results suggest that the WSNs formed by many low-range and high-rate transmission links have substantially higher capacity than the ones with fewer long-range and low-rate transmission links.

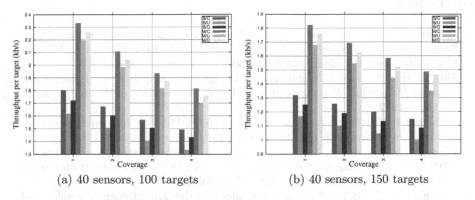

(a) 40 sensors, 100 targets   (b) 40 sensors, 150 targets

Fig. 3. Impact of Q_coverage on throughput

Finally, we illustrate in Fig 3a and 3b the impact of coverage level on throughput. Again by solving identical scenarios, and letting the coverage q to vary from 1 to 4, we observe that throughput per target decreases linearly when q increases. Regardless on reliability and fidelity issues that may impose to monitor each target by multiple sensors, the results indicate that more data packets about a given target are delivered to the sink in a unit time period by monitoring this one by a single sensor rather than by several ones. This is non-trivial and implies in fact that there exists a trade-off to consider between throughput and fidelity when it comes to specify the number of sensors monitoring each target.

5 Conclusion

We addressed in this paper the problem of achieving maximal capacity in TDMA wireless sensor networks by considering a physical interference model. We considered sensors that can adjust dynamically their transmission power to obtain efficient data rates while mitigating interference. Furthermore, the considered transmission links are interrelated with the traffic patterns resulted from target coverage requirements and multi-path routing. We formulated this problem as a mixed-integer linear program using a column generation scheme. With regard to power control, we considered three alternatives, namely, continuous, discrete, and uniform transmission power levels. Experimental results showed that multi-rate capabilities and sensor density increase significantly the overall throughput. Besides, results suggest that capacity could be enhanced if more discrete power levels are available on the sensor radio. We showed also that coverage requirements involve some trade-off between performance and sensing fidelity. As a future work, we will consider solving efficiently the pricing problem by mean of heuristics-based approaches.

Acknowledgement. This work is supported by TASSILI research program 11MDU839 between the University of Oran (ALGERIA) and the University of Nice (France).

References

1. Baronti, P., Pillai, P., Chook, V.W.C., Chessa, S., Gotta, A., Fun Hu, Y.: Wireless Sensor Networks: A Survey on the State of the Art and the 802.15.4 and ZigBee Standards. Computer Communications 30(7), 1655–1695 (2007)
2. Shi, Y., Hou, Y., Liu, J., Kompella, S.: Bridging the Gap between Protocol and Physical Models for Wireless Networks. IEEE Transactions on Mobile Computing 99, 1 (2012)
3. Chlamtac, I., Kutten, S.: A spatial reuse TDMA/FDMA for mobile multi-hop radio networks. In: Proceedings of IEEE INFOCOM 1985, pp. 389–394 (1985)
4. Wang, Y., Wang, W., Li, X., Song, W., Li, I.: Interference-aware joint routing and TDMA link scheduling for static wireless networks. IEEE Transactions on Parallel and Distributed Systems 19(12), 1709–1726 (2008)
5. Kar, K., Luo, X., Sarkar, S.: Delay guarantees for throughput-optimal wireless link scheduling. In: Proceedings of IEEE INFOCOM 2009, pp. 2331–2339 (2009)
6. Hou, Y., Shi, Y., Sherali, H.: Optimal spectrum sharing for multi-hop software defined radio networks. In: Proceedings of IEEE INFOCOM 2007, pp. 1–9 (2007)
7. Borbash, S., Ephremides, A.: Wireless link scheduling with power control and SINR constraints. IEEE Transactions on Information Theory 52(11), 5106–5111 (2006)
8. Tang, J., Xue, G., Chandler, C., Zhang, W.: Link scheduling with power control for throughput enhancement in multihop wireless networks. IEEE Transactions on Vehicular Technology 55(3), 733–742 (2006)
9. Wang, Q., Wu, D.O., Fan, P.: Delay-Constrained Optimal Link Scheduling in Wireless Sensor Networks. IEEE Transactions on Vehicular Technology 59(9), 4564–4577 (2010)
10. Uddin, M.F., Assi, C.: Joint Routing and Scheduling in WMNs with Variable-Width Spectrum Allocation. IEEE Transactions on Mobile Computing 99(PrePrints) (2012)
11. Chipcon: CC2420 2.4 GHz IEEE 802.15.4/ZigBee-ready RF Transceiver (2004), http://www.chipcon.com
12. IBM: IBM ILOG CPLEX Optimization Studio Getting Started with CPLEX, Version 12 Release 4 (2011)
13. Lanzisera, S., Mehta, A.M., Pister, K.S.J.: Reducing Average Power in Wireless Sensor Networks through Data Rate Adaptation. In: ICC 2009: Proceedings of IEEE International Conference on Communications, pp. 1–6 (June 2009)
14. Moscibroda, T., Wattenhofer, T.: The Complexity of Connectivity in Wireless Networks. In: INFOCOM 2006: Proceedings of 25th IEEE International Conference on Computer Communications, pp. 1–13 (April 2006)

Estimating Time Complexity
of Rumor Spreading in Ad-Hoc Networks

Dariusz R. Kowalski[1,*] and Christopher Thraves Caro[2,**]

[1] Department of Computer Science, University of Liverpool, Liverpool L69 3BX, UK
[2] GSyC, Universidad Rey Juan Carlos, Spain

Abstract. Rumor spreading is a fundamental communication process: given a network topology modeled by a graph and a source node with a message, the goal is to disseminate the source message to all network nodes. In this work we give a new graph-based formula that is a relatively tight estimate of the time complexity of rumor spreading in ad-hoc networks by popular PUSH&PULL protocol. We demonstrate its accuracy by comparing it to previously considered characteristics, such as graph conductance or vertex expansion, which in some cases are even exponentially worse than our new characterization.

Keywords: Rumor spreading, conductance, vertex expansion, synchronous model, asynchronous model, PUSH&PULL protocol.

1 Introduction

Rumor spreading is one of the fundamental problems in many areas of computer science, including networks [17], distributed systems [11], and distributed databases [8]. It has been widely studied from different perspectives (c.f. [16]), but still a number of open problems remain. In this work, we focus on one of them: estimating time performance of simple epidemic PUSH&PULL protocol in ad-hoc networks.

The input of the problem is an undirected graph $G = (V, E)$ and a *source* node in V that initially holds a rumor. The network is ad-hoc in the sense that its topology is not a priori known to the nodes, except of direct neighborhood. The goal is to spread the rumor to every node in the graph. The rumor is transmitted by means of a *communication protocol* (or simply protocol). We say that a node is *informed* at a given moment if it has previously received the rumor; otherwise it is *uninformed*. A communication protocol is the way the rumor passes from informed to uninformed nodes.

We restrict our attention to the *fairly simple* PUSH&PULL protocol introduced in [8]. In the PUSH part of the protocol, an informed node chooses uniformly at random one neighbor to communicate with and spread the rumor to it (*push*

* Supported by the Polish National Science Centre grant DEC-2012/06/M/ST6/00459.
** Supported by Spanish MICINN grant Juan de la Cierva, Comunidad de Madrid grant S2009TIC-1692 and Spanish MICINN grant TIN2008-06735-C02-01.

J. Cichoń, M. Gębala, and M. Klonowski (Eds.): ADHOC-NOW 2013, LNCS 7960, pp. 245–256, 2013.
© Springer-Verlag Berlin Heidelberg 2013

action). On the other hand, in the PULL part of the protocol, an uninformed node chooses uniformly at random one neighbor to communicate with and asks for the rumor, which is then immediately transferred to the requesting node (in the same step) if the requested node is informed (*pull action*). Throughout this document, we consider discrete time steps, and we assume that one time step is sufficiently large so that communication between two nodes completes.

We consider two models of analyzing PUSH&PULL protocol: *asynchronous* and *synchronous* models. In an asynchronous execution of PUSH&PULL, at each step only one node performs an action. At each step, such a node is chosen uniformly at random among the n nodes of the graph. If the chosen node is informed, it performs a push action. Otherwise, it performs a pull action. In a synchronous execution of PUSH&PULL, at each step every node performs an action. If a node is informed, it performs a push action, otherwise, it performs a pull action (c.f. [1]). Sometimes, for simplicity, we will call asynchronous and synchronous executions of PUSH&PULL by asynchronous and synchronous PUSH&PULL, respectively.

In order to measure the *time complexity*, or *runtime*, of rumor spreading, we measure the number of steps required so that every node is informed, regardless which node initially has the rumor (worst case analysis). Since in an asynchronous execution of PUSH&PULL only one node performs an action per step, while in a synchronous execution of PUSH&PULL every node performs an action per step, in order to compare them fairly we consider the concept of a *round*, which is a single step in case of synchronous executions (i.e., round and step are equivalent in synchronous setting) and n consecutive steps in case of asynchronous ones. In this work we continue the quest for finding accurate mathematical formulas to estimate the runtime of PUSH&PULL protocol.

1.1 Previous Results

Previous results focused on estimating the rumor spreading runtime via graph characteristics such as *conductance* denoted by Φ (which is roughly speaking the smallest possible ratio of the number of edges in a cut and the total number of edges on the "smaller" side of the cut) and *vertex expansion* denoted by α (which is roughly speaking the smallest possible ratio of the number of nodes adjacent to the border of a set and the total size of the set).

We start by discussing papers that relate runtime of rumor spreading to vertex expansion. Sauerwald and Stauffer [20] proved that for an arbitrary graph G with vertex expansion α, and a constant $\epsilon > 0$ sufficiently small, PUSH&PULL protocol informs all nodes in time $O((1/\alpha)n^{1-\epsilon})$ with high probability (whp). (In this document, by "with high probability (whp)" we mean with probability $1 - O(n^{-c})$, for an arbitrary constant $c > 2$.) Giakkoupis and Sauerwald [15] pushed forward the understanding of runtime with respect to vertex expansion in general graphs providing an almost tight bound. Their main result says that for any graph with vertex expansion at least α, PUSH&PULL protocol informs all nodes in $O((1/\alpha) \log^{2.5} n)$ steps whp.

Runtime for PUSH&PULL protocol has been bounded for different graphs via conductance. For instance, under complete graphs, it is known that $\theta(\log n)$ rounds are sufficient to spread the rumor [8,17,12,9]. Chierichetti et al. [7]

studied rumor spreading restricted to social networks modeled by Preferential Attachment model. They proved that PUSH&PULL protocol informs all nodes in this kind of graphs within $O(\log^2 n)$ rounds whp. Going further, Doerr et al. improved that result by proving that $\theta(\log n)$ rounds are sufficient for PUSH&PULL protocol in Preferential Attachment graphs. Furthermore, if PUSH&PULL protocol is slightly modified by allowing each node to choose uniformly at random among all neighbors but the last contacted node, the required time reduces to $\theta(\frac{\log n}{\log \log n})$, which is optimal since $\theta(\frac{\log n}{\log \log n})$ is the diameter of those graphs [10]. Rumor spreading in random graphs with a power law degree distribution with an arbitrary exponent is studied in [13]. The authors show that when the exponent ranges in the open interval $(2,3)$, the rumor is spread to almost all nodes in $\Theta(\log \log n)$ rounds whp. On the other hand, if the exponent is strictly larger than 3, then $\Omega(\log n)$ rounds are required.

Results that connect runtime of rumor spreading with conductance in general graphs follow an interesting progression. Chierichetti, Lattanzi and Panconesi contributed with two improvements. First in [6], the authors proved that given any graph G and any source node, PUSH&PULL protocol broadcasts the message within $O(\Phi^{-6}(G) \log^4 n)$ many rounds. The same authors pushed forward their bound in [5] proving a tighter bound of $O(\frac{\log^2 1/\Phi}{\Phi} \log n)$ many rounds for PUSH&PULL protocol, regardless of the source. Giakkoupis in [14] presented a tight bound for rumor spreading via conductance. His main result says that for any graph with conductance at least Φ, PUSH&PULL protocol informs all nodes in $O((1/\Phi) \log n)$ rounds, whp.

On the other hand, Censor-Hillel and Shachnai [4] slightly modified the random protocol by adding some determinism in it, so that bottlenecks are *detected* and information is sent across them. They defined *weak conductance* Φ_c. The authors proved that for any $c \geq 1$ and δ in the interval $(0, 1/(3c))$, their protocol informs all nodes in $O(c(\frac{\log n + \log \delta^{-1}}{\Phi_c}) + c)$ rounds with probability at least $1 - 3c\delta$. Recently, Censor-Hillel et al. [2] modified PUSH&PULL protocol in order to solve the information dissemination problem with no dependence on conductance. Indeed, they proved that via their new version of PUSH&PULL protocol, the rumor spreading problem is solved in at most $O(D + \text{polylog}(n))$ rounds in a graph of diameter D. However, their protocol is not the classical PUSH&PULL protocol.

All the results mentioned previously hold in the synchronous model. In the asynchronous model, Sauerwald [19] studied PUSH protocol, where he proved that the expected number of rounds for rumor spreading in this model is asymptotically equivalent to the expected number of rounds in the synchronous model for PUSH protocol. Then, he introduced a new measure that considers for each value $1 \leq k \leq n-1$ the subset S of k nodes that minimizes $\sum_{v \in S} d_{S^c}(v)/d(v)$ (denoted Φ_k). Using this characterization, Sauerwald proved that the expected runtime time of PUSH protocol in the asynchronous model is at most $\sum_{k=1}^{n-1} 1/\Phi_k$.

1.2   Room for Improvement: Three Classes of Graphs

We observe that there is still room for substantial improvement in this line of research. Even if conductance and vertex expansion have shown to be related with

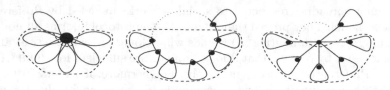

Fig. 1. This figure shows a graphic representation of the three families described in the text. From left to right, first it shows a graphic representation of a siameses-cliques, necklace and cliques-star. In the three subfigures, *big* dots represent central nodes or connector nodes of the graphs. On the other hand, cliques are represented by closed curves. As well, in all the three figures a dashed close curve shows graphically the partition that we use to bound the conductance and vertex expansion for these families; in the three cases half of the cliques are enclosed by the dashed line.

the runtime for rumor spreading, this relations match only in some cases. In this section, we present families of graphs whose runtime under PUSH&PULL protocol shows a gap with respect to the previous bounds predicted via conductance and vertex expansion. The main components of these families are equally sized complete graphs (cliques). The way in which those cliques are glued together describes each family. The size of the cliques and the amount of cliques of a graph fully determine the graph. We use letters b and c to denote the amount of cliques in a graph and the size of the cliques, respectively. Note that $bc = \Theta(n)$, where n is the number of nodes.

- *Siameses-cliques*: In this family, cliques are connected by a central node that belongs to every clique. We denote by $\mathcal{S}_{(b,c)}$ the siameses-cliques with b cliques, where each clique has size c.
- *Necklace*: A (b,c)-*necklace* graph consists in a cycle of length b (the necklace) and b cliques of size c (the jewels), where each clique is glued to one node of the cycle via one node of the clique. A (b,c)-necklace is denoted by $\mathcal{N}_{(b,c)}$.
- *Cliques-star*: In each graph of this family, there exists a central node that does not belong to any clique. For every clique, there exists an edge that connects one node of the clique with the central node. We denote by $\mathcal{ST}_{(b,c)}$ the cliques-star with b cliques, where each clique has size c.

Fig. 1 shows a graphical representation of $\mathcal{S}_{(b,c)}$, $\mathcal{N}_{(b,c)}$ and $\mathcal{ST}_{(b,c)}$. Even though these families may look like very specific cases, similar types of topologies are common in real networks, which makes our study widely applicable. The methods of estimating rumor spreading runtime developed for these three classes could be extended to other types of graphs having them as their components.

In Table 1, for each of the defined families of graphs, we show a lower bound for the time complexity for PUSH&PULL protocol, the best possible time complexity predicted via conductance, the best possible time complexity predicted via vertex expansion, and the time complexity predicted via the measure γ introduced in this work. All these bounds are in terms of expected number of synchronous and asynchronous rounds; this is in view of Theorem 2 relating synchronous and

Table 1. Time complexity predictions

Graph	PUSH&PULL runtime	Φ-prediction	α-prediction	γ-prediction
$\mathcal{S}_{(b,c)}$	$\Omega(\log(bc))$	$O(c\log(bc))$	$O(bc\log^{2.5}(bc))$	$O(\min\{b\log c; c\log b\})$
$\mathcal{N}_{(b,c)}$	$\Omega(bc)$	$O(bc^2\log(bc))$	$O(bc\log^{2.5}(bc))$	$O(bc)$
$\mathcal{ST}_{(b,c)}$	$\Omega(\log c+$ $\min\{c\log c, b\log b\})$	$O(c^2\log(bc))$	$O(bc\log^{2.5}(bc))$	$O(\min\{b\log c; c\log b\}+$ $\lceil\frac{b}{c}\rceil\log c)$

asynchronous executions of PUSH&PULL and the fact that each of the bounds is at least logarithmic in the number of nodes $n = \Theta(bc)$.

From these numbers, we can extract some conclusions. The first conclusion is the existence of graphs whose conductance prediction has a huge gap with respect to the runtime for PUSH&PULL protocol. For instance, conductance predicts an $O(n\log n)$ time complexity for PUSH&PULL in any graph $\mathcal{S}_{(b,c)}$ with c in $\Theta(n)$, while time complexity for PUSH&PULL in those graphs is in $O(\log n)$. Or even worse, conductance predicts an $O(n^2\log n)$ time complexity for PUSH&PULL in any graph $\mathcal{N}_{(b,c)}$ with c in $\Theta(n)$, while time complexity for PUSH&PULL in those graphs is in $O(n)$. In the case of $\mathcal{ST}_{(b,c)}$ graphs with c in $\Theta(n)$, conductance predicts an $O(n^2\log n)$ time complexity for PUSH&PULL, while time complexity for PUSH&PULL in those graphs is in $O(\log n)$.

We conclude that there is still room to search for measures better correlated with the time complexity of PUSH&PULL. Particularly, we see in the table that the γ measure introduced in this work improves considerably the estimate in the cases when conductance and vertex expansion fail. If we compare with the examples mentioned previously, γ predicts $O(\log n)$ time complexity in any graph $\mathcal{S}_{(b,c)}$ with c in $\Theta(n)$, or $O(n)$ time complexity in any graph $\mathcal{N}_{(b,c)}$ with c in $\Theta(n)$. These numbers make us conjecture that γ measure deserves to be studied.

1.3   Our Results

Our work resumes the search on accurate mathematical characterization of rumor spreading properties, or in general, properties of communication tasks. We introduce a new measure γ, prove that it is an upper bound on the expected time complexity of an asynchronous execution of PUSH&PULL protocol in any graph (Theorem 1 in Section 3), and study its further properties. In particular:

- We show that, for *almost regular* graphs, it gives estimates asymptotically not worse than those given by the previously considered conductance measure (Theorem 3 in Section 3).
- We prove that when the maximum degree of a graph is in $O(\log^{1.5} n)$, the new measure gives estimates asymptotically not worse than those given by the previously considered vertex expansion measure (Theorem 4 in Section 3).
- We show that for some families of graphs, as defined in Section 1.2, the new measure provides considerably better estimates for the expected time complexity of rumor spreading. In particular, for some types of cliques-star graphs, the new estimates are even exponentially better than the previous ones (Theorems 5, 6 and 7 in Section 4)

We also compare synchronous and asynchronous executions of PUSH&PULL:

- The number of rounds in an execution of asynchronous PUSH&PULL, for any graph, is asymptotically not bigger than the number of steps in a synchronous execution of PUSH&PULL protocol plus $\log n$, whp. (Theorem 2 in Sec. 3).

Therefore, the expected numbers of rounds made by PUSH&PULL protocol in synchronous and asynchronous executions for a given graph are asymptotically the same, up to a logarithmic additive component.

2   Technical Preliminaries

Some notation is presented before introducing *conductance* and *vertex expansion*. Let us denote with $n = |V|$ the number of nodes in the graph. For any subset of nodes $S \subseteq V$, vol(S) denotes the *volume* of S and it is defined as $\sum_{v \in S} d(v)$, where $d(v)$ is the degree of node v. Given a graph G, we define its maximum and minimum degree as follows: $d_{max} = \max\{d(v) : v \in V\}$ and $d_{min} = \min\{d(v) : v \in V\}$, respectively. For any set of nodes $S \subseteq V$ and any node v, we denote by $d_S(v)$ the degree of v restricted to S. For any two sets of nodes $S, T \subseteq V$, we define $E(S, T)$ to be the set of edges between nodes in S and nodes in T. For any set of nodes $S \subseteq V$, we define the complement of S, denoted by S^c, as $V \setminus S$. For any subset of nodes $S \subseteq V$, $\partial(S)$ denotes the set of nodes from S^c such that each of them has at least one edge to some node in S.

Conductance. In the literature it is possible to find more than one way to define it. Indeed, conductance was introduced in the context of Markov Chains, and it was used to bound the convergence rate of a Markov Chain that represents a random walk in a graph. A traditional definition of conductance can be found in Sinclair's book [21]. It is also possible to find a definition of conductance especially contextualized for rumor spreading, we refer the reader to Mosk-Aoyama et al. [18] and Censor-Hillel et al. [3,4]. On the other hand, authors such as Giakkoupis [14] and Sauerwald and Stauffer [20] have used a definition of conductance that measures how *well-connected* the graph is. In this document, we use this definition of conductance, basically, due to its *graph theoretic* nature. The *conductance* of a graph $G = (V, E)$ is: $\Phi(G) := \min_{S \subseteq V; \text{vol}(S) \leq \text{vol}(V)/2} \frac{|E(S, S^c)|}{\text{vol}(S)}$.

Vertex Expansion. A second important notion of expansion is *vertex expansion*. The *vertex expansion* of a graph $G = (V, E)$ is: $\alpha(G) := \min_{S \subseteq V : 1 \leq |S| \leq n/2} \frac{|\partial(S)|}{|S|}$. We remark that, in the case of regular graphs, conductance and vertex expansion are related. For any graph with maximum degree d, it holds that $\alpha/d \leq \Phi \leq \alpha$.

3   New Measure

In this section we introduce γ measure; a new measure to estimate time complexity of rumor spreading. This measure might be seen as a PUSH&PULL version of the measure introduced by Sauerwald in [19]. Sauerwald's measure defined for the PUSH protocol considers $\sum_{k=1}^{n-1} \Phi'_k$, where Φ'_k minimizes the value of $\sum_{v \in S^c} d_S(v)/d(v)$ over subsets S of k nodes, for $1 \le k \le n-1$. Our PUSH&PULL oriented extension of this measure assures that for every value k we would be always considering the worst case among all possible sets of k nodes, where for each set the measure considers the best between PUSH and PULL in order to estimate runtime for PUSH&PULL protocol. We remark that this is not just the minimum running time of two separate executions of protocols PUSH and PULL respectively. This is to reflect the key property of the PUSH&PULL protocol, which is that whenever one of the component protocol does not perform well in some part of the PUSH&PULL execution then the other usually does. This design differentiates γ measure from a trivial extension of Sauerwald's measure.

For instance, it was proved by Chierichetti et al. in [7] that when the network is a preferential attachment graph, both PUSH and PULL protocols require polynomially many steps to spread the rumor. While, Doerr et al. in [10] proved that the combination of both protocols, i.e., PUSH&PULL protocol, requires only $O(\log n)$ steps in the same graphs. Such a synergy cannot be obtained by running the protocols separately (i.e., without taking into account the progress done by the other algorithm) or by measuring (combined) PUSH&PULL protocol using the minimum of the Sauerwald's measures for PUSH and for PULL protocol. In order to overcome this type of a drawback, not only in preferential attachment graphs, we consider a more sophisticated extension of the measure introduced by Sauerwald.

We propose the following new parameter for a given graph G:

$$\beta_k(G) = \min_{S:S \subseteq V, |S|=k} \max\left\{ \sum_{v \in S} \frac{d_{S^c}(v)}{d(v)}, \sum_{w \in S^c} \frac{d_S(w)}{d(w)} \right\}.$$

This parameter captures the worst-case bi-directional expansion through a cut of one border of size k. The new measure, is defined as follows:

$$\gamma(G) = \sum_{k=1}^{n-1} \frac{1}{\beta_k(G)}.$$

We may skip parameter G from the above parameters $\gamma(G), \beta_k(G)$ if graph G is clearly understood from the context.

Theorem 1. *For any given graph G, the asynchronous expected round complexity of* PUSH&PULL *protocol is at most $\gamma(G)$.*

Proof. We partition an asynchronous execution of PUSH&PULL protocol into consecutive phases as follows: phase k contains steps in which exactly k nodes are informed. Note that some phases may be empty, and the partition into phases is related with the partition into asynchronous rounds, though the former depends on the random choices made in asynchronous steps while the latter is fixed (i.e., each round contains exactly n asynchronous steps).

Observe that the expected number of steps in phase k is at most $1/\beta_k(G)$. Indeed, let W be the set of k informed nodes during phase k. The probability that a single push operation delivers a message from some node in W to some node in W^c is $\frac{1}{n}\sum_{v\in W}\frac{d_{W^c}(v)}{d(v)}$, while the probability that a single pull operation delivers a message from some node in W to some node in W^c is $\frac{1}{n}\sum_{w\in W^c}\frac{d_W(w)}{d(w)}$. Therefore, the probability that at least one of these operations succeeds in delivering the message to some uninformed node, and thus the phase will terminate, is at least $\frac{1}{n}\max\left\{\sum_{v\in W}\frac{d_{W^c}(v)}{d(v)},\sum_{w\in W^c}\frac{d_{W^c}(w)}{d(w)}\right\}\geq\frac{\beta_k(G)}{n}$. Because these operations are applied independently in consecutive steps, the expected time for termination of the phase is at most $\frac{n}{\beta_k(G)}$. Hence, the expected time of computation is the sum of expected numbers of steps, over phases $1\leq k\leq n-1$, which gives at most $\sum_{k=1}^{n-1}\frac{n}{\beta_k(G)}$ steps, which in turn is equal to $\gamma(G)$ asynchronous rounds.   □

Theorem 2. *For any graph G, the expected number of rounds in a synchronous execution of* PUSH&PULL *protocol on graph G is* $O(\gamma(G)+\log n)$.

Proof. Consider an asynchronous execution of PUSH&PULL protocol. Partition asynchronous steps into consecutive (asynchronous) rounds. Consider the following modification of the execution: for each asynchronous round and each node, the node skips all its activities (i.e., push and pull operations) done in this round except the first one. By definition, the number of steps (and thus the number of asynchronous rounds), needed for rumor dissemination in the modified execution is not smaller than the number of steps (resp., asynchronous rounds) resulting from the original PUSH&PULL execution. We apply this modification to any asynchronous execution of PUSH&PULL protocol, and the distribution of the obtained modified executions we call a *modified asynchronous* PUSH&PULL *protocol*. Note that the modified protocol may not correspond to any set of executions resulted from a distributed protocol, but its round complexity is (stochastically) not smaller than the round complexity of PUSH&PULL.

In order to compare the round complexity of the modified asynchronous PUSH&PULL protocol with the round complexity of a synchronous execution of PUSH&PULL protocol, let us fix a sequence of random bits used by nodes in a synchronous execution of PUSH&PULL protocol, i.e., bits that are used by push and pull operations in order to select a random neighbor. Let G be a given n-node graph. Consider a node v and the path from the source of the rumor to node v through which the rumor is delivered to v in the considered execution of PUSH&PULL with the fixed bits (i.e., to each node on this path, the first delivered message comes through its predecessor on this path). Now consider an asynchronous execution of the modified protocol for the same set of random bits. Note that we fix only bits used for push and pull operations, while bits used to generate an asynchronous order of active nodes are still random. The probability that the number of rounds needed for delivering the rumor to node v through the same path in asynchronous execution of the modified protocol is asymptotically different from the number of rounds used in the corresponding synchronous execution (i.e., for the same fixed sequence of random bits for push and pull) by $\omega(\log n)$, is polynomially small, and drops exponentially with the growth of the constant hidden in the asymptotic notation. (Here the probability distribution is over random bits used for selecting an asynchronous order of nodes.) To see this, observe that the probability of not skipping a round by a node on the path during the execution of the modified algorithm (due to not being selected to do a push-pull activity in that round) is at least $1 - (1 - 1/n)^n = \Omega(1)$. By Chernoff bound, the number of skipped rounds is within a constant factor from the number of rounds in the fixed synchronous execution (plus additional $O(\log n)$ if that number is smaller than $\log n$), with probability at least polynomially small and exponentially dropping with the growth of the constant factor.

In order to complete the proof, observe that the modified protocol in asynchronous environment completes rumor spreading on graph G in $O(\gamma(G))$ rounds, in expectation. Indeed, we can enhance the proof of Theorem 1 in a way to capture the skipped steps made by the modified protocol: a push-pull step is skipped with a constant probability, as the expected number of repetitions of nodes within a round is a constant. This probability introduces an additional constant in front of all the formulas in the original proof of Theorem 1, which does not change the asymptotic result of $O(\gamma(G))$ rounds, in expectation. Next we apply the relation between the round complexity of a synchronous execution of PUSH&PULL and an asynchronous execution of the modified protocol, proved in the previous paragraph. This implies the expected $O(\gamma(G) + \log n)$ round complexity of synchronous PUSH&PULL protocol.   □

Let us consider now almost regular graphs, and see how γ prediction compares with respect to the prediction given by conductance. We say that a graph G is *almost regular (a.r.)* if $d_{max}(G)/d_{min}(G) = O(1)$.

Theorem 3. *For every a.r. graph G with n nodes, it holds: $\gamma(G) = O(\frac{\log n}{\Phi(G)})$.*

Proof. Observe that for a given almost regular graph G of n nodes, in which $d_{max}/d_{min} = c = O(1)$, and for any subset of nodes S of size k, we have: $k \cdot d_{min}(G) \le vol(S) \le k \cdot d_{max}(G) \le c \cdot vol(S)$ Therefore, the following holds:

$$\beta_k(G) = \min_{S:S\subseteq V,|S|=k} \max\left\{\sum_{v\in S}\frac{d_{S^c}(v)}{d(v)}, \sum_{w\in S^c}\frac{d_S(w)}{d(w)}\right\} \ge \min_{S:S\subseteq V,|S|=k}\sum_{v\in S}\frac{d_{S^c}(v)}{d(v)}$$

$$\ge \min_{S:S\subseteq V,|S|=k}\frac{\sum_{v\in S}d_{S^c}(v)}{d_{max}(G)} \ge \frac{k}{c}\cdot\min_{S:S\subseteq V,|S|=k}\frac{|E(S,S^c)|}{vol(S)} \ge \frac{k}{c}\cdot\Phi(G) .$$

Note that the last inequality might be not true since $vol(S)$ might be larger than $vol(V)/2$. In that case, it holds that $vol(S^c) \le vol(V)/2$. Hence, the same analysis developed here holds using $\sum_{w\in S^c}\frac{d_S(w)}{d(w)}$ in the first inequality. Finally, we get $\gamma(G) = \sum_{k=1}^{n-1}\frac{1}{\beta_k(G)} \le c\cdot\sum_{k=1}^{n-1}\frac{1}{k\cdot\Phi(G)} = c\cdot\frac{H_{n-1}}{\Phi(G)}$.   □

It follows that the estimate of rumor spreading time based on our new characteristic γ is asymptotically not worse than the previous estimate based on conductance derived in [14], in case of almost regular graphs. We show now that γ prediction is also asymptotically not worse than the best previous estimate based on vertex expansion derived in [15], when the graph has bounded node degree by $\frac{d_{max}^2}{d_{min}} = O(\log^{1.5} n)$.

Theorem 4. *For any graph G with n nodes and such that its maximum degree satisfies $\frac{d_{max}^2}{d_{min}} = O(\log^{1.5} n)$, it holds that $\gamma(G)$ is in $O(\frac{\log^{2.5} n}{\alpha(G)})$.*

Proof. For any graph G, it holds that $\Phi(G) \geq \alpha(G)/d_{max}$. Then, if we repeat the same analysis that we did for almost regular graphs with respect to conductance (c.f., the proof of Theorem 3) we obtain $\beta_k(G) \geq \frac{k \cdot d_{min}}{d_{max}} \cdot \Phi(G)$. Now, if we plug in the above mentioned relation between conductance and vertex expansion, we obtain that $\beta_k(G) \geq \frac{k \cdot d_{min} \alpha(G)}{d_{max}^2}$. Consequently, it holds that $\gamma(G)$ is in $O(\frac{d_{max}^2}{d_{min}} \cdot \alpha^{-1}(G) \log n)$. In the case when $\frac{d_{max}^2}{d_{min}}$ is in $O(\log^{1.5} n)$, $\gamma(G)$ is in $O(\frac{\log^{2.5} n}{\alpha(G)})$. ☐

4   Case Study

In this section, we use the previously introduced measure γ to estimate time complexity of rumor spreading in the three families defined in Section 1.2. Moreover, in this section we explain how to obtain the bounds shown in Table 1.

Siameses-cliques. Runtime of rumor spreading in siameses-cliques with n nodes is at least $\Omega(\log n)$. That can be explained because, no matter the source node is, the central node receives the rumor in $O(\log c)$ steps, whp. Once the central node is informed, every other node receives the rumor in $\Omega(\log n)$ steps, whp. Therefore, in total the rumor needs at least $\Omega(\log n)$ steps to be spread in siameses-cliques with n nodes. On the other hand, conductance in a siameses-cliques with n nodes is smaller than $\frac{b(c-1)/2}{(c-1)^2 b/2} = \frac{1}{c-1}$. Such an upper bound on the conductance comes from the partition where $b/2$ cliques are in one set and the rest of the cliques plus the central node are in the opposite set (Fig. 1 shows a graphic representation of this partition). Therefore, the best estimate for rumor spreading runtime in $\mathcal{S}_{(b,c)}$ guaranteed by conductance is of order at least $c \log(bc)$, according to [14]. Consider the same partition of the the the set of nodes of $\mathcal{S}_{(b,c)}$ in order to compute its vertex expansion. Then it holds: $\alpha(\mathcal{S}_{(b,c)}) \leq \frac{1}{(c-1)b/2-1/2}$. Concluding that, the best estimate for rumor spreading runtime in $\mathcal{S}_{(b,c)}$ guaranteed by vertex expansion is of order at least $bc \log^{2.5}(bc)$, according to [15]. As we show, the γ measure gives more accurate upper bound.

Theorem 5. *For a Simeses-cliques with $n = b(c-1)+1$ nodes, it holds that $\gamma(\mathcal{S}_{(b,c)})$ is in $O(\min\{b \log c; c \log b\})$.*

Necklaces. In the case of a necklace with $n = bc$ nodes, rumor spreading runtime is at least $\Omega(bc)$. Note that if the source node is not part of the ring, the corresponding node that is part of the ring shall be informed in $O(\log c)$ steps. Once a node in the ring is informed, it requires at least $\Omega(c)$ steps to transmit the rumor to a neighbor in the ring. That is because the probability to transmit the rumor to one neighbor in the ring via a PUSH action is $1/c$, and the probability to transmit the rumor to one neighbor in the ring via a PULL action is also $1/c$. Therefore, at least c steps are required so that the rumor goes from one node in the ring to another node in the ring, whp. Since there are b such rumor-relays needed, we obtain the sought lower bound for rumor spreading runtime in $\mathcal{N}_{(b,c)}$. Conductance for $\mathcal{N}_{(b,c)}$ is at most $\frac{2}{(c-1)^2 b/2}$. This upper bound comes from the partition (depicted in Fig. 1), where $b/2$ cliques belong to one set and the other $b/2$ cliques belong to the other set of the partition. Therefore, the best prediction via conductance (according to [14]) is at least $O(bc^2 \log(bc))$. The same partition gives an upper bound for the vertex expansion of $\frac{2}{(c-1)b/2}$. Hence, the best vertex-expansion-based upper estimate (according to [15]) is at least $O(bc \log^{2.5}(bc))$. We prove that γ measure considerably improves these estimations.

Theorem 6. *For any b and any c, it holds that $\gamma(\mathcal{N}_{(b,c)})$ is $O(bc)$.*

Cliques-stars. Time complexity for rumor spreading in a cliques-star with $n = bc + 1$ nodes is at least $\Omega(\log c + \min\{b \log b; c \log c\})$. The $\log c$ part comes from the fact that, if the source node is not the central node of $\mathcal{ST}_{(b,c)}$ neither a connector, then the rumor needs $O(\log c)$ steps so that a connector receives it. Once a connector is informed, the central node receives the rumor with probability $1/c$ via PUSH and $1/c$ via PULL. Therefore, the coupon collector's problem assures that the central node receives the rumor whp in at least $c \log c$ steps. Using the same arguments, once the central node is informed, every connector receives the rumor in at least $b \log b$ steps, whp. Hence, all in all, PUSH&PULL runtime in $\mathcal{ST}_{(b,c)}$ is at least $\Omega(\log c + \min\{b \log b; c \log c\})$. Conductance for $\mathcal{ST}_{(b,c)}$ is at most $\frac{b/2}{(c-1)^2 b/2}$, according to the partition where $b/2$ cliques belong to one set and $b/2$ cliques plus the central node belong to the opposite set (see Fig. 1 for a graphical example). Therefore, the best upper bound on PUSH&PULL runtime based on conductance is at least $O(c^2 \log(bc))$, according to [14]. The same partition gives $\alpha(\mathcal{ST}_{(b,c)}) \leq \frac{1}{bc}$. Hence, according to [15], the best estimate given by vertex expansion is at least $O(bc \log^{2.5}(bc))$. We prove that γ measure improves these estimates.

Theorem 7. *For a cliques-star with $n = bc + 1$ nodes, when $b \leq c$, it holds that $\gamma(\mathcal{ST}_{(b,c)}) = O(\min\{b \log c; c \log b\} + \lceil \frac{b}{c} \rceil \log c)$.*

5   Conclusions and Future Work

The main contribution of this work is the introduction of a new γ measure that estimates the expected time complexity for asynchronous PUSH&PULL protocol in a more accurate way than the previously considered measures. However, the quest towards accurate characterization of random communication processes in ad-hoc networks is still far from being completed. In particular, as we see especially in example of siameses-cliques, there are still some topologies for which we do not know how to accurately characterize fundamental communication processes such as PUSH&PULL rumor spreading. Further research involves searching for accurate characterizations of performances of other types of algorithms and communication tasks.

References

1. Avin, C., Borokhovich, M., Censor-Hillel, K., Lotker, Z.: Order optimal information spreading using algebraic gossip. In: Proceedings of PODC 2011, pp. 363–372 (2011)
2. Censor-Hillel, K., Haeupler, B., Kelner, J.A., Maymounkov, P.: Global computation in a poorly connected world: fast rumor spreading with no dependence on conductance. In: Proceedings of STOC 2012, pp. 961–970 (2012)
3. Censor-Hillel, K., Shachnai, H.: Partial information spreading with application to distributed maximum coverage. In: Proceedings of PODC 2010, pp. 161–170 (2010)
4. Censor-Hillel, K., Shachnai, H.: Fast information spreading in graphs with large weak conductance. In: Proceedings of SODA 2011, pp. 440–448 (2011)
5. Chierichetti, F., Lattanzi, S., Panconesi, A.: Almost tight bounds for rumour spreading with conductance. In: Proceedings of STOC 2010, pp. 399–408 (2010)
6. Chierichetti, F., Lattanzi, S., Panconesi, A.: Rumour spreading and graph conductance. In: Proceedings of SODA 2010, pp. 1657–1663 (2010)
7. Chierichetti, F., Lattanzi, S., Panconesi, A.: Rumor spreading in social networks. Theor. Comput. Sci. 412(24), 2602–2610 (2011)
8. Demers, A., Greene, D., Houser, C., Irish, W., Larson, J., Shenker, S., Sturgis, H., Swinehart, D., Terry, D.: Epidemic algorithms for replicated database maintenance. SIGOPS Oper. Syst. Rev. 22, 8–32 (1988)
9. Doerr, B., Fouz, M.: Asymptotically optimal randomized rumor spreading. In: Aceto, L., Henzinger, M., Sgall, J. (eds.) ICALP 2011, Part II. LNCS, vol. 6756, pp. 502–513. Springer, Heidelberg (2011)
10. Doerr, B., Fouz, M., Friedrich, T.: Social networks spread rumors in sublogarithmic time. In: Proceedings of STOC 2011, pp. 21–30 (2011)
11. Dolev, S., Schiller, E., Welch, J.L.: Random walk for self-stabilizing group communication in ad hoc networks. IEEE Trans. Mob. Comput. 5(7), 893–905 (2006)
12. Elsässer, R.: On the communication complexity of randomized broadcasting in random-like graphs. In: Proceedings of SPAA 2006, pp. 148–157 (2006)
13. Fountoulakis, N., Panagiotou, K., Sauerwald, T.: Ultra-fast rumor spreading in social networks. In: Proceedings of SODA 2012, pp. 1642–1660 (2012)
14. Giakkoupis, G.: Tight bounds for rumor spreading in graphs of a given conductance. In: Proceedings of STACS 2011, pp. 57–68 (2011)
15. Giakkoupis, G., Sauerwald, T.: Rumor spreading and vertex expansion. In: Proceedings of SODA 2012, pp. 1623–1641 (2012)
16. Hromkovic, J., Klasing, R., Pelc, A., Ruzicka, P., Unger, W.: Dissemination of Information in Communication Networks: Broadcasting, Gossiping, Leader Election, and Fault-Tolerance. Texts in Theoretical Computer Science. An EATCS Series. Springer (2010)
17. Karp, R.M., Schindelhauer, C., Shenker, S., Vocking, B.: Randomized rumor spreading. In: Proceedings of FOCS 2000, pp. 565–574 (2000)
18. Mosk-Aoyama, D., Shah, D.: Fast distributed algorithms for computing separable functions. IEEE Transactions on Information Theory 54(7), 2997–3007 (2008)
19. Sauerwald, T.: On mixing and edge expansion properties in randomized broadcasting. Algorithmica 56, 51–88 (2010)
20. Sauerwald, T., Stauffer, A.: Rumor spreading and vertex expansion on regular graphs. In: Proceedings of SODA 2011, pp. 462–475 (2011)
21. Sinclair, A.: Algorithms for random generation and counting - a Markov chain approach. Progress in theoretical computer science. Birkhäuser (1993)

Strong Connectivity of Wireless Sensor Networks with Double Directional Antennae in 3D

Evangelos Kranakis[1,*], Fraser MacQuarrie[1],
Izabela Karennina Travizani Maffra[2,**], and Oscar Morales Ponce[3]

[1] Carleton University, School of Computer Science, Ottawa, ON K1S 5B6, Canada
[2] Federal University of Minas Gerais, Computer Science Department, Brazil
[3] Chalmers University, Department of Computing, S-412 96 Goeteborg, Sweden

Abstract. Using directional antennae in forming a wireless sensor network has many advantages over omnidirectional, including improved energy efficiency, reduced interference, increased security, and improved routing efficiency. We propose using double (Yagi) directional antennae in 3D space: for a given spherical angle such antennae transmit from their apex simultaneously directionally along two diametrically opposing cones in 3D. We study the resulting network formed by such directional sensors. We design a new algorithm to address strong connectivity of the resulting network and compare its hop-stretch factor with the three-dimensional omnidirectional model. We also obtain a lower bound on the minimum range required to ensure strong connectivity for sensors with double antennae. Further, we present simulation results comparing the diameter of a traditional sensor network using omnidirectional and one using directional antennae.

Keywords: Antennae, Diameter, Directional, Range, Sensor Network, Stretch Factor, Strong Connectivity, Yagi Antenna.

1 Introduction

Most studies of wireless sensor networks (WSNs) assume that sensors employ omnidirectional antennae to communicate. For sensors with identical transmission range, this has lead to the so-called UDG (Unit Disk Graph) model (also known as *protocol model*) whereby sensors are able to communicate with each other if and only if the distance between them is less than or equal to the transmission range of the two sensors. In this paper we consider sensors in 3D and adopt a different model, where the sensors use *directional antennae*. This offers many possible advantages such as: reduced energy consumption, lowered interference, tighter security, and improved routing efficiency. In 2D, the directional antenna model originates in the work of [2]. However our work is mainly motivated by the work in [4], which explored for the first time the use of single directional antennae in 3D space.

More specifically, we address *the orientation problem for strong connectivity* and the *stretch factor problem* for double directional antennae in three-dimensional space.

---

* Supported in part by NSERC grant.
** Research supported by MITACS Globalink and CAPES scholarship.

J. Cichoń, M. Gębala, and M. Klonowski (Eds.): ADHOC-NOW 2013, LNCS 7960, pp. 257–268, 2013.
© Springer-Verlag Berlin Heidelberg 2013

Both problems are concerned with finding an orientation of the antennae. However, in the former case we merely want to ensure the resulting network is strongly connected, while in the latter for a given angle of the antenna we want to determine the minimum possible range (of the antenna) so as to guarantee that a directional c-hop-spanner is induced, for some constant $c > 0$. This guarantees that the length of shortest paths in the resulting directional graph are at most c times the length of shortest paths in the underlying sensor network of omnidirectional antennae.

Several communication models are possible depending on how sensors send (directional or omnidirectional transmissions) and receive (directional or omnidirectional receptions) messages. We adopt the model where transmissions occur directionally while receptions occur omnidirectionally. This is the simplest and most intuitive communication model but it has the additional benefit that it simplifies and illustrates the underlying complexity issues of the resulting directed graph.

Although our work presents mainly theoretical results, we note that directional antennae have proven in real-world applications and their use is emerging in various settings where energy, interference, security, etc, are of primary concern (see [2] and [5] for extensive discussions and references concerning these issues). We also note that papers such as [1,8] discuss electronic beam steering, which relies on and uses the advantages of directional antennae in its implementation.

1.1 Preliminaries and Notation

In this section we define several concepts and ideas that will be used in the main results. Throughout this paper, we assume that sensors are located at points in 3D space. The closed ball centered at the point p with radius r is denoted by $\mathbb{B}[p, r]$. Given a set S of sensors, define the Unit Ball Graph $U := UBG(S)$ as the graph whose set of vertices consists of S and two vertices u, v are connected by an edge if and only if $d(u, v) \leq 1$. We denote the hop distance between two sensors u and v in a graph G by $d_G(u, v)$ (When the graph is easily understood from the context we omit the subscript G so as to simplify notation.) Our sensor network is formed from directional antennae which replace corresponding omnidirectional antennae. We measure the quality of the resulting graph of directional antennae with the stretch factor which compares the length of shortest paths in the two kinds of graphs. If G is the graph resulting by orienting the antennae on the set S of points, the stretch factor can be defined by $\tau_G(S) = \max_{\forall u,v \in P} \frac{d_G(u,v)}{d_U(u,v)}$, where $d_G(u, v), d_U(u, v)$ is the hop-distance between u and v in the graphs G, U, respectively. To measure the stretch factor of the resulting directional graph it will be convenient to define the concept of k-coverage as follows (see [6]). Define a k-orientation over a set S of sensors as the orientation of the sensors of S such that $\forall s \in S$: 1) $\mathbb{B}[s, 1]$ is covered by S, and 2) $\forall p \in \mathbb{B}[s, 1]$, the shortest path from s to a sensor covering p has length at most $k - 1$. This means that those sensors which are reachable by 1 hop in the $UBG(S)$ are still reachable by a constant number of hops k in the new orientation.

The *solid angle* subtended by a surface is defined as the ratio between the area of the spherical cap and the square of the radius of the sphere of which it is part. It is usually denoted by Ω. The *apex angle* of a spherical cone, denoted by 2θ, is defined as the maximum planar angle between any two generatrices of the spherical cone.

The apex angle 2θ and the solid angle Ω are related by the following well-known relation due to Archimedes:

$$\Omega = 2\pi \cdot (1 - \cos\theta). \tag{1}$$

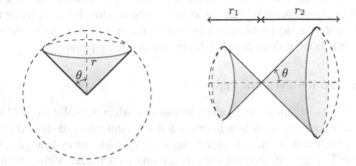

Fig. 1. Left: Illustration of a single antenna of apex angle 2θ and radius r. Right: Illustration of a Yagi double antenna of apex angle 2θ and radii r_1 and r_2.

A *single antenna* is modelled as a spherical cone characterized by its apex angle (which also determines the solid angle) and its range. A $(2\theta, r)$ *double antenna* is modelled as two spherical cones diametrically opposed, of apex angle 2θ and radii both equal to r. A more general model, is based on the *Yagi antenna*, in which the opposing spherical cones do not necessarily have the same range, formally denoted as $(2\theta, r_1, r_2)$ antenna, where r_1 and r_2 are the radii of the two cones of the antenna. A three-dimensional Yagi antenna is depicted in Figure 1. Given a set of points S in three-dimensional space that form a Unit Ball Graph U, the *optimal range* is defined as the length of the longest edge of the minimum spanning tree of U and is denoted by $r_{MST}(S)$. It is clear that any range lower than r_{MST} will fail to provide a strongly connected graph since the MST will be disconnected.

A packing problem that relates to the problems addressed in this paper is the *Tammes Radius* [10], which can be described as the maximum length r of the radius of n equal circles placed in the surface of a sphere, without overlapping and it is denoted by R_n. Another related problem is the *Kissing Number*, which can be described for the three-dimensional scenario as the maximum number of disjoint unit spheres such as every sphere is tangent to a given unit sphere [7].

1.2 Related Work

Many advantages of networks using directional antennae have been explored, and there are many works which examine the topological changes inherent to the use of directional antennae. A comprehensive survey can be found in [5].

A three-dimensional scenario was first adopted in [4] where a lower bound on the solid angle necessary to ensure strong connectivity with optimal range is calculated as $\frac{18\pi}{5}$. A proof is given showing that the problem of determining the existence of an orientation which achieves strong connectivity with optimal range is NP-complete for sensors of solid angle $\Omega < \pi$. An algorithm is also presented which finds such an orientation for antennae with beam width $\frac{18}{5}\pi \leq \Omega \leq 4\pi$. Finally, simulation results are provided which show how the stretch factor of the directional model compares with the omnidirectional model.

Other works closely related to our paper are [3] and [6]. The former presents algorithms to orient directional single antennae with constant stretch factor, while the latter deals with the connectivity problem and the stretch factor problem for double-antennae. Unlike our work, both of these papers refer to sensors in the plane.

1.3 Outline and Results of the Paper

In Section 2, we show how to orient a single antenna, when the solid angle Ω is greater than 2π. In Section 3, we show how to orient a double-antenna and show that the solid angle Ω for which it is possible to attain connectivity with optimal range is bounded below by $\Omega \geq \frac{25}{13}\pi$. In both sections, the orientations are achieved with constant stretch factor.

Table 1. Summary of results

Solid Angle	Antenna	Antenna Radius	Stretch Factor	Proof
$\Omega \geq 2\pi$	Single	$\max(1, 2\sin\theta)$	2	Theorem 1
$2\pi \leq \Omega < \frac{18\pi}{5}$	Single	$r_{MST}(S) \cdot \frac{\sqrt{\Omega \cdot (4\pi - \Omega)}}{\pi}$	N/A	[4]
$\Omega \geq \frac{18\pi}{5}$	Single	$r_{MST}(S)$	N/A	[4]
$\Omega \geq \frac{25\pi}{13}$	Double	r_{MST} (both)	N/A	Theorem 2
$\pi \leq \Omega < 2\pi$	Double	$4\cdot\sin(\frac{\pi}{4} + \theta)$ and 2	4	Theorem 3

In Section 4, we present the results of our simulation, which evaluated how the use of directional double-antenna impacts the diameter of the resulting graph, compared to the original UBG. Table 1 summarizes our results, along with other existing results.

2   Orienting Single Antennae with Constant Stretch Factor

In this section we show how to orient small groups of sensors with apex angle $2\theta \geq \pi$, so as to form a strongly connected directed graph with constant stretch factor.

Using [6][Lemmas 4 and 6] and the results presented in [4], it is possible to derive the following lemmas, for the three-dimensional case.

Lemma 1. *Given two sensors u, v, in three-dimensional space, with apex angle $2\theta > \pi$, if the Euclidean distance between them is δ, there exists a 2-orientation of u and v with transmission range $\max\{1, \delta, \sqrt{1 + \delta^2 - 2\delta\cos(2\theta)}\}$.*

Proof. Orient each antenna in such a way that the straight line which connects u and v contains one edge of the apex angle of v, and vice-versa, one orientation being clockwise and the other one counterclockwise. The farthest point which lies in $\mathbb{B}[u, 1]$ must be covered by v is at a distance d equal to $\sqrt{1 + \delta^2 - 2\delta \cos(2\theta)}$ (by the law of cosines), or equal to δ, in the case of large values of 2θ. Furthermore, in the case of $\delta < 1$ and large values of 2θ, the distance d might be less than 1, but u and v need to cover points which are within a distance of 1 from itself and are not covered by the other node. Therefore, a range $r = \max\{1, \delta, \sqrt{1 + \delta^2 - 2\delta \cos(2\theta)}\}$ is required to ensure that $\mathbb{B}[u, 1]$ and $\mathbb{B}[v, 1]$ are covered. As u covers v directly, and vice-versa, this is a 2-orientation.

Lemma 2. *Given a set S of $n \geq 3$ sensors in three-dimensional space, with apex angle $2\theta > \pi$. Suppose there exists $s \in S$ such that the maximum distance between s and every other sensor in S is δ. If all the sensors in $S \setminus \{s\}$ are contained within a solid spherical sector centered at s with apex angle 2θ, then there is a 2-orientation of S with transmission range $r = \max\{1, \delta, \sqrt{1 + \delta^2 - 2\delta \cos(2\theta)}\}$.*

Proof. Consider sensor s and any sensor $t \in S \setminus \{s\}$. Due to Lemma 1, there is a 2-orientation of s and t with range $\max\{1, \delta, \sqrt{1 + \delta^2 - 2\delta \cos(2\theta)}\}$. Assume that t is on the edge of the apex angle of the spherical sector containing all the sensors $x \in S \setminus \{s\}$, i.e, t forms an angle of at least $2\pi - 2\theta$ with another sensor $u \in S \setminus \{s\}$, such that the spherical sector defined by t, u is empty. Since all the sensors $x \in S \setminus \{s\}$ exist in a spherical sector of apex angle 2θ, t certainly exists.

Therefore, there is an orientation of s and t so that they form a 2-orientation, and so that s covers every sensor $x \in S \setminus \{s\}$. Each of the other sensors x can be oriented to cover s and the portion of $\mathbb{B}[x, 1]$ not covered by s. By the law of cosines, the farthest point from s which is covered by it is at a distance $\max\{\delta, \sqrt{1 + \delta^2 - 2\delta \cos(2\theta)}\}$. Therefore, this is the range required to cover $\mathbb{B}[x, 1]$. The resulting orientation is strongly connected and any sensor can be reached in 2 hops, which means it is a 2-orientation.

Following [6][Lemmas 4 and 6], paper [6] presents the theorem which creates a connected graph with stretch factor $\tau_G(S) \leq 2$. The three-dimensional version uses the same ideas, since the lemmas are adaptable to three dimensions, as well as the algorithm for finding the convex hull, which is used in the proof.

Theorem 1. *Given a a set of sensors S in three-dimensional space, each with one directional antenna of apex angle $2\theta \geq \pi$ and let $U(S) = UBG(S)$. Then there exists an antenna orientation of S with range $\max\{1, 2\sin(\theta)\}$ which creates a directed spanner G_θ with stretch factor $\tau_{G_\theta} \leq 2$.*

Proof. Define $C(G)$ as the set of the vertices from the union of the convex hulls of all connected components of a graph G. Let Q be a hierarchical structure defined as follows: $Q_0(V_0, E_0) = U(S)$ and $Q_{k+1}(V_{k+1}, E_{k+1}) = Q_k[V_k - C(Q_k)]$. In other words, at each iteration the components of the convex hull of each connected component are taken away, which means that every iteration is a proper subset of the previous one. Using this hierarchical structure, it is possible to prove by induction that an orientation can be found for the unit ball graph $U(S)$ on S. This is done by maintaining the invariant that in each iteration every sensor is either *locally convex* or *oriented*.

Definition 1 (Local convexity). *A sensor s is locally convex in a graph G if it is a member of the convex hull of the set of s and its 1-hop neighbours in G.*

Throughout the proof we will use the term *convex* interchangeably to mean locally convex. Since all sensors have apex angles $2\theta \geq \pi$, if a sensor s is convex, it is possible to orient s so that it covers all its neighbours in G. Consider now the iteration Q_i such that $Q_{i-1} \neq \emptyset$ and $Q_i = \emptyset$, i.e., the first iteration in which we get an empty set. Since Q_i is empty, Q_{i-1} must have only convex sensors - so the invariant holds immediately. The iteration Q_{i-2} is the first which may contain sensors which are not convex. We must orient these non-convex sensors in order to satisfy the invariant. We do so as follows: each non-convex sensor t requests to orient with one of its neighbours in $C(Q_{i-2})$ (t must have at least one such neighbour, otherwise it would have been non-convex in Q_{i-1})). Now, for each sensor in $C(Q_{i-2})$, there are three possibilities:

- No request to orient is received.
- A single request to orient is received: The convex sensor and the requestor orient themselves according to Lemma 1, forming a 2-orientation.
- Multiple requests to orient are received: The convex sensor and all requestors orient themselves according to Lemma 2, also forming a 2-orientation.

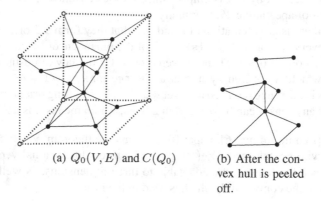

(a) $Q_0(V, E)$ and $C(Q_0)$

(b) After the convex hull is peeled off.

Fig. 2. One iteration of the construction of Q for a given $UBG(V)$. $C(Q_0)$ is denoted by hollow points.

Consider now any iteration Q_k. Assuming Q_{k+1} has a valid orientation, the sensors in Q_k either are convex, already oriented, or can be oriented as explained above, which creates a valid orientation for Q_k. As a valid orientation for the basis case Q_{i-1} was already shown, then by induction there is a valid orientation $G_\theta(S)$ for $Q_0 = U(S)$. The way this orientation was constructed always guaranteed that a sensor can reach one of its neighbours in $U(S)$ in at most 2 hops. As proved in [6][Lemma 1], when groups are merged, this property still holds. So, $G_\theta(S)$ is a connected graph, with $\tau_{G_\theta} \leq 2$. The detailed algorithm is as follows.

Algorithm 1. Orienting sensors with single antenna when apex angle is greater than π.

$Q_0 \leftarrow UBG(S)$
$i \leftarrow 0$
while $Q_i \neq Q_{i+1}$ **do**   ▷ Will actually happen when they are empty sets
    $Q_{i+1} \leftarrow Q_i[V_k - C(Q_k)]$ ▷ Peel off the convex hull of previous iteration
end while
for all $q \in Q$ **do**
    for all sensors $\in q$ **do**
        if sensor is convex **then**
            continue
        else
            request_to_orient(sensor)   ▷ Sensor requests to orient with one of its neighbours
        end if
    end for
    for all sensors $\in q$ **do**
        if sensor was requested to orient **then**
            orient(sensor)   ▷ Orient according to Lemmas (1) or (2)
        end if
    end for
end for

The pseudocode for this orientation is given in Algorithm 1. The proof of Theorem 1 is now complete.

3 Orienting Double Antennae

In this section, we will show a lower bound on the solid angle of the antennae for which connectivity can be achieved with optimal range. An algorithm will also be presented for orienting double antennae with constant stretch factor.

Theorem 2. *Given a set S of points in three-dimensional space and a spherical angle $\Omega \geq \frac{25\pi}{13}$, there exists a polynomial time algorithm that computes a strong orientation of three-dimensional double-antennae of spherical sector with solid angle Ω and having optimal range.*

Proof. Consider a Euclidean minimum spanning tree (MST) T of S, in which $r_{MST}(S)$ is its longest edge. It will be determined how to orient the antennae at each point $p \in S$.

Let B_p be the sphere centered at each point p, with the minimum radius r_p such that all the neighbours of p in T are covered. This implies that $r_p \leq r_{MST}(S)$. For each neighbour u of p in T, let u' and u'' be the intersection points of B_p with the straight line containing the segment defined by u and p. Let $N_{B_p}(p)$ be the set of points projected on the surface of B_p. The maximum degree of a Euclidean MST is in general bounded by the *Kissing Number*, which means that in three dimensions it is bounded by 12. Since p has maximum degree 12 in T, $|N_{B_p}(p)| \leq 24$.

Let DT_p be the Delaunay Triangulation of $N_{B_p}(p)$ (in 3D) on the surface of B_p. The number of triangles of a complete triangulation of n points on a sphere is $2n - 4$. This

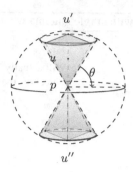

Fig. 3. Projections of three of the neighbours of p that form the greatest triangle. The projections of neighbours u, u' and u'', are highlighted.

can be proved easily by induction on the number n of points. Since $n \leq 24$ this gives at most $2 \cdot 24 - 4 = 44$ triangles. By the way the points were projected on B_p, we know that for each triangle there is a symmetric and identical triangle diammetrically opposed to it. Consider the two largest opposing triangles, t_p and t'_p (a tie can be broken arbitrarily). Orient the antennae at p with range r_p in such a way that the circles circumscribed to t_p and to t'_p are the only part not covered. This can be easily achieved by orienting each antenna towards the normal line to the straight line containing the centers of t_p and t'_p, that passes through the center of B_p.

To prove the lower bound on the solid angle at each point p, observe that, by the pigeonhole principle, the radius of the circumscribed circles of the greatest triangles will have length at least equal to the Tammes' Radius R_{44} [10] and the planar angle θ at the center of the sphere B_p is at least $\arcsin(R_{44})$. From [9], we find the optimal value for 2θ: $2\theta = 31.9834230°$. Therefore, the solid angle of the antennae can be calculated, using Archimedes' Equation as follows: $\Omega = 2\pi - 2\pi(1 - \cos(\theta)) = 2\pi(1 - (1 - \cos(\theta))) = 2\pi \cdot \cos(\theta) < \frac{25}{13}\pi$, where the last inequality is obtained after numerical calculation. It is easy to see that the resulting transmission graph is strongly connected, since T is connected and all edges of T are covered by exactly two antennae at opposite endpoints. This completes the proof of Theorem 2.

Next we study the necessary range to orient the network with constant stretch factor. The algorithm consists of maximally partitioning the vertices of the graph into triples, in such a way that at least one vertex is a neighbour of the other two in the UBG. After this step, it is necessary to determine the necessary range to orient the vertices in the triples to cover each other, as well as to cover vertices that are not part of any triple.

Lemma 3. *When maximally partiotioning the set of vertices of the UBG, the vertices which are not part of any triple are at distance at most 2 of the closest triple.*

Proof. Assume there is a vertex v which is at distance greater than 2. As the UBG is connected, there must be two more unmatched vertices, w and x that would connect v to the closest triple in the UBG. But if those exist, they could form a triple and the partition would not be maximal. Then, by contradiction, it is evident that v must be at distance of at most 2.

Next it will be shown that with infinite range, it is possible to orient antennae in each triple to cover the whole space.

Lemma 4. *Consider three points A, B, C in the space, forming a triangle. Three identical double antennae of apex angle $2\theta \geq \pi/2$ and infinite range can be oriented so as to cover the whole space.*

Proof. Consider the greatest angle of the triangle to be α. The orientation will depend on the value of α.

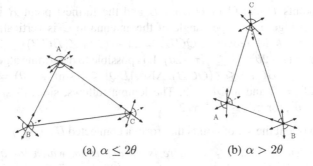

(a) $\alpha \leq 2\theta$ (b) $\alpha > 2\theta$

Fig. 4. Orientation of three double antennae with infinite range

(i) $\alpha \leq 2\theta$: Without loss of generality assume that BC is horizontal and A is above BC. Orient the antennae at as depicted in Figure 4(a) so that the antennae cover all three-dimensional subspace delimited by the triangle and its projections which are normal to the paper. It is easy to see that all the 3-dimensional space is covered.

(ii) $\alpha > 2\theta$: Without loss of generality, assume that AB is the second smallest edge in the triangle, AB is horizontal and C is above AB. Orient the antennae as depicted in Figure 4(b), so that the one antenna wedge of the apex angle of C is vertical and the wedge of the apex angle of the antennae at A and B are on AB. To prove the orientation covers the "whole space", observe that the antennae at A and B only leave uncovered the 3-dimensional subspace below the plane containing AB, that is normal to the paper. However, the antenna at C covers this subspace.

Lemma 5. *Let A, B, C be three points in the 3-dimensional space, such that $d(A, B) \leq 1$ and $d(A, C) \leq 1$. Assume that $\frac{\pi}{2} \leq 2\theta \leq \pi$. We can orient three $(2\theta, r, 2)$-double antennae of apex angle 2θ at A, B, C so that every point at distance at most two from one of these points is covered by at least one of the three antennae, where $r \leq 4 \cdot \sin(\frac{\pi}{4} + \theta)$.*

Proof. Consider the balls $\mathbb{B}[A, 2]$, $\mathbb{B}[B, 2]$, $\mathbb{B}[C, 2]$, of radius 2 centered at A, B and C, respectively. Let $\mathcal{D} = \mathbb{B}[A, 2] \cup \mathbb{B}[B, 2] \cup \mathbb{B}[C, 2]$. Observe that each antenna covers one spherical sector of angle $\pi \leq \Omega \leq 2\pi$, relating to θ as described in Archimedes' Equation, with range two. It remains to prove that a range $r \leq 4 \cdot \sin(\frac{\pi}{4} + \theta)$ is sufficient to cover the remaining area. Two cases are to be considered:

(i) $\alpha \leq 2\theta$: Observe that the area of $\mathcal{D}$ that A covers is at most at distance three. Let us assume, without loss of generality, that $|AB| \leq |AC|$. Therefore, $\angle(BCA) \leq \angle(CBA) \leq \pi - 2\theta$. Hence, we only need to consider the case for C. Let P the farthest point of C in the coverage area $\mathcal{D}$. Observe that $\angle(PBC) = (\pi - 2\theta) + \angle(CBA)$. Therefore, a range of $4 \sin(\frac{\angle(PBC)}{2})$ is always sufficient to cover $\mathcal{D}$, since $|BC| \leq 2$ and $|BP| = 2$.

(ii) $\alpha > 2\theta$: From the orientation of Lemma 4, $\alpha = \angle(BAC)$ is the largest angle and $\beta = \angle(CBA)$ is the smallest angle in the triangle. Therefore, $\beta \leq \frac{\pi - 2\theta}{2}$. Observe that the farthest point of $\mathcal{D}$ that A covers is at distance three. Then, consider the farthest points P and Q in $\mathcal{D}$ from B and the farthest point R in $\mathcal{D}$ from C. Since the wedge of the apex angle of the antenna in C is vertical, $\angle(BCP) = \pi/2 + \beta \leq \pi - \theta$. Moreover, $\angle(QCB) = 2\pi - ((2\theta + \angle(BCP)) \leq 3\pi/2 - 2\theta$ and $\angle(CBR) = (\pi - 2\theta) + \beta \leq \frac{3}{2}(\pi - 2\theta)$. It is possible to determine algebraically that $\angle(CBR) \leq \angle(BCP) \leq \angle(QCB)$. Also, $|BQ| \leq 4 \cdot \sin(\frac{3\pi}{4} - \theta) = 4 \cdot \sin(\frac{\pi}{4} + \theta)$, since $|BC| = 2$ and $|CQ| = 2$. The lemma follows, since $4 \cdot \sin(\pi - 2\theta) \leq 4 \cdot \sin(\frac{\pi}{4} + \theta)$, for $\pi/4 \leq \theta \leq \pi/2$.

It is possible to orient the set of points that form a connected $UBG(P)$.

Theorem 3. *Given $\frac{\pi}{2} \leq 2\theta \leq \pi$, there is an algorithm which for any connected $UBG(P)$ on a set P of points in the space, orients $(2\theta, 4 \cdot \sin(\frac{\pi}{4} + \theta), 2)$-double antenna so that the resulting graph is also connected and has stretch factor four. Furthermore, it can be done in linear time.*

Proof. Let $\mathcal{T}$ be any partition of the UBG with a maximal number of triples in such a way that the triangle defined by the three sensors has at least two edges of length at most one. This partition can be constructed in linear time, by selecting a node which is not yet in the partition and then trying to select two of its neighbours in a such a way that the criteria mentioned before is met. For any triangle T in $\mathcal{T}$, the antennae is oriented as shown in Lemma 5. For each sensor, the antenna covering an internal angle of the triangle have radius $4 \cdot \sin(\frac{\pi}{4} + \theta)$, while the opposite antenna have range 2. The remaining sensors, which are not in a triple, must be oriented towards its nearest triangle, which will be at distance at most two.

Let G be the directed spanner induced by the antennae. It is easy to see that it will be strongly connected, by the way the orientations are done. It remains to prove that for each edge u, v there is a directed path P from u to v and also a directed path from v to u of hop-length no more than 4 hops. Let T and T' be two different triangles in the partition $\mathcal{T}$. We consider three cases, depending on the location of sensors u and v:

(i) $u, v \in T$: Then $|P| \leq 2$ and $|P'| \leq 2$;

(ii) $u \in T$ and $v \in T'$: Since $d(u, v) \leq 1$, v is in the coverage area of the triangle T. Therefore, v is reachable by u in at most three hops, which means $|P| \leq 3$. An analogous argument shows that $|P'| \leq 3$;

(iii) At least one of u and v is not unmatched, i.e., is not in any triangle of the partition. Assume without loss of genreality that u is unmatched. Observe that there exists a triangle $T \in \mathcal{T}$ at distance at most two from u. Otherwise, $\mathcal{T}$ would not be maximal. Therefore, u can reach v through T in at most four hops, i.e., $|P| \leq 4$. Similarly, we can prove that $|P'| \leq 4$.

4   Simulation Results

In this section we use simulation results to analyze how replacing omnidirectional antennae with directional antennae in three-dimensional space impacts the diameter of the graph. The diameter $D(G)$ of a graph G is defined as the length of the maximum shortest path between any two nodes of the graph. For each simulation, a random set of points S was generated and the corresponding UBG was constructed. If the UBG was not connected, the set of points was discarded and a new one was generated, until a connected UBG was obtained. A directed spanner of S was constructed using the algorithm from Theorem 3. The construction of the triples was executed in a greedy and random manner.

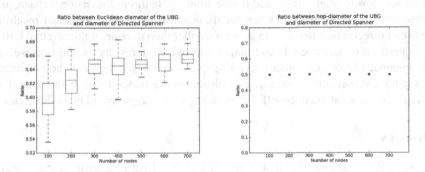

Fig. 5. Left: Boxplot comparing the Euclidean diameter of the UBG and the directed spanner, when varying the number of nodes. Right: Comparison of the hop diameter of the UBG and the directed spanner, when varying the number of nodes.

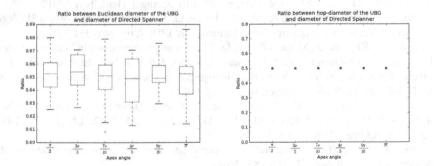

Fig. 6. Left: Boxplot comparing the Euclidean diameter of the UBG and the directed spanner, when varying the number of nodes. Right: Comparison of the hop diameter of the UBG and the directed spanner, when varying the apex angle.

We compared the hop-diameter of both graphs, as well as the Euclidean diameter. In the first simulation, the apex angle 2θ was fixed to $\frac{\pi}{2}$ and the number of nodes n

varied from 400 to 1000, in increments of 100. For each n, the simulation ran 30 times. Figure 5 shows the results. In the second simulation, the number of nodes was fixed to 500 and the apex angle varied from $\frac{\pi}{2}$ to π in increments of $\frac{\pi}{10}$. The simulation was run 30 times for each angle. Figure 6 shows the results. Both simulations show that the diameter of the directed spanner is in general smaller than the one of the UBG, with the hop-diameter of the directed spanner being half the diameter of the UBG. This advantage is most probably due to the increased range of communication present in the directed spanner.

5 Conclusion

We discussed how to orient single and doube antennae in three-dimensional space, and also observed with the simulations that the diameter of the directed spanner resulting by the use of directional antenna is in general smaller then the one of the original UBG. Several questions remain open. In addition to improving our results for single as well as double antennae (Table 1), another interesting question is concerned with how to orient sensors in three-dimensional space when each sensor is equipped with k antennae, $1 < k \leq 12$, as well as what the trade-offs between angle and range would be in these cases.

References

1. Capsalis, P.K.: Electronic beam steering using switched parasitic smart antenna arrays. Progress in Electromagnetics Research 36, 101–119 (2002)
2. Caragiannis, I., Kaklamanis, C., Kranakis, E., Krizanc, D., Wiese, A.: Communication in wireless networks with directional antennas. In: Proceedings of the Twentieth Annual Symposium on Parallelism in Algorithms and Architectures, pp. 344–351. ACM (2008)
3. Eftekhari Hesari, M., Kranakis, E., MacQuarie, F., Morales-Ponce, O., Narayanan, L.: Strong Connectivity of Sensor Networks with Double Antennae. In: Even, G., Halldórsson, M.M. (eds.) SIROCCO 2012. LNCS, vol. 7355, pp. 99–110. Springer, Heidelberg (2012)
4. Kranakis, E., Krizanc, D., Modi, A., Morales-Ponce, O.: Connectivity Trade-offs in 3D Wireless Sensor Networks Using Directional Antennae. In: IPDPS, pp. 345–351 (2011)
5. Kranakis, E., Krizanc, D., Morales-Ponce, O.: Maintaining Connectivity in Sensor Networks Using Directional Antennae. In: Nikoletseas, S., Rolim, J. (eds.) Theoretical Aspects of Distributed Computing in Sensor Networks. Monographs in Theoretical Computer Science. An EATCS Series, pp. 59–84. Springer, Heidelberg (2011)
6. Kranakis, E., MacQuarrie, F., Morales-Ponce, O.: Stretch Factor in Wireless Sensor Networks with Directional Antennae. In: Lin, G. (ed.) COCOA 2012. LNCS, vol. 7402, pp. 25–36. Springer, Heidelberg (2012)
7. Pfender, F., Ziegler, G.: Kissing numbers, sphere packings and some unexpected proofs. Notices Amer. Math. Soc. 51, 873–883 (2004)
8. Sievenpiper, D., Schaffner, J., Loo, B., Tangonan, G., Harold, R., Pikulski, J., Garcia, R.: Electronic beam steering using a varactor-tuned impedance surface. In: IEEE Antennas and Propagation Society International Symposium, vol. 1, pp. 174–177 (2001)
9. Sloane, N.J.: Spherical Codes: Nice arrangements of points on a sphere in various dimensions, http://www2.research.att.com/~njas/packings/index.html#I
10. Tammes, P.M.L.: On the origin of number and arrangement of the places of exit on the surface of pollen-grains. Rec. Trav. Bot. Neerl. 27, 1–84 (1930)

A Graph Parameter That Matches the Resilience of the Certified Propagation Algorithm*

Chris Litsas, Aris Pagourtzis, and Dimitris Sakavalas

School of Electrical and Computer Engineering
National Technical University of Athens, 15780 Athens, Greece
chlitsas@central.ntua.gr, pagour@cs.ntua.gr, sakaval@corelab.ntua.gr

Abstract. We consider the Secure Broadcast problem in incomplete networks. We study the resilience of the Certified Propagation Algorithm (CPA), which is particularly suitable for *ad hoc* networks. We address the issue of determining the maximum number of corrupted players $t_{\max}^{\text{CPA}}$ that CPA can tolerate under the t-locally bounded adversary model, in which the adversary may corrupt at most t players in each player's neighborhood. For any graph G and dealer-node D we provide upper and lower bounds on $t_{\max}^{\text{CPA}}$ that can be efficiently computed in terms of a graph theoretic parameter that we introduce in this work. Along the way we obtain an efficient 2-approximation algorithm for $t_{\max}^{\text{CPA}}$. We further introduce two more graph parameters, one of which matches $t_{\max}^{\text{CPA}}$ exactly.

Keywords: Distributed Protocols; Ad Hoc Networks; Secure Broadcast; Byzantine Faults; t-Locally Bounded Adversary Model.

1 Introduction

A fundamental problem in distributed networks is Secure Broadcast, in which the goal is to distribute a message correctly despite the presence of Byzantine faults. In particular, an adversary may control several nodes and be able to make them deviate from the protocol arbitrarily by stopping, rerouting, or even altering a message that they should normally relay intact to certain nodes. In general, agreement problems have been primarily studied under the threshold adversary model, where a fixed upper bound t is set for the number of corrupted players and broadcast can be achieved if and only if $t < n/3$, where n is the total number of players. The Broadcast problem has been extensively studied in complete networks under the threshold adversary model mainly in the period from 1982, when it was introduced by Lamport, Shostak and Pease [9], to 1998, when Garay and Moses [4] presented the first fully polynomial Broadcast protocol optimal in resilience and round complexity.

* Work supported by ALGONOW project of the Research Funding Program THALIS, co-financed by the European Union (European Social Fund – ESF) and Greek national funds through the Operational Program "Education and Lifelong Learning" of the National Strategic Reference Framework (NSRF).

J. Cichoń, M. Gębala, and M. Klonowski (Eds.): ADHOC-NOW 2013, LNCS 7960, pp. 269–280, 2013.
© Springer-Verlag Berlin Heidelberg 2013

The case of a threshold adversary in incomplete networks has been studied to a much lesser extent [1–3, 8], mostly through protocols for Secure Message Transmission which, combined with a Broadcast protocol for complete networks, yield Broadcast protocols for incomplete networks. Naturally, connectivity constraints are required to hold in addition to the $n/3$ bound.

In the case of an honest dealer the impossibility threshold of $n/3$ does not hold; for example, in complete networks the problem becomes trivial. However, in incomplete networks the situation is different. A small number of traitors (corrupted players) may manage to block the entire protocol if they control a critical part of the network, e.g. if they form a separator of the graph. It therefore makes sense to define criteria depending on the structure on the graph (graph parameters), in order to bound the number or restrict the distribution of traitors that can be tolerated.

An approach in this direction is to consider topological restrictions on the adversary's corruption capacity. The importance of local restrictions comes, among others, from the fact that they may be used to derive local criteria which the players can employ in order to achieve Broadcast in *ad hoc* networks. Such an example is the *t-locally bounded adversary model*, introduced in [6], in which at most t-corruptions are allowed in the neighborhood of every node.

Koo [6] proposed a simple, yet powerful protocol for the t-locally bounded model, the *Certified Propagation Algorithm* (CPA) (a name coined by Pelc and Peleg in [10]), and applied it to networks of specific topology. In 2005 Pelc and Peleg [10] considered the t-locally bounded model in generic graphs and gave a sufficient topological condition for CPA to achieve Broadcast in such graphs. They also provided an upper bound on the number of corrupted players t that can be locally tolerated in order to achieve Broadcast by any protocol, in terms of an appropriate graph parameter; they left the deduction of tighter bounds as an open problem. To this end, Ichimura and Shigeno [5] proposed an efficiently computable graph parameter which implies a tighter, but not exact, characterization of the class of graphs on which CPA achieves Broadcast.

1.1   Our Results

In this paper we study the behavior of CPA in generic (incomplete) networks, with an honest dealer. Our main contribution is the exact determination of the maximum number of corrupted players $t_{\max}^{\text{CPA}}$ that can be locally tolerated by CPA, for any graph G and dealer D. We do this by developing three graph parameters:

- $\mathcal{K}(G, D)$ is determined via an appropriate *level-ordering* of the nodes of the graph. We show that $t < \mathcal{K}(G, D)/2$ is a sufficient condition for CPA to be t-locally resilient and that $t < \mathcal{K}(G, D)$ is a necessary condition, implying that $\lceil \mathcal{K}(G, D)/2 \rceil - 1 \leq t_{\max}^{\text{CPA}} < \mathcal{K}(G, D)$. We prove that our parameter coincides with the parameter $\widetilde{\mathcal{X}}(G, D)$ of [5]. We further propose an efficient algorithm for computing $\mathcal{K}(G, D)$ which is faster than the algorithm for computing $\widetilde{\mathcal{X}}(G, D)$ proposed in [5]. Note that this immediately gives an asymptotic 2-approximation for t_{max}^{CPA}; we provide an example that shows that the ratio of this algorithm is tight.

- $\mathcal{M}(G, D, t)$, depending also on a value t, is a parameter that immediately reveals whether CPA is t-locally resilient for graph G and dealer D, by simply checking whether $\mathcal{M}(G, D, t) \geq t + 1$. Therefore, via this parameter, we provide a *necessary and sufficient condition* for CPA to be t-locally resilient. Such a condition was not known until very recently, when a necessary and sufficient condition was independently given in [11]. However, the way in which the condition of [11] is defined implies a superexponential time algorithm to check it (actually no algorithm is given in [11]). On the other hand, we will see that even a naïve algorithm to compute $\mathcal{M}(G, D, t)$ would need single exponential time.

- $\mathcal{T}(G, D) = \max\{t \in \mathbb{N} \mid \mathcal{M}(G, D, t) \geq t + 1\}$, gives the maximum number of corrupted players that CPA can tolerate in every node's neighborhood, hence exactly determining t_{max}^{CPA}.

1.2   Problem and Model Definition

We will now formally define the adversary model and the CPA algorithm; both notions were developed in [6]; the term t-locally bounded is due to [10]. We will also define basic notions and terminology that we will use throughout the paper. We refer to the participants of the protocol by using the notions *node* and *player* interchangeably.

Secure Broadcast with Honest Dealer. We assume the existence of a designated honest player, called the *dealer*, who wants to broadcast a certain value x_D to all players. We say that a distributed protocol achieves Secure Broadcast if by the end of the protocol every honest player has *decided on* x_D, i.e. has been able to deduce that x_D is the value originally sent by the dealer and output it as its own decision.

The above problem is trivial in complete networks and we will consider the case of incomplete networks here. In the sequel we will refer to the problem as the Broadcast problem.

t-Locally Bounded Adversary Model. We consider a network where nodes may be corrupted, but at most t-corruptions are allowed in the neighborhood of every node. A corruption set with the above property is called *t-local set*. An algorithm which achieves Broadcast in the t-locally bounded adversary model is called *t-locally resilient*.

The previously mentioned Certified Propagation algorithm uses only local information and thus is particularly suitable for *ad hoc* networks. CPA is probably the only Broadcast algorithm known up to now for the t-locally bounded model, not requiring knowledge of the network topology.

Certified Propagation Algorithm

1. The dealer D sends its initial value x_D to all of its neighbors, decides on x_D and terminates.
2. If a node is a neighbor of the dealer, then upon receiving x_D from the dealer, decides on x_D, sends it to all of its neighbors and terminates.
3. If a node is not a neighbor of the dealer, then upon receiving $t + 1$ copies of a value x from $t + 1$ distinct neighbors, it decides on x, sends it to all of its neighbors and terminates.

Definition 1 (Max CPA Resilience). *For a graph G and dealer-node D, $t_{max}^{CPA}(G, D)$ is the maximum t such that CPA is t-locally resilient.*

Whenever G and D are implied by the context, we will simply write t_{max}^{CPA}.

Bounds vs Conditions. Let us now make a simple but useful observation: for a graph-theoretic parameter X, showing that $t < X$ is a sufficient topological condition for CPA to be t-locally resilient provides a lower bound of $\lceil X \rceil - 1$ on t_{max}^{CPA}. Respectively, necessary conditions of similar form imply upper bounds on t_{max}^{CPA}. We will often use this relation between bounds and conditions throughout the paper.

2   Lower Bounds on Max CPA Resilience

Pelc and Peleg [10] were the first to present a graph-theoretic parameter $\mathcal{X}(G, D)$ that associates the maximum tolerable number of local corruptions with the topology of the graph. This parameter represents the maximum number b such that every node v has at least b neighbors with distance to D smaller than that of v. They give a sufficient condition for CPA resilience, namely $\mathcal{X}(G, D) \geq 2t + 1$, which implies that the nodes of graph G can be arranged in levels w.r.t. their distance from D, the first level being the neighborhood of D, and every node in level k having at least $2t + 1$ neighbors in level $k - 1$. This, in turn implies that every node in distance k from D (level k) decides in the k-th round, because it will certainly receive at least $t + 1$ correct values from honest nodes in level $k - 1$. However, as shown in the same paper, this condition is not necessary, because a node in level k may collect correct values from neighbors in level k or $k + 1$ also, thus completing the necessary number of $t + 1$ identical values. In other words, $\lceil \mathcal{X}/2 \rceil - 1$ is a lower bound for Max CPA Resilience but not a tight one.

2.1   A New Parameter for Bounding Max CPA Resilience

In order to derive tighter bounds on t_{max}^{CPA} we introduce the notion of *minimum k-level ordering* of a graph which generalizes the level ordering that was implicit in [10]. Intuitively, a minimum k-level ordering is an arrangement of nodes into disjoint levels, such that every node has at least k neighbors in previous levels and belongs to the minimum level for which this property is satisfied for this node. Formally:

Definition 2. *A* Minimum k-Level Ordering $\mathcal{L}_k(G, D)$ *of a graph* $G = (V, E)$ *for a given dealer-node* D *is a partition* $V \setminus \{D\} = \bigcup_{i=1}^{m} L_i$, $m \in \mathbb{N}$ *s.t.*

$$L_1 = \mathcal{N}(D), \quad L_i = \{v \in V \setminus \bigcup_{j=1}^{i-1} L_j : |\mathcal{N}(v) \cap \bigcup_{j=1}^{i-1} L_j| \geq k\}, 2 \leq i \leq m$$

We next define the relaxed k-level ordering notion which will be useful for our proofs, by dropping the level minimality requirement for nodes.

Definition 3. *A* Relaxed k-Level Ordering *of a graph* $G = (V, E)$ *for a given dealer-node* D *is a partition* $V \setminus \{D\} = \bigcup_{i=1}^{m} L_i$, $m \in \mathbb{N}$ *s.t.*

$$L_1 = \mathcal{N}(D), \quad \forall v \in L_i : |\mathcal{N}(v) \cap \bigcup_{j=1}^{i-1} L_j| \geq k$$

Properties of k-Level Orderings. Note that while there may exist several relaxed k-level orderings of a graph, the minimum k-level ordering is unique, as can be shown by an easy induction. Let us also observe that a relaxed k-level ordering may be easily transformed to the unique minimum k-level ordering; to show this we will use a new notion: Given a relaxed k-level ordering $\mathcal{L}'$: $V = \bigcup_{i=1}^{m} L_i$, $m \in \mathbb{N}$ we will refer to a player $u \in L_h \in \mathcal{L}'$ as *delayed node* in $\mathcal{L}'$ if $\exists d$ with $1 < d < h \leq m$ s.t. $|\mathcal{N}(u) \cap \bigcup_{j=1}^{d-1} L_j| \geq k$. Now, given any relaxed k-level ordering $\mathcal{L}'$ we can construct a minimum k-level ordering $\mathcal{L}_k$ simply by repeatedly moving every delayed node to the lowest level such that the partition remains a relaxed k-level ordering. It is not hard to see that a relaxed k-level ordering with no delayed nodes is actually a minimum k-level ordering. Therefore, the following holds (proof omitted):

Proposition 1. *Given a graph* G *and dealer* D, *for every* $k \in \mathbb{N}$, *if there exists a Relaxed k-Level Ordering for G, D then there exists a unique Minimum k-Level Ordering for G, D.*

Definition 4 (Parameter $\mathcal{K}$). *For a graph* G *and dealer* D,

$$\mathcal{K}(G, D) \stackrel{def.}{=} \max\{k \in \mathbb{N} \mid \exists \text{ a Minimum } k\text{-Level Ordering } \mathcal{L}_k(G, D)\}$$

Theorem 1 (Sufficient Condition). *For every graph G, dealer D and $t \in \mathbb{N}$, if $t < \mathcal{K}(G, D)/2$ then CPA is t-locally resilient.*

Proof. Observe that $2t < \mathcal{K}(G, D)$ implies the existence of a minimum $(2t + 1)$-level ordering $\mathcal{L}_{2t+1}(G, D)$. Let $\mathcal{L}_{2t+1}(G, D)$ be the partition $\{L_1, \ldots, L_m\}$ of V, i.e. $V = \bigcup_{i=1}^{m} L_i$. It suffices to show that for $1 \leq i \leq m$, every honest player $v \in L_i$ decides on the dealer's value x_D. By strong induction on i:

Every honest player $v \in L_1 = \mathcal{N}(D)$ decides on the dealer's value x_D due to the CPA steps 1 and 2. If all honest players $u \in L_i, 1 \leq i \leq h$, decide on x_D at some round, then every honest player $v \in L_{h+1}$ receives $|\bigcup_{j=1}^{h} L_j \cap \mathcal{N}(v)| \geq 2t+1$ messages from its decided neighbors in previous levels and at least $t + 1$ of them are honest. Thus v decides on x_D. $\square$

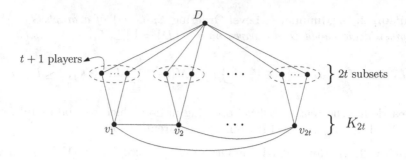

Fig. 1. Graph with $\mathcal{K}(G, D) = t + 1$, for which CPA is t-locally resilient

Corollary 1 (Lower Bound). *For any graph G and dealer D it holds that $t_{\max}^{\mathrm{CPA}} \geq \lceil \mathcal{K}(G, D)/2 \rceil - 1$*

2.2 Non-tightness of the Lower Bound

In Theorem 1 we proved that $t < \mathcal{K}(G, D)/2$ is sufficient for CPA to be t-locally resilient; we next prove that it is not a necessary condition. Intuitively, the reason is that the topology of the graph may prevent the adversary from corrupting t players in *each* player's neighborhood, hence some players will correctly decide by executing CPA even if they have only $t + 1$ neighbors in previous levels.

Proposition 2. *There exists a family of instances (G, D), such that CPA is $(\mathcal{K}(G, D) - 1)$-locally resilient for (G, D).*

Proof. Figure 1 provides such an instance for each value of t. In this instance the neighborhood of D consists of $2t^2 + 2t$ nodes, nodes $v_1, \ldots, v_{2t}$ form a clique of size $2t$ and are connected with $\mathcal{N}(D)$ as shown in the figure. We can easily check that $t = \mathcal{K}(G, D) - 1$. If we run CPA on G then any player $v_i \in \{v_1, \ldots, v_{2t}\}$ receives M correct messages, with

$$M = M_A + M_B \tag{1}$$

where, M_A = number of messages received from $\mathcal{N}(D)$ and
M_B = number of messages received from $B = \{v_1, \ldots, v_{2t}\} \setminus \{v_i\}$.

Let $T_i = T \cap \mathcal{N}(D) \cap \mathcal{N}(v_i)$ be the set of traitors that are common neighbors of D and v_i. Then

$$M_A = |\mathcal{N}(D) \cap \mathcal{N}(v_i) \setminus T_i| = t + 1 - |T_i| \tag{2}$$

In order to compute the number of correct messages that v_i receives from players in B, we define the sets:

$$C_{B_1} = \{v \in B \mid v \text{ receives at most } t \text{ messages from } \mathcal{N}(D) \}$$
$$C_{B_2} = \{v \in B \mid v \text{ is corrupted} \}$$
$$C_B = C_{B_1} \cup C_{B_2}$$

We observe that C_{B_1} becomes maximum in cardinality if the adversary corrupts exactly one player in every set $\mathcal{N}(v_j) \cap \mathcal{N}(D), \forall v_j \in B$. Therefore $\max\limits_{T:t\text{-local set}} |C_{B_1}| = \max\limits_{T:t\text{-local set}} |T \cap (\mathcal{N}(D) \setminus \mathcal{N}(v_i))| = t - |T_i|$. Also $|C_{B_2}| \leq t - |T_i|$ because B and $\mathcal{N}(v_i) \cap \mathcal{N}(D)$ form the neighborhood of v_i where the corruptions can be at most t. Next we compute an upper bound on C_B.

$$|C_B| = |C_{B_1} \cup C_{B_2}| \leq |C_{B_1}| + |C_{B_2}| \leq (t - |T_i|) + (t - |T_i|) = 2t - 2|T_i|$$

and thus,

$$M_B = 2t - 1 - |C_B| = 2t - 1 - 2t + 2|T_i| = 2|T_i| - 1 \qquad (3)$$

Finally we can compute the total number of messages M,

$$(1),(2),(3) \Rightarrow M = M_A + M_B \geq t + 1 - |T_i| + 2|T_i| - 1 =$$
$$= t + |T_i|$$

For any v_i, if $|T_i| > 0$ then $M \geq t + 1$. Otherwise $|T_i| = 0$ and v_i receives $t + 1$ correct messages from $\mathcal{N}(D)$. Thus CPA successfully achieves Broadcast on (G, D).

□

3   An Upper Bound on Max CPA Resilience

In the previous section we have shown that $t_{\max}^{\text{CPA}} \geq \lceil \mathcal{K}(G, D)/2 \rceil - 1$; we have also demonstrated cases in which $\mathcal{K}(G, D) - 1$ traitors are locally tolerated by CPA. In this section we will show that the latter is the best possible: $\mathcal{K}(G, D) - 1$ is an upper bound on the number of local traitors for any G and D. We do this by proving a necessary condition for CPA to be t-locally resilient.

Theorem 2 (Necessary Condition). *For any graph G, dealer D and $t \geq \mathcal{K}(G, D)$, CPA is not t-locally resilient.*

Proof. Assume that CPA is t-locally resilient, with $t \geq \mathcal{K}(G, D)$. Since, by assumption, CPA is t-locally resilient there must be a positive integer, let s, so that the algorithm terminates after s steps in G. Consider now the operation of CPA on graph G in terms of sets. Let L_i denote the set of nodes that decide in the i-th round. Since every node in L_i decides at the i-th round we get that it has at least $t + 1$ neighbors in sets $L_1, \ldots, L_{i-1}$. That is,

$$\forall v \in L_i \Rightarrow |\mathcal{N}(v) \cap \bigcup_{j=1}^{i-1} L_j| \geq t + 1.$$

Observe that the above sequence is a relaxed $(t+1)$-level ordering for G, D. From the above observation and according to the Proposition 1 we get that there must be a minimum $(t + 1)$-level ordering for G, D. But this is a contradiction since we assumed that $t \geq \mathcal{K}(G, D)$. □

Corollary 2 (Upper bound on $t_{\max}^{\text{CPA}}$). *For any graph G and dealer D it holds that $t_{\max}^{\text{CPA}} < \mathcal{K}(G, D)$*

3.1   Comparison with the Ichimura-Shigeno Parameter

In [5], Ichimura and Shigeno introduce a graph theoretic parameter $\widetilde{\mathcal{X}}(G, D)$ which can be used to obtain a sufficient condition for CPA resilience. For a graph $G = (V, E)$ and dealer D, they consider a total ordering $\sigma = (v_1, v_2, \ldots)$ of the set $V \setminus (\mathcal{N}(D) \cup D)$, and use $\delta(W_i, v)$ to denote the number of neighbors that v has in the set $\mathcal{N}(D) \cup \{v_1, \ldots, v_{i-1}\}$. The total ordering σ has the property that $\forall i, j,$ with $1 \leq i < j \leq |V \setminus (\mathcal{N}(D) \cup D)|$ it holds that $\delta(W_{i-1}, v_i) \geq \delta(W_{i-1}, v_j)$. This ordering is also referred to as *max-back ordering*. They define parameter $\widetilde{\mathcal{X}}(G, D) = \min\{\delta(W_{i-1}, v_i) \mid i = 1, 2, \ldots\}$. and prove that it is unique, i.e., is the same for all max-back orderings. They essentially prove that,[1]

$$\lceil \widetilde{\mathcal{X}}(G, D)/2 \rceil - 1 \leq t_{\max}^{\text{CPA}} < \widetilde{\mathcal{X}}(G, D). \tag{1}$$

Hence, their parameter gives similar bounds as ours. We next show that there is a good reason for this coincidence: despite the different way of defining the parameters $\mathcal{K}(G, D)$ and $\widetilde{\mathcal{X}}(G, D)$, they prove to be equal.

Proposition 3. $\mathcal{K}(G, D) = \widetilde{\mathcal{X}}(G, D)$

Proof. Consider the max-back ordering $\sigma = (v_1, v_2, \ldots)$. Then the sequence $\{L_1 = \mathcal{N}(D), L_2 = \{v_1\}, L_3 = \{v_2\}, \ldots\}$ is trivially a relaxed $\widetilde{\mathcal{X}}(G, D)$-level ordering, because the minimum connectivity between a level and its predecessors is $\widetilde{\mathcal{X}}(G, D)$. Thus, due to Proposition 1, there exists a minimum $\widetilde{\mathcal{X}}(G, D)$-level ordering, therefore $\mathcal{K}(G, D) \geq \widetilde{\mathcal{X}}(G, D)$. Thus, combining the last inequality with inequality (1) we get the following:

$$t_{\max}^{\text{CPA}} < \widetilde{\mathcal{X}}(G, D) \leq \mathcal{K}(G, D)$$

Since Proposition 2 implies that there is a graph for which CPA is $(\mathcal{K}(G, D) - 1)$-locally resilient the above relation yields the equality of $\mathcal{K}(G, D)$ and $\widetilde{\mathcal{X}}(G, D)$, since $\widetilde{\mathcal{X}}(G, D) < \mathcal{K}(G, D)$ would lead to $t_{\max}^{\text{CPA}} < \mathcal{K}(G, D) - 1$, a contradiction.   □

Although the two parameters $\mathcal{K}(G, D)$ and $\widetilde{\mathcal{X}}(G, D)$ are equal, the fact that $\mathcal{K}(G, D)$ is defined in a completely different way leads to an improved complexity of computing it, as we will see in the next section.

4   Approximation of Max CPA Resilience

Let us now consider the approximability of computing the *Max CPA Resilience*; we will give an efficient 2-approximation algorithm. We first show how to check if there exists a minimum m-level ordering, for a graph G and dealer D, using a slight variation of the standard BFS algorithm. Subsequently, we obtain the approximation by simply computing $\mathcal{K}(G, D)$, using the above check. The ratio follows immediately, by combining Corollaries 1 and 2.

---

[1] Note that the condition $t \leq \widetilde{\mathcal{X}}(G, D)$ was given as necessary in [5]; however their proof can be easily modified to show the tighter bound $t < \widetilde{\mathcal{X}}(G, D)$, implying the right part of (1).

Existence check of a minimum m-level ordering for (G, D)
On input (G, D, m) do the following:

1. Assign a zero counter to each node.
2. Enqueue the dealer and every one of its neighbors.
3. Dequeue a node and increase the counters of all its neighbors. Enqueue a neighbor only if its counter is at least m.
4. Repeat Step 3 until the queue is empty.
5. If all nodes have been enqueued then output *'True'* (a minimum m-level ordering exists); otherwise, output *'False'*.

Note that the above algorithm can be modified to compute the minimum m-level ordering $\mathcal{L}_m(G, D)$.

2-Approximation of $t_{\max}^{\text{CPA}}$

1. Compute $\mathcal{K}(G, D)$: since $\mathcal{K}(G, D) < \min\limits_{v \in V \setminus (\mathcal{N}(D) \cup D)} \deg(v) = \delta$, the exact value of $\mathcal{K}(G, D)$ is computed by $\log \delta$ repetitions of the existence check, by simple binary search.
2. Return $\lceil \mathcal{K}(G, D)/2 \rceil - 1$

Since $t \geq \mathcal{K}(G, D) \Rightarrow$ CPA is not t-locally resilient, it holds that $t_{\max}^{\text{CPA}} < \mathcal{K}(G, D)$, consequently, the returned value is at least $\lceil t_{\max}^{\text{CPA}}/2 \rceil - 1$.

A tight example for the approximation ratio of the algorithm is in fact given by the instance in Figure 1 in which we present a graph for which $\mathcal{K}(G, D) = t+1$ and CPA is t-locally resilient.

The complexity of the above approximation algorithm is obviously given by the complexity of the computation of $\mathcal{K}(G, D)$. As explained above the algorithm requires at most $\log \delta$ executions of the existence check. The latter requires $O(|E|)$ time (same complexity as BFS). Altogether, we get that the time complexity of the algorithm is $O(|E| \log \delta)$, which significantly improves upon the complexity bound for the equivalent parameter $\widetilde{\mathcal{X}}(G, D)$ given in [5]; the complexity stated there is $O(|V|(|V| + |E|))$.

5   Determining $t_{\max}^{\text{CPA}}$ Exactly

In this section we present a procedure to compute the exact value of $t_{\max}^{\text{CPA}}$. To this end, we introduce two new graph parameters.

For a corruption set (t-local set) T and graph $G = (V, E)$ we will denote with $G_{\bar{T}} = (V \setminus T, E')$ the node induced subgraph of G on the node set $V \setminus T$.

Definition 5. *For any graph G, dealer D and positive integer t, the t-safety threshold is the quantity* $\mathcal{M}(G, D, t) = \min\limits_{T:\ t\text{-local set}} \mathcal{K}(G_{\bar{T}}, D)$.

Theorem 3 (Necessary and Sufficient Condition). *For a graph $G = (V, E)$ and dealer D, CPA is t-locally resilient iff $\mathcal{M}(G, D, t) \geq t + 1$.*

Proof. ($\Leftarrow$) Assume $\mathcal{M}(G, D, t) \geq t + 1$ and let $T \subseteq V \setminus D$ be any t-local corruption set. It must hold that $\mathcal{K}(G_{\bar{T}}, D) \geq t + 1$. Hence, there exists a minimum $(t+1)$-level ordering $\mathcal{L}_{t+1}(G_{\bar{T}}, D) = \{L_1, \ldots, L_m\}$. Therefore every honest player v has at least $t + 1$ honest neighbors in previous levels of $\mathcal{L}_{t+1}(G_{\bar{T}}, D)$; by a simple induction we can show that v will decide on the dealer's value x_D.

($\Rightarrow$) If CPA is t-locally resilient then for any t-local corruption set, T, we have that every honest player in $G_{\bar{T}}$ decides on x_D and let the total number of rounds for the termination of the protocol is $m \in \mathbb{N}$. Define the sequence of sets $L_i = \{v \in V \setminus T \mid v$ decides in round i of CPA $\}, i \in \{1, \ldots, m\}$. Then we will show by induction that the sequence $(L_i)_{i=1}^m$ is the (unique) minimum $(t+1)$-level ordering on graph $G_{\bar{T}}$ with dealer D. Note first that $L_1 = \mathcal{N}(D) \setminus T$ because the players that decide in round 1 are exactly the neighborhood of the dealer. For the induction basis, we observe that $L_2 = \{v \in V \setminus T \mid \mathcal{N}(v) \cap L_1 \geq t + 1\}$ because the players that decide in round 2 are exactly those who will receive $t + 1$ identical messages from decided players in round 1. Assuming now that $L_k = \{v \in V \setminus \{T \cup \bigcup_{j=1}^{k-1} L_j\} : |\mathcal{N}(v) \cap \bigcup_{j=1}^{k-1} L_j| \geq t + 1\}$ it turns out that $L_{k+1} = \{v \in V \setminus \{T \cup \bigcup_{j=1}^{k} L_j\} : |\mathcal{N}(v) \cap \bigcup_{j=1}^{k} L_j| \geq t + 1\}$ due to the fact that the players that decide in round $k + 1$ are exactly the players who receive at least $t + 1$ messages from previously decided players. Since the above hold for any T, the claim follows. ☐

For exactly determining the maximum CPA resilience $t_{\max}^{\text{CPA}}$ we need the parameter,

$$\mathcal{T}(G, D) = \max\{t \in \mathbb{N} \mid \mathcal{M}(G, D, t) \geq t + 1\}$$

It should be clear by the above discussion that $\mathcal{T}(G, D)$ is exactly the maximum CPA resilience:

Corollary 3. $t_{\max}^{\text{CPA}} = \mathcal{T}(G, D)$

A simple algorithm to compute the t-safety threshold requires exponential time (consider all the t-local corruption sets and compute $\mathcal{K}(G_{\bar{T}}, D)$ as in Section 4). Note that a different necessary and sufficient condition for CPA to be t-locally resilient was independently given in [11]. However, a superexponential time to check that condition is implicit (no algorithm is given in [11]).

Moreover, for computing $t_{\max}^{\text{CPA}} = \mathcal{T}(G, D)$ it suffices to perform at most $\log \delta$ $\mathcal{M}(G, D, t)$ computations, where δ is the minimum degree of any node in $V \setminus (\mathcal{N}(D) \cup D)$.

6   Conclusions

In this paper we developed three new graph parameters, depending on graph G and dealer-node D, for bounding the maximum resilience $t_{\max}^{\text{CPA}}$ of CPA, i.e., the maximum number of corrupted players in each node's neighborhood that CPA can tolerate. The first parameter, $\mathcal{K}(G, D)$, can be efficiently computed and can be used for approximating $t_{\max}^{\text{CPA}}$ within a factor of 2. The t-safety threshold,

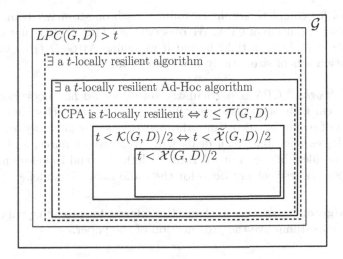

Fig. 2. Overview of conditions related to the existence of t-locally resilient algorithms. Parameters $LPC(G, D)$ and $\mathcal{X}(G, D)$ are defined in [10] and $\widetilde{\mathcal{X}}(G, D)$ is from [5]. Continuous lines show strict inclusions.

$\mathcal{M}(G, D, t)$, may be used as a test to check whether CPA is t-locally resilient for a certain graph G and dealer D. The third parameter, $\mathcal{T}(G, D)$, coincides with $t_{\max}^{\text{CPA}}$ and thus provides a necessary and sufficient condition for CPA to be t-locally resilient, namely $t \leq t_{\max}^{\text{CPA}} = \mathcal{T}(G, D)$. An overview of conditions (sufficient and/or necessary) for CPA resilience together with the corresponding classes of graphs is depicted in Figure 2.

Efficiency of Computing $t_{\max}^{\text{CPA}}$. A trivial approach to compute $t_{\max}^{\text{CPA}}$ via the parameter $\mathcal{T}(G, D)$ requires exponential time. It is therefore interesting to find an efficient algorithm or alternatively a hardness proof for this problem. Another direction would be to define another efficiently computable parameter yielding more tight bounds than $\mathcal{K}(G, D)$ in order to obtain a better approximation algorithm for $t_{\max}^{\text{CPA}}$.

CPA Uniqueness Conjecture. We believe that having a new necessary and sufficient condition in terms of the new parameter $\mathcal{T}(G, D)$ may help as a starting point for proving or disproving the *CPA uniqueness conjecture* of Pelc and Peleg [10] for *ad hoc* networks, which states that no algorithm can locally tolerate more traitors than CPA.

Dealer Corruption. It is well known that CPA works under the assumption that the dealer is honest. In order to address the case in which the dealer is corrupted one may observe that if the total number of traitors is strictly less than $n/3, n = |V|$, and the number of traitors in each node's neighborhood is bounded by $\mathcal{T}(G, D)$ then we can achieve Secure Broadcast by simulating any protocol for complete graphs as follows: each one-to-many transmission is replaced by an execution of CPA. We observe that $\mathcal{T}(G, D)$ may not be tight in this case. We can obtain a tight bound if we define $\mathcal{M}(G, D, t)$ by considering only corruption sets of size strictly less than $n/3$.

Wireless Networks. CPA is particularly suited for *ad hoc* networks, however it does not deal with radio network collisions. Only few articles have addressed the problem of secure broadcast in radio networks so far and only for restricted graph topologies (e.g. [6], which deals with Byzantine failures, and [7], which studies the problem in the fault-tolerant model). It would therefore make sense to develop locally resilient protocols for the radio network model.

Acknowledgment. We wish to thank the referees for providing valuable comments that helped improve the presentation of the paper.

References

1. Dolev, D.: The byzantine generals strike again. J. Algorithms 3(1), 14–30 (1982)
2. Dolev, D., Dwork, C., Waarts, O., Yung, M.: Perfectly secure message transmission. J. ACM 40(1), 17–47 (1993), http://doi.acm.org/10.1145/138027.138036
3. Franklin, M., Wright, R.N.: Secure communication in minimal connectivity models. Journal of Cryptology 13, 9–30 (2000), http://dx.doi.org/10.1007/s001459910002
4. Garay, J.A., Moses, Y.: Fully polynomial byzantine agreement for $n > 3t$ processors in $t + 1$ rounds. SIAM J. Comput. 27(1), 247–290 (1998)
5. Ichimura, A., Shigeno, M.: A new parameter for a broadcast algorithm with locally bounded byzantine faults. Inf. Process. Lett. 110(12-13), 514–517 (2010)
6. Koo, C.Y.: Broadcast in radio networks tolerating byzantine adversarial behavior. In: Chaudhuri, S., Kutten, S. (eds.) PODC, pp. 275–282. ACM (2004)
7. Kranakis, E., Krizanc, D., Pelc, A.: Fault-tolerant broadcasting in radio networks. Journal of Algorithms 39(1), 47–67 (2001), http://www.sciencedirect.com/science/article/pii/S0196677400911477
8. Kumar, M.V.N.A., Goundan, P.R., Srinathan, K., Rangan, C.P.: On perfectly secure communication over arbitrary networks. In: Proceedings of the Twenty-First Annual Symposium on Principles of Distributed Computing, PODC 2002, pp. 193–202. ACM, New York (2002), http://doi.acm.org/10.1145/571825.571858
9. Lamport, L., Shostak, R.E., Pease, M.C.: The byzantine generals problem. ACM Trans. Program. Lang. Syst. 4(3), 382–401 (1982)
10. Pelc, A., Peleg, D.: Broadcasting with locally bounded byzantine faults. Inf. Process. Lett. 93(3), 109–115 (2005)
11. Tseng, L., Vaidya, N.H., Bhandari, V.: Broadcast using certified propagation algorithm in presence of byzantine faults. CoRR abs/1209.4620 (2012)

Beacon-Less Mobility Assisted Energy Efficient Georouting in Energy Harvesting Actuator and Sensor Networks

Nathalie Mitton[1], Enrico Natalizio[2], and Riaan Wolhuter[3],*

[1] Inria, France
nathalie.mitton@inria.fr
[2] Heudiasyc Lab, UTC, France
enrico.natalizio@hds.utc.fr
[3] Stellenbosch University, South Africa
wolhuter@sun.ac.za

Abstract. In the next years, wireless sensor networks are expected to be more and more widely deployed. In order to increase their performance without increasing nodes' density, a solution is to add some actuators that have the ability to move. However, even actuators rely on batteries that are not expected to be replaced. In this paper, we introduce MEGAN (Mobility assisted Energy efficient Georouting in energy harvesting Actuator and sensor Networks), a beacon-less protocol that uses controlled mobility, and takes account of the energy consumption and the energy harvesting to select next hop. MEGAN aims at prolonging the overall network lifetime rather than reducing the energy consumption over a single path. When node s needs to send a message to the sink d, it first computes the "ideal" position of the forwarder node based on available and needed energy, and then broadcasts this data. Every node within the transmission range of s in the forward direction toward d will start a backoff timer. The backoff time is based on its available energy and on its distance from the ideal position. The first node whose backoff timer goes off is the forwarder node. This node informs its neighborhood and then moves toward the ideal position. If, on its route, it finds a good spot for energy harvesting, it will actually stop its movement and forward the original message by using MEGAN, which will run on all the intermediate nodes until the destination is reached. Simulations show that MEGAN reduces energy consumption up to 50% compared to algorithms where mobility and harvesting capabilities are not exploited.

1 Introduction

Wireless sensor networks (WSNs) are intended to be deployed in harsh environments. Therefore, it is expected that a large number of cheap sensor devices

---

* This work was partially supported by CPER NPC/FEDER CIA and the LIRIMA PREDNET.

J. Cichoń, M. Gȩbala, and M. Klonowski (Eds.): ADHOC-NOW 2013, LNCS 7960, pp. 281–292, 2013.

will be randomly scattered over a region of interest. These devices are powered by batteries and have limited processing and memory capabilities. When batteries deplete, sensors stop covering their area and being part of the underlying communication network. One solution is to add actuators able to move to areas where resources are most needed in order to efficiently route packets between two given nodes. This has been shown [21] that deploying mobile devices in a network can provide the same performance as increasing the nodes' density.

Energy efficient data routing strategies for WSNs have mainly aimed to increase network lifetime [8,13], but none of the most widespread solutions consider energy harvesting. Ambient energy harvesting as a power solution has gained momentum in recent years, especially with significant progress in the functionality of low power embedded electronics. By generating power from environmental energy, the dependency on batteries can be reduced or even eliminated [19].

In this work, we introduce MEGAN (Mobility assisted Energy efficient Georouting in energy harvesting Actuator and sensor Networks) that uses controlled mobility and takes account of the energy consumption (for sending data and moving) and the energy harvesting to select next hop. MEGAN aims at prolonging the overall network lifetime rather than reducing the energy consumption over a single path. Indeed, even if a path selection tries to optimize energy consumption like in [4, 8], the same nodes might be selected at each iteration causing a quick depletion of their batteries. In MEGAN, the selection of nodes is based on their current available energy, which could increase or decrease at any time due to energy harvesting. MEGAN has the following characteristics:

- **Localized:** a node is aware only of its location and that of the destination.
- **Memory-less:** no information has to be stored at the node or in the message.
- **Loop free:** nodes choose forwarder among the neighbors which reside in the forwarding direction toward the destination.
- **Beacon-less:** a node does not need to keep neighborhood tables up to date.
- **Energy efficient:** MEGAN optimizes the energy consumption along a routing path but also balances the remaining energy over nodes by taking into account the energy consumption for sending and moving, and the harvestable energy. Simulation results show that MEGAN reduces energy consumption up to 50% compared to some algorithms where mobility and harvesting are not exploited.

The rest of this paper is organized as follows. Section 2 reviews literature works. Notations and models used in this paper are detailed in Section 3. Our contribution is detailed in Section 4. Performance evaluation is conducted in Section 5. Finally, Section 6 concludes this work by presenting future works.

2   Related Work

In this section, we first discuss works about the optimal node placement and movement to allow optimal node positions. Then, we recall algorithms of geographic routing, and finally scan the literature considering energy harvesting.

In this work we are mainly interested in the energy efficiency of the node placement, even though other works focused on other network parameters. The optimal placement of a fixed number of nodes has been mathematically determined in the case of a monodirectional [15] and bidirectional [14] data flow to extend the path lifetime when nodes on the path have different residual energies. All the cited works consider the deployment of static nodes, whereas in our work nodes are mobile and can reach a different and possibly more convenient location in the field in terms of energy efficiency. The movement of nodes involves an energy expenditure that can negatively affect the overall energy budget, therefore algorithms and heuristics for energy-efficient movements have been proposed [7]. The main difference between our work and the mentioned works is that MEGAN bases the node selection on the energy expenditure for movement and transmission, as well as the harvestable energy.

Unlike MTPR [20] or MMBCR [16] which pursue similar objectives but are on-demand routing protocols (they require a route discovery step), MEGAN is a geographic routing protocol. The basic principle of geographic routing is that each node is aware of its position, the positions of its neighbors and that of the destination. The basic greedy routing has been extended to provide energy efficiency [11], to guarantee delivery [3] and to both provide energy efficiency and guarantee delivery [4]. Nevertheless, all these solutions require a neighbor discovery protocol proved to be very energy-consuming because of required message exchanges [2]. EBGRES [10] is beacon-less like BOSS [18] or BRAVE [1] but differs from these approaches since it takes benefits of controlled mobility. In EBGRES, each node sends out the data packet first. The neighbor selection is processed only among those neighbors that successfully received it. EBGRES uses a 3-way handshake, whereas MEGAN considers a 2-way handshake. In addition MEGAN adapts power transmission to save energy and reduce interferences. These works consider networks composed of static nodes. Current solutions for routing in mobile sensor networks adopt existing routing protocols to find an initial route, and iteratively move each node to the midpoint of its upstream and downstream nodes on the route. However, existing routing protocols may not be efficient. Moving strategy in [5] may cause useless zig-zag movements. In MobileCOP [12], next hop on the path is selected in a cost-over-progress (COP) [11] fashion and then moved to a straight line connecting the source to the destination. Once the first routing has succeeded, nodes on the path are moved and placed equidistantly on the line. Such move may disconnect the network, induces a memory overhead on nodes and a latency. CoMNET [8] was the first solution to propose considering the moving cost into the routing decision and to ensure that node mobility does not disconnect the network. Nevertheless, CoM-NET implies a neighborhood discovery, unlike MEGAN, and do not consider the energy that could be harvested. In addition, all these works aim at optimizing the energy consumption along a path and not the consumption of the whole network.

3   Notations and Models

3.1   Notation and Network Model

We model the network as a graph $G = (V, E)$ where V is the set of nodes, E is the set of edges, and $uv \in E$ if nodes u and v are in transmission range one of each other.

Let $N(u)$ be the neighborhood of node u, $N(u) = \{v \mid uv \in E\}$ and $N_D(u)$ the set of neighbors of node u in the forwarding direction toward the destination node D $(N_D(u) = \{v \in N(u) \mid \|\boldsymbol{v} - \boldsymbol{D}\| \leq \|\boldsymbol{u} - \boldsymbol{D}\|\}$ where u denotes the identity of node u when $\boldsymbol{u}$ is used for node u position. $\|\boldsymbol{u} - \boldsymbol{v}\|$ is the Euclidean distance between nodes u and v.

We assume that every node can adapt its transmission range between 0 and R_{max} by steps with regards to distance to be reached. Each node u has its own energy level E_u such that $0 \leq E_u \leq E_{max}$ where E_{max} is the maximum energy level that a node can have (same for all nodes). The energy level of node u, E_u evolves along time since it can decrease because of sending/receiving message and because of moving. It can also increase based on energy it can harvest from its environment. We will refer to the energy available at node u through the time variable t, $E_u(t)$. Note that MEGAN is model-independent but for the sake of evaluation, we use some common energy models computed as follows.

3.2   Energy Models

Energy Model for Communication. The most common energy model [17] is such that $E_{com}(d) = d^{\alpha} + c$ if $d \neq 0$ or 0 otherwise, where d is the distance separating two neighboring nodes; α is a real constant $(1 < \alpha)$ that represents the path loss; c is a distance-independent term that takes account of signal processing overhead at both the transmitter and the receiver (phase-locked loops, voltage-controlled oscillators, bias currents, etc.). As in [9] we assume that the energy consumption for this overhead is the same at both sides of the communication.

Energy Model for Movement. To the best of our knowledge, so far, there is no accurate model for defining the energy consumed to move nodes. Therefore, in the following, we use the model adopted in the literature [5, 12]: $E_{mov_u}(d) = ad$ where d is the covered distance by node u and a a constant to be defined. It only considers the *kinetic friction* that nodes have to win in order to move, the *static friction* can be considered simply by adding another constant value.

Energy Model for Harvesting. The amount of energy harvested from the environment can be very different from node to node due to the diversity of harvesters, the locations of the nodes, the deployment policy and the rate of harvesting, etc. The energy model used in this paper is for a solar based harvesting sensor node as defined in [10].

We denote $R_{harv_u}(\boldsymbol{u}, t)$ the rate of energy harvested by node u at time t in position $\boldsymbol{u}$. $R_{harv_u}(\boldsymbol{u}, t)$ is a deterministic value issued from a previous study of the environment. Node u can operate at time t if its residual energy is greater than the energy needed for communicating or moving at time t. If $R_{harv_u}(\boldsymbol{u}, t)$ is greater that the rate of the energy consumed $R_{cons_u}(t)$, then $R_{harv_u}(\boldsymbol{u}, t) - R_{cons_u}(t)$ could be wasted. In these cases we assume that the surplus energy is used to recharge the node battery or can be stored in a capacitor for a later use. In the rest of the paper we will refer to $R_{harv_u}(\overrightarrow{ug_u}, t)$ as the harvested energy measured by node u on the path between $\boldsymbol{u}$ and $\boldsymbol{g_u}$ and averaged over time t. We assume that at $t = 0$, the battery of the node u is completely charged $E_u = E_{max}$, and that the energy available for node u at any time is given by:

$$E_u(t) = \min\{E_u(t-1) + \int_{t-1}^{t} R_{harv_u}(\boldsymbol{u}, t)dt - \int_{t-1}^{t} R_{cons_u}(t)dt, E_{max}. \quad (1)$$

4 Contribution

The idea of MEGAN is to combine the benefits brought by energy harvesting and controlled mobility. MEGAN takes account of several energy components, *i.e.* the energy spent for sending a message, the energy spent for moving, the energy that can be harvested and the residual energy available at a node to take routing decision. We assume that the transmitter adjusts its transmitting power in order to deliver the minimum required power for a correct reception at the receiver. This allows energy saving. Unlike previous geographical protocols, MEGAN does not assume that its neighborhood is known *a priori* and thus does not rely in any neighborhood discovery scheme. The idea is the following. We illustrate MEGAN with Fig. 1 and the whole process is summed up by Algo. 1. We assume that source node S needs to send a packet to destination node D, the generic node u chosen by the routing protocol first estimates the optimal position g_u of the next forwarder. We will come back on the optimal position computing in the following section.

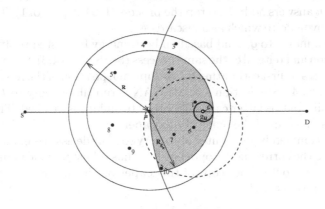

Fig. 1. Example of MEGAN

As a source node, node S runs Algo. 1 only from L. 8. Unlike EBGRES [10], MEGAN takes account of not only the sending energy but also the energy spent for moving, the energy that can be harvested and the residual energy available at node u at current time. Indeed, our goal is not only to have the minimum energy consumption over the routing path but to balance the remaining energy on nodes considering that some of them might harvest more energy than others.

Once g_u is computed, node u sends its message M containing g_u. u adjusts its range to save energy so that every node located in the circle of radius $R_{g_u} = \|u - g_u\| + \epsilon$ around u can be reached. Each generic node v that receives M runs Algo. 1. If v is located in the forwarding direction uD (Dashed area in Fig. 1, nodes 1, 2, 6, 7 and 10), it computes a backoff time T_v, otherwise, it discard M.

Node v prepares its ACK and waits the backoff time to transmit it. If during the waiting time, v receives an ACK from another node, it discards M (L5-9, Algo. 1). Otherwise, node v sends the ACK, to make all other waiting nodes discard M. The ACK is used to (i) avoid collisions, since every other candidate will discard the message and (ii) notify u that the greedy routing has succeeded. Node v is the selected forwarder node. T_v has been computed in such a way that if v is the first to answer (T_v is the smallest value), it means that v is the closest node to g_u (and thus is the one that needs less energy to move to g_u) or if there exists a node w closer to g_u, its residual energy after the movement would be smaller than the residual energy of v (See Sect. 4.1). Node u waits for an answer from other nodes for a time T_{max} calculated according to the furthest allowed position (R_{g_u}). If two nodes v and w, which are not in the transmitting range of each other, both transmit their ACK to u. Since the best forwarder node is the first one that sent the ACK (since it has the lowest backoff time), u advertises all other ones to discard M (L.18-26, Algo. 1).

For instance on Fig. 1, if node 6 is the first one to answer to u, it sends an ACK with range $\|u - 6\|$ (dash circle). Nodes 1 and 7 overhear the ACK and thus stop the process. But, this is not the case for nodes 2 and 10 that do not lie in the range of this ACK. Nodes 2 and 10 continue decreasing their back-off time and node 2 sends at its turn an ACK to node u. Since this is the second ACK received by u within T_{max}, u answers node 2 to stop the process (L.24-26, Algo. 1). This ACK is also received by node 10 which also discards M.

Node v then moves to g_u and based on best energy harvested position adjusts its final position and reiterates the same process (See Section 4.2) Node v reiterates the whole process by first estimating g_v, the optimal position of the next forwarder, by running Algo. 4. (See Section 4.3) MEGAN is run till delivery to D or till the greedy step fails. In this latter case, a recovery technique is invoked. The design of the recovery step is out of the scope of this paper.

MEGAN is completely distributed and every routing decision is made when needed with regards to the current node topology. It is thus robust to node failure. Dealing with multi-flows is out of the scope of this paper but in such cases, an extension of [6] could be investigated.

4.1   Back-Off Computing

The back-off time should be computed such that the node with more residual energy after the movement to g_u be the first node that answers to the routing request of node u. Thus, the back-off computation needs to take account of the residual energy of v at the current time t, $E_v(t)$, and the energy that the same node will spend for the movement E_{mov_v}, in order to compute its residual energy after the movement, $E_v(t + k)$, where k is the number of time units spent to move to the new location. The energy spent for the movement is computed as detailed

---

Algorithm 1. MEGAN (v, D, g_u) - Run at node v upon reception of a message from u to D for position g_u.

1: **If** $v \notin N_d(u)$ **then**{Discard Message} Exit **end if**
2: $t \leftarrow$ ComputeBackOff(g_u)
3: **while** $t - -$ **do**
4:    **If** v receives ACK from another node **then** Exit **end if**
5: **end while**
6: Send ACK and Move to g_u and Measure $R_{harv_v}(\overrightarrow{vg_u}, t)$
7: AdjustPosition$(R_{harv_v}(g_u, t))$
8: $g_v \leftarrow$ ComputeOptimalPosition(v, D)
9: $Send(D, g_v)$
10: $t \leftarrow T_{max}; OK \leftarrow$ FALSE
11: **while** $t - -$ **do**
12:    **If** OK=FALSE $\wedge$ v receives ACK **then** OK$\leftarrow$ TRUE **end if**
13:    **if** OK=TRUE AND v receives ACK from w **then**
14:       Advertise w to discard; {Better node than w is selected, w must give up.}
15:    **end if**
16: **end while**
17: **If** OK=FALSE **then** Triggers recovery process **end if**

---

in Sec. 3. Once the future residual energy is computed such that $E_v(t + k) = E_v(t) - E_{mov_v}$, the back-off time, for the node to wait before sending its ACK, is simply the ratio between the maximum energy level E_{max} and this value. In fact, since $0 < E_v(t + k) \leq E_{max}$ then the back-off time will be $1 \leq \frac{E_{max}}{E_v(t+k)} < \infty$ [ms]. Nodes that calculate a back-off larger than a given threshold can just discard the routing request and avoid replying to u. Let us assume on Fig. 1 that node u sends its routing request in a radius R_{g_u}. All nodes v, such that $||u - v||$, (nodes $1, 2, 5, 6, 7, 8, 9, 10$) receive it. Only nodes in the forwarding direction of D, (nodes $1, 2, 6, 7, 10$) triggers a back-off time. Let the following values for the energies at time t: $E_1(t) = 4, E_6(t) = 7, E_{com}(||1-g_u||) = 2$ and $E_{com}(||6-g_u||) = 4$. After k time units needed for the physical movement, we would have $E_1(t+k) = 4-2 = 2$ and $E_6(t + k) = 7 - 4 = 3$. So, even if 6 is further to g_u than 1, it will be chosen. This favors the balance of energy consumption over nodes.

Algorithm 2. ComputeBackOff(g_u) - Run at node v.

1: Compute $E_v(t + k) \leftarrow E_v(t) - E_{mov_v}(||v - g_u||)$
2: $T_v \leftarrow \frac{E_{max}}{E_v(t+k)}$
3: Return T_v

4.2 Position Adjustment

When node v (that is in position v before the movement) moves to the estimated optimal position g_u, it can instantaneously measure the energy harvesting rate in its new position. During its movement, it has retrieved information about the

$R_{harv_v}(\overrightarrow{vg_u}, t)$ in all the intermediate positions between its original position v and position g_u. Through all this information, when v reaches g_u, it can decide to move of ϵ in one of the following four directions (as shown on Fig. 2):

1. $\overrightarrow{g_u u}$, i.e. toward u;
2. $\overrightarrow{ug_u}$, i.e. against u;
3. $\overrightarrow{vg_u}$, i.e. it keeps going in the current travel direction;
4. $\overrightarrow{g_u v}$, i.e. it backtracks.

Fig. 2. Position adjustment

Intuitively, in ideal conditions of node density and residual energy, $\overrightarrow{ug_u}$ would be directed as the vector $\overrightarrow{SD}$ and $\overrightarrow{vg_u}$ would be perpendicular to $\overrightarrow{SD}$. Therefore, the four mentioned directions would represent the extreme points of two perpendicular diameters of the circle having g_u as the center and ϵ as the radius, as shown in Fig. 1. This means that the position adjustment algorithm, in ideal conditions, can take $360°$ around g_u into account. Direction 1 is chosen when the energy harvested in u is significantly larger than that harvested in g_u and the gradient of the R_{harv_v} along $\overrightarrow{vg_u}$ is not significantly different from 0. Direction 2 is chosen when the energy harvested in u is significantly smaller than that harvested in g_u and the gradient of the R_{harv_v} along $\overrightarrow{vg_u}$ is not significantly different from 0. Direction 3 is chosen when the gradient of the R_{harv_v} along $\overrightarrow{vg_u}$ is significantly positive and the energy harvested in u is not significantly different from that harvested in g_u. Direction 4 is chosen when the gradient of the R_{harv_v} along $\overrightarrow{vg_u}$ is significantly negative and the energy harvested in u is not significantly different from that harvested in g_u. For the fourth case, a further optimization would make node v stop its movement in the direction $\overrightarrow{vg_u}$, instead of making it reach g_u and then backtracks.

When the gradient of the R_{harv_v} along $\overrightarrow{vg_u}$ is zero and the energy harvested in u is not significantly different from that harvested in g_u, then v does not move from g_u. On the contrary if both the gradient and the mentioned difference are significantly high then the node will move simultaneously along both the axes. In order to quantify the differences between the energy harvested along the direction $\overrightarrow{SD}$ and that perpendicular to $\overrightarrow{SD}$ we introduce two threshold values, H_{par} and H_{per}, respectively. For sake of simplicity in Algo. 3 we omitted the time variable.

4.3   Computing of Optimal Position for Forwarder Node

In order to estimate the optimal position for next node in the path towards the destination node D, we assume that node v is aware of E_{max}, its energy at current time t, $E_v(t)$, its energy consuming and harvesting rates, R_{cons_v} and R_{harv_v}, respectively.

Algorithm 3. AdjustPosition($R_{harv_v}(\boldsymbol{g_u}, t)$) - Run at node v.

1: Move of:
2: $(|R_{harv_v}(\boldsymbol{u})\text{-}R_{harv_v}(\boldsymbol{g_u})| > H_{par}) \cdot \epsilon \cdot \frac{R_{harv_v}(\boldsymbol{u})-R_{harv_v}(\boldsymbol{g_u})}{|R_{harv_v}(\boldsymbol{u})-R_{harv_v}(\boldsymbol{g_u})|}$ along $\overrightarrow{ug_u}$
3: $(|\text{Grad}(R_{harv_v}(\overrightarrow{vg_u}))| > H_{per}) \cdot \epsilon \cdot \frac{\text{Grad}(R_{harv_v}(\overrightarrow{vg_u}))}{|\text{Grad}(R_{harv_v}(\overrightarrow{vg_u}))|}$ along $\overrightarrow{vg_u}$

By multiplying the energy consuming and harvesting rates for a certain number of Δt time units, k, node v can estimate its future residual energy, $E_v(t + k\Delta t)$, available after k time units. If this value does not decrease in respect of the current residual energy $(E_v(t + k\Delta t) \geq E_v(t))$, then the ideal position, computed from the optimal transmission radius, is given by:

$$h_v = v + d^* \cdot \frac{D - S}{\|D - S\|} \tag{2}$$

In this case, h_v is also the optimal position $\boldsymbol{g_v}$ for next relay ($\boldsymbol{g_v} = h_v$). Otherwise (if $E_v(t+k\Delta t) \leq E_v(t)$)) an intermediate position will be considered as the estimated optimal position. Notice that several approaches could be considered in order to estimate next forwarder optimal position, depending on the criteria of optimality. The definition of the criteria of optimality is out of the scope of this work. Hence, we introduce two simple approaches depicted in Algo. 4 based on energy computation: a *conservative* approach that keeps the forwarder node close to $\boldsymbol{v}$, and an *optimistic* that pushes it close to position $vectg_u$ when $E_v(t + k) \geq E_v(t)$). We thus have:

$$\boldsymbol{g_v} = \begin{cases} (\frac{E_v(t)-E_v(t+k\Delta t)}{E_{max}}) \cdot h_v & \text{if conservative approach} \\ (1 - \frac{E_v(t)-E_v(t+k\Delta t)}{E_{max}}) \cdot h_v & \text{if optimistic approach} \end{cases} \tag{3}$$

Algorithm 4. ComputeOptimalPosition($v, \boldsymbol{D}$) - Run at node v.

1: Measure $E_v(t)$, $R_{cons_v}(\Delta t)$ and $R_{harv_v}(\Delta t)$
2: $E_v(t + k\Delta t) \leftarrow E_v(t) + R_{harv_v}(k\Delta t) - R_{cons_v}(k\Delta t)$, $\boldsymbol{h_v} \leftarrow \boldsymbol{v} + d^* \cdot \frac{D-S}{\|D-S\|}$
3: **If** $E_v(t + k\Delta t) \geq E_v(t)$ **then** $g_v \leftarrow h_v$
4: **else** $g_v \leftarrow \frac{E_v(t+k\Delta t)}{E_{max_v}} \cdot h_v$ (*conservative approach*)
   or: $g_v \leftarrow (1 - \frac{E_v(t)-E_v(t+k\Delta t)}{E_{max}}) \cdot h_v$ (*optimistic approach*)
5: **end if**
6: Return g_v

---

5   Simulation Results

To evaluate the performance of MEGAN, we run the simulation in OMNET++ with a CSMA/CA MAC layer. We assume a free space propagation model and packet collisions to carry out the tests, and that when a node dies it cannot be resuscitated. Sensor nodes are uniformly spread throughout a square area of 1000 ×

$1000m^2$ and they can adapt their transmission range between 0 and 100m. Only connected networks are considered in the results. The DATA message payload is set to 128 bytes, the size of the ACK message is 25 bytes. We use the cost models described in Section 3 with the propagation loss exponent α set to 2 and $c = 50nJ/bit$. We use two values of a for the cost model related to movement : 0.1 and 10, to make the energy to send prevail over the energy to move and vice versa, respectively and we simulate two scenarios accordingly. Each sensor node is initialized with 2J of battery energy.

The transmission range depends on the energy available at the node and the transmission delay . We are mainly interested in evaluating the total energy consumption for sensor-to-sink packet delivery independently from the MAC layer used. We conducted 50 replications of each simulation scenario to obtain statistically significant output. In each replication, each of the 20 nodes, which are randomly chosen as sources, generates 40 data packets to be transmitted to a randomly chosen destination. The simulation terminates when all the data packets generated in the network are delivered. We study the evolution of the cumulative energy spent to send (and receive) all the data and control packets in the network, when both the transmission range and the nodes' density vary.

We compare MEGAN with EBGRES and COMNET, because those are the two conceptually closest algorithms to MEGAN, which is both a beacon-less energy harvesting routing scheme, as EBGRES, and a mobility based routing scheme, as COMNET.

Energy Consumption versus Transmission Range. Fig. 3 shows the total energy expenditure of the network for both the scenarios (sending cost higher than moving cost, and vice versa), when the transmission range varies. In both cases, MEGAN outperforms the other protocols. We can see that, for all the algorithms, the energy expenditure decreases exponentially when the transmission range increases, because a smaller number of intermediate nodes is needed to cover the source-destination path.

Both MEGAN and COMNET are mobility based algorithms, thus they perform best in cases that favor mobility (Fig. 3 (a)). MEGAN consumes less energy than COMNET because it is a beacon-less approach and has less message handshakes needed to set up the transmission path. Furthermore, in comparison with MEGAN, COMNET uses a flooding approach for neighbor discovery and is not reactive. When sending data is more convenient than moving nodes (Fig. 3 (b)), COMNET consumes too much energy to move to the optimal position, whereas MEGAN moves in a more intelligent way and saves enough energy to perform better than EBGRES to send the same amount of data.

Energy Consumption versus Nodes' Density. Fig. 4 shows the total energy expenditure of the network for both the scenarios, when the transmission range is set to 100 m, and the nodes' density varies. We can see that, for all the algorithms, the energy expenditure decreases linearly when the nodes' density increases, because, for all the simulated algorithms, the selection procedure has more chances to find a relay node close to the optimal position. We can see from Fig. 4 (a) that when sending data is more costly than moving nodes, EBGRES and COMNET perform

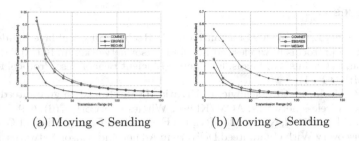

(a) Moving < Sending   (b) Moving > Sending

Fig. 3. Cumulative energy consumption versus transmission range

almost identically. Whereas, from Fig. 4 (b), we can see that when moving nodes is more costly than sending data EBGRES performs better than COMNET. However, in both the scenarios, MEGAN spends less energy than the other protocols, due to the mentioned reasons of using mobility in a smarter way than COMNET and of requiring less control messages than EBGRES as well as of finding better positions than EBGRES to forward data. An interesting remark is that when moving nodes is "expensive" the energy saved by finding a better position in MEGAN for each intermediate node is compensated by the additional energy needed to move to that position. In fact, EBGRES and MEGAN almost overlap in this scenario.

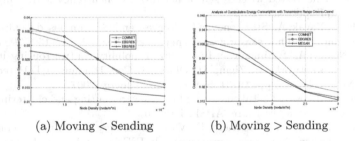

(a) Moving < Sending   (b) Moving > Sending

Fig. 4. Cumulative energy consumption versus node density

6   Conclusion and Future Works

In this paper, we introduced MEGAN, the very first beacon-less geographic routing that takes advantage of controlled mobility and energy harvested from the environment to dynamically adapt the energy consumption. Simulation results show that MEGAN outperforms literature protocols by reducing up to 50% the energy consumption under some parameters.

As a next step, we intend to investigate recovery mechanisms that also takes account of these hardware characteristics to improve the routing performances.

References

1. Stojmenovic, I., Khan, A., Constantinou, C.: Realistic physical layer modelling for georouting protocols in wireless ad-hoc and sensor networks. In: ICUMT (2009)
2. Amadou, I., Chelius, G., Valois, F.: Energy-Efficient Beacon-less Protocol for WSN. In: PIMRC (2011)
3. Bose, P., Morin, P., Stojmenovic, I., Urrutia, J.: Routing with guaranteed delivery in ad-hoc wireless networks. ACM/Kluwer Wireless Networks 7(6), 609–616 (2001)
4. Elhafsi, E.H., Mitton, N., Simplot-Ryl, D.: End-to-End Energy Efficient Geographic Path Discovery With Guaranteed Delivery in Ad hoc and Sensor Networks. In: IEEE PIMRC (2008)
5. Goldenberg, D.K., Lin, J., Morse, A.S.: Towards mobility as a network control primitive. In: Mobihoc, pp. 163–174 (2004)
6. Gouvy, N., Hamouda, E., Zorbas, D., Mitton, N.: Energy Efficient Multi-Flow Routing in Mobile Sensor Networks. In: IEEE WCNC, Shanghai, China (2013)
7. Guerriero, F., Violi, A., Natalizio, E., Loscri, V., Costanzo, C.: Modelling and solving optimal placement problems in wireless sensor networks. Applied Mathematical Modelling 35(1) (2011)
8. Hamouda, E., Mitton, N., Simplot-Ryl, D.: Energy efficient mobile routing in actuator and sensor networks with connectivity preservation. In: Frey, H., Li, X., Ruehrup, S. (eds.) ADHOC-NOW 2011. LNCS, vol. 6811, pp. 15–28. Springer, Heidelberg (2011)
9. Heinzelman, W.B., Chandrakasan, A.P., Balakrishnan, H.: An application-specific protocol architecture for wireless microsensor networks. IEEE Trans. on Wireless Com. 1(4) (2002)
10. Jumira, O., Wolhuter, R., Zeadally, S.: Energy-efficient beacon-less geographic routing in energy harvested wireless sensor networks. Concurrency and Computation: Practice and Experience (2012)
11. Kuruvila, J., Nayak, A., Stojmenovic, I.: Progress and location based localized power aware routing for ad hoc sensor wireless networks. IJDSN 2, 147–159 (2006)
12. Liu, H., Nayak, A., Stojmenović, I.: Localized mobility control routing in robotic sensor wireless networks. In: Zhang, H., Olariu, S., Cao, J., Johnson, D.B. (eds.) MSN 2007. LNCS, vol. 4864, pp. 19–31. Springer, Heidelberg (2007)
13. Momma, S., Mikoshi, T., Takenaka, T.: Power aware routing and clustering scheme for wireless sensor networks. In: APSITT (2010)
14. Natalizio, E., Loscri, V., Guerriero, F., Violi, A.: Energy spaced placement for bidirectional data flows in wireless sensor network. IEEE Com. Letters 13(1) (2009)
15. Natalizio, E., Loscri, V., Viterbo, E.: Optimal placement of wireless nodes for maximizing network lifetime. IEEE Com. Letters 12(5) (2008)
16. Nishat, H., Srinivasa Rao, D., Balaswamy, C.: Energy efficient routing protocols for mobile adhoc networks. Int. Journal on Computer Applications 26(2) (2011)
17. Rodoplu, V., Meng, T.: Minimizing energy mobile wireless networks. IEEE JSAC 17(8), 1333–1347 (1999)
18. Sánchez, J.A., Marín-Pérez, R., Ruiz, P.M.: Boss: Beacon-less on demand strategy for geographic routing in wireless sensor networks. In: MASS (2007)
19. Vigorito, C.M., Ganesan, D., Barto, A.G.: Adaptive control of duty cycling in energy harvesting wireless sensor networks. In: SECON (2007)
20. Vir, D., Agarwal, S.K., Imam, S.A., Mohan, L.: Performance analysis of MPTR routing protocol in power deficient node. Int. Journal on AdHoc Networking Systems (IJANS) 2(4) (2012)
21. Wang, W., Srinivasan, V., Chua, K.-C.: Extending the lifetime of wireless sensor networks through mobile relays. IEEE/ACM Trans. Netw. 16(5), 1108–1120 (2008)

Localization of Submerged Sensors Using Radio and Acoustic Signals with Single Beacon

Anisur Rahman, Vallipuram Muthukkumarasamy, and Elankayer Sithirasenan

School of Information and Communication Technology, Griffith University
Gold Coast, QLD 4222, Australia
{anis.rahman,v.muthu,e.sithirasenan}@griffith.edu.au

Abstract. This paper proposes a simple mechanism to accurately determine the distance between nodes using radio and acoustic signals in underwater wireless sensor networks (UWSN). It also delineates a new method of determining the coordinates of sensors with a single beacon. As the knowledge of precise coordinates of the sensors is as important as the collected data in UWSN, the accurate distance measurement between the nodes is the prime factor for better accuracy. Mostly, in range free method, received signal strength indicator (RSSI) is used to determine the propagation loss from which the inter nodes distances are calculated. However, in underwater, RSSI of acoustic signal is heavily affected by multipath fading that eventually leads to erroneous coordinates. The proposed mathematical model of coordinate-determination has better immunity from multipath fading as well as synchronization in distance determination resulting precise location of the sensors. Moreover, a single beacon is used to determine the coordinates of the sensor nodes where none of them has *a priori* knowledge about its location.

Keywords: localization, submerged sensors, radio, acoustic, coordinates.

1 Introduction

UWSN are envisioned to enable applications for oceanographic data collection, pollution monitoring, offshore exploration, disaster prevention, assisted navigation and tactical surveillance applications [1]. Despite varieties of application of UWSN, idea of submerged wireless communication may still seem far-fetched and got attention of researchers since last decades. As positions of submerged sensors play a vital role for the significance of validity of data, determining the coordinate precisely is very crucial. There are many distance measurement techniques for terrestrial applications based on signal strength but those algorithms are not suitable for underwater environment because of numerous factors associated with it including multipath effects.

However, in the recent years, there has been a growing interest to explore and fulfil the needs of a multitude of underwater applications. In addition to underwater sensors, the network may also comprise surface stations and autonomous underwater vehicles and regardless of the type of deployment, the location of sensors needs to be determined for meaningful interpretation of the sensed data [2]. In UWSNs, acoustic

J. Cichoń, M. Gębala, and M. Klonowski (Eds.): ADHOC-NOW 2013, LNCS 7960, pp. 293–304, 2013.

channels are naturally employed, and range measurement using acoustic signals are much more accurate than using radio [3, 4]. However, multipath phenomenon and time synchronization have been the most challenging factors.

In this paper, we propose to use both radio and acoustic signals; radio's limited propagation restricts our problem domain for few hundred meters depth as carefully chosen frequency can propagate up to 323m [5]. As radio signal's speed in underwater is many times faster than sound speed, we have utilized this merit of radio signal for synchronization the beacon and sensor nodes.

In [6], Duff and Muller proposed a method to solve the multilateration equations by means of nonlinear least square optimization when no positions are known. The algorithm is based on a degree-of-freedom analysis – which says enough measurement from different positions will provide enough equations to solve the problem. In [7], same technique is used incorporating extended Kalman filter. However, the degree-of- freedom analysis does not guarantee a unique solution in a system of nonlinear equations, such as trilateration, when the only data available is the distance measured between the nodes [8]. In [9], Guevara et al. introduced a new closed-form solution where no position information of any nodes is required to determine the positions of multiple static beacon nodes, the only information they used is the distance measurement between static beacons and mobile node.

Having analyzed the various studies discussed above, in this paper we propose a closed-form solution to determine the coordinates of the submerged sensors having only one beacon node at the surface using both radio and acoustic signals. The precise conditions for obtaining initial subsets of nodes were justified using rigidity theory in [8]. Section 2 describes solvable configuration and multipath and Section 3 explains the proposed distance measurement technique. Section 4 explains the proposed theoretical mechanism to determine the sensors coordinates. Analysis is given in Section 5. Simulation results and discussions are reported in Section 6 and finally conclusions in Section 7.

2   Solvable Configuration and Multipath

2.1   Problem Domain

In the proposed method we need at least three sensors and a floating beacon; a boat or a buoy can be used as a beacon and sensors are deployed at the bottom of the water. While measuring the distances between the beacon and sensors using radio and acoustic signals as described in Section 3, it is assumed that sensors and beacon are in parallel plane. For the sake of simplicity, in this paper we also assume that the submerged sensors are static for the time of computation – the time that is required to measure the distances from different positions of the beacon. A solvable configuration of one beacon with three submerged sensors is denoted in Fig. 2. Our proposed mechanism exploits the advantage of both radio and acoustic signal propagation;   in sea water radio can propagate as far as 323m [5]. Since most of the marine explorations take place in shallow water, our proposed model has wide ranging practical applications.

2.2   The Effect of Multipath Fading

Multipath formation in the ocean is governed by two factors: sound reflection at the surface and bottom, and sound refraction in the water. The latter is a consequence of sound speed variation with depth, which is mostly evident in deep water channels.

Typical frequencies associated with underwater acoustics are between 10 Hz and 1 MHz. The propagation of sound in the ocean at frequencies lower than 10 Hz is usually not possible without penetrating deep into the seabed, whereas frequencies above 1 MHz are rarely used because they are absorbed very quickly. So for any appropriate acoustic frequency that is used in underwater to measure the propagation loss is affected by multipath fading. Fig. 1 shows the obvious present of multipath fading; the impulse response of an acoustic channel is influenced by the geometry of the channel and its reflection properties, which determine the number of significant propagation paths, their relative strengths and delays [10] . Strictly speaking, as there are infinitely many signal echoes, the resultant signal at any point would be $\sum_{i=0}^{\infty} f_i$; but those that have undergone multiple reflections and lost much of the energy can be discarded, leaving only a finite number of significant paths.

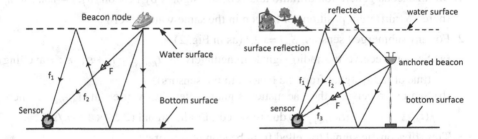

Fig. 1. Multipath for different configurations

However, in our model, the multipath fading phenomenon is minimal because arrival time difference of radio and acoustic signals is used for distance computation, not the signal strength. Shortest Euclidean distance will be travelled earliest and that is our focal point. Even though the radio signal does not propagate as of acoustic signal in underwater, but the right frequency of radio signal will give certain distance of propagation, which is enough to cover the shallow water [5]. The tremendous speed of radio signal compared to that of acoustic has been used for timing purposes. Section 3 elaborates on the mechanism in details.

3   Proposed Distance Determination

Despite the underwater limitations of both radio and acoustic signals, we propose to use each of its merit in our method to minimize the distance measurement error. In our method, radio signal will be used to measure the flight time of the acoustic signal. Even though the speed of radio signal is little less than that of in the vacuum, considering the problem domain, the reduced speed will not affect much of the proposed localization method. Moreover, the speed of acoustic signal, which varies due to different factors, is the only variable that we need to use for coordinates determination.

3.1   Determination of Flight Time of Acoustic Signal

Assumptions:

- The beacon can generate radio and acoustic signals simultaneously.
- The environmental factors that affect the acoustic signal will be considered while measuring inter-node distances.
- Sensor nodes are stationary during the short measurement period.
- Beacon (boat or buoy - at the water level) and sensor nodes (at the bottom) are in parallel plane.
- Each sensor node has a unique ID.

Steps:

1. Simultaneous generation of radio and acoustic signals by beacon $S_j, j = 4,5....$ at t_0 (here, S_j :different positions of beacon in the same water level)
2. For any submerged sensors $S_i, i = 1,2,3$ (as in Fig. 2)
   (a) Sensors receive the radio signal immediately at $t_{Ra(rec)} = t_0 + \varepsilon$ (ε : travelling time of radio signal from the beacon to the sensors)
   (b) Sensor receives the acoustic signal after a while at $t_{Ac(rec)}$; here $(t_{Ac(rec)} - t_0) \gg (t_{Ra(rec)} - t_0)$ due to speed of radio signal ($2.25 \times 10^6 m/s$).
3. Time of acoustic signal travelled from beacon to sensors:
$$T_{ij(Travel), i=1,2,3; j=4,5,6....} = t_{Ac(rec)} - t_{Ac(tra)} = t_{Ac(rec)} - t_{Ra(tra)} \quad \because t_{Ac(tra)} = t_{Ra(tra)}$$
$$\therefore T_{ij(Travel)} \approx t_{Ac(rec)} - t_{Ra(rec)} \quad \because t_{Ra(rec)} = t_0 + \varepsilon \approx t_{Ra(tra)}$$
4. Sensor nodes send back the time $T_{ij(Travel)}$ with individual sensor's ID back to the beacon using acoustic signal.
5. Beacon node computes the distance between the beacon and sensors: $d_{ij} = v_A \times T_{ij(travel)}$ (here, v_A is average speed of acoustic signal for the water column)

3.2   Determination of Average Speed of Acoustic Signal

A typical speed of acoustic waves near the ocean surface is about 1500m/s, more than four times faster than the speed of sound in air. However, the speed of sound in underwater is affected prominently by temperature, salinity and depth. As these parameters are variable and different from place to place in the water; so as the speed. The speed of sound v at any point in water can be calculated according to the following Mackenzie equation [11].

$$v = 1448.96 + 4.591T - 5.304 \times 10^{-2}T^2 + 2.374 \times 10^{-4}T^3 + 1.340 (S-35) + 1.630 \times 10^{-2}D + 1.675 \times 10^{-7}D^2 - 1.025 \times 10^{-2}T(S - 35) - 7.139 \times 10^{-13}TD^3 = f(T, D, S)$$

where T, D and S are temperature, depth and salinity and constrains for these variables are 2-30°C, 0-8000m and 25-40ppt respectively. Mackenzie equation gives
sound speed for any specific value of T, D and S whereas for our problem domain we
need average speed of sound through the water column where all three variables
change dynamically. It should be noted that in shallow water temperature predomi-
nates whereas depth and salinity have very minimal effect on sound speed as we can
see in simulation. Here, the pressure at the bottom by the sensors will be converted
to depth D following equation stated in [12].

Our proposed average speed of sound from the surface to bottom for the problem
domain can be calculated as (1). (A : area created by the limit of T, D and S)

$$\bar{v} = f_{avg}(T,D,S) = \frac{1}{A} \int\limits_{S_i}^{S_f} \int\limits_{0}^{D_f} \int\limits_{T_i}^{T_f} f(T,D,S)dTdDdS \tag{1}$$

4   Coordinates Computation

The objective of localization algorithms is to obtain the coordinates of all the sensors.
Only values available here to compute is the distance measurements and typically it is
considered as optimization problem where objective functions to be minimized have
residuals of the distance equations. The variables of any localization problem is the coordinates of the nodes; in principle more distance equations than number of variables are
required to solve this kind of problem. However, this approach known as degree-of-
freedom analysis may not guarantee the unique solution in a nonlinear system.

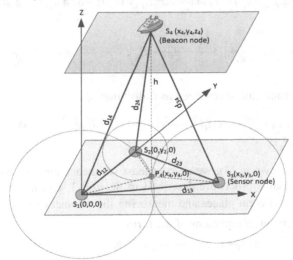

Fig. 2. Subset of nodes and coordinates determination

Trilateration or multilateration techniques are nonlinear system usually used to
determine location or coordinates of the sensors in partial or full. According to

Guevara et al. [13] the convergence of optimization algorithms and Bayesian methods depend heavily on initial conditions used and they circumvent the convergence problem by linearizing the trilateration equations.

Fig. 2 shows the initial subset composed of the beacon node $S_j, j=4$ and three sensor nodes $S_i, i = 1,2,3$. Without loss of generality, a coordinate system can be defined using one of the sensor nodes $S_i, i = 1,2,3$ as the origin $(0,0,0)$ of the coordinate system. Now the trilateration equations can be written as a function of two groups of distance measurements: the distance between beacon and sensors d_{14}, d_{24}, d_{34}.... which are measured data, and inter-sensor distances d_{12}, d_{13}, d_{23} and the volume of tetrahedron V_t (formed by the beacon and sensors), which are unknown.

Based on the local positioning system (LPS) configuration of Fig. 2, we need to write equation that will includes all known and unknown distances. For that matter, we express the volume of tetrahedron V_t using Cayley-Menger determinant as following:

$$288 \, V_t^2 = \begin{vmatrix} 0 & 1 & 1 & 1 & 1 \\ 1 & 0 & d_{12}^2 & d_{13}^2 & d_{14}^2 \\ 1 & d_{12}^2 & 0 & d_{23}^2 & d_{24}^2 \\ 1 & d_{13}^2 & d_{23}^2 & 0 & d_{34}^2 \\ 1 & d_{14}^2 & d_{24}^2 & d_{34}^2 & 0 \end{vmatrix} \tag{2}$$

By expanding and grouping known-unknown variables of (2), we get

$$d_{34}^2\left(d_{12}^2 - d_{23}^2 - d_{13}^2\right) + d_{14}^2\left(\frac{d_{23}^4}{d_{12}^2} - d_{23}^2 - \frac{d_{13}^2 d_{23}^2}{d_{12}^2}\right) + d_{24}^2\left(\frac{d_{13}^4}{d_{12}^2} - \frac{d_{13}^2 d_{23}^2}{d_{12}^2} - d_{13}^2\right) - \left(d_{14}^2 d_{24}^2 + d_{14}^2 d_{34}^2 - d_{24}^2 d_{34}^2 - d_{14}^4\right)\frac{d_{23}^2}{d_{12}^2}$$

$$- \left(d_{34}^2 d_{24}^2 - d_{14}^2 d_{34}^2 + d_{14}^2 d_{24}^2 - d_{24}^4\right)\frac{d_{13}^2}{d_{12}^2} + \left(144\frac{V_t^2}{d_{12}^2} + d_{13}^2 d_{23}^2\right) = \left(d_{24}^2 d_{34}^2 - d_{34}^4 + d_{14}^2 d_{34}^2 - d_{14}^2 d_{24}^2\right)$$

Here,

$\left(d_{12}^2 - d_{23}^2 + d_{13}^2\right)$, $\left(\frac{d_{23}^2}{d_{12}^2} - d_{23}^2 - \frac{d_{13}^2 d_{23}^2}{d_{12}^2}\right)$, $\left(\frac{d_{13}^4}{d_{12}^2} - \frac{d_{13}^2 d_{23}^2}{d_{12}^2} - d_{13}^2\right)$, $\frac{d_{23}^2}{d_{12}^2}$, $\frac{d_{13}^2}{d_{12}^2}$ and $\left(144\frac{V_t^2}{d_{12}^2} + d_{13}^2 d_{23}^2\right)$ are

unknown terms.

The above expansion can be rewritten as following:

$$d_{14}^2 X_1 + d_{24}^2 X_2 + d_{34}^2 X_3 - \left(d_{14}^2 - d_{34}^2\right)\left(d_{24}^2 - d_{14}^2\right)X_4 - \left(d_{24}^2 - d_{14}^2\right)\left(d_{34}^2 - d_{24}^2\right)X_5 + X_6$$
$$= \left(d_{24}^2 - d_{34}^2\right)\left(d_{34}^2 - d_{14}^2\right) \tag{3}$$

As we have six unknown in (3), we need at least six measurements, which could be done following the same procedure described in Section 3 steering the beacon node S_j in six $j = 4,...,9$ different places and measuring the distances in the vicinity of S_4. Eventually we get m linear equations of the form

$$a_{11}x_1 + a_{12}x_2 + \cdots + a_{1n}x_n = b_1,$$
$$a_{21}x_1 + a_{22}x_2 + \cdots + a_{2n}x_n = b_2,$$
$$\vdots$$
$$a_{m1}x_1 + a_{m2}x_2 + \cdots + a_{mn}x_n = b_m, \tag{4}$$

If we omit reference to the variables, then system (4) can be represented by an array of all coefficients known as the augmented matrix where the first row represents the first linear equation and so on; that could be expressed in $AX = b$ linear form.

After doing so for our system:

$$A = \begin{bmatrix} d_{14}^2 & d_{24}^2 & d_{34}^2 & -\left(d_{14}^2 - d_{34}^2\right)\left(d_{24}^2 - d_{14}^2\right) & -\left(d_{24}^2 - d_{14}^2\right)\left(d_{34}^2 - d_{24}^2\right) & 1 \\ d_{15}^2 & d_{25}^2 & d_{35}^2 & -\left(d_{15}^2 - d_{35}^2\right)\left(d_{25}^2 - d_{15}^2\right) & -\left(d_{25}^2 - d_{15}^2\right)\left(d_{35}^2 - d_{25}^2\right) & 1 \\ \vdots & \vdots & \vdots & \vdots & \vdots & \vdots \\ d_{19}^2 & d_{29}^2 & d_{39}^2 & -\left(d_{19}^2 - d_{39}^2\right)\left(d_{29}^2 - d_{19}^2\right) & -\left(d_{29}^2 - d_{19}^2\right)\left(d_{39}^2 - d_{29}^2\right) & 1 \end{bmatrix}$$

$$X = \begin{bmatrix} \left(\dfrac{d_{23}^4}{d_{12}^2} - d_{23}^2 - \dfrac{d_{13}^2 d_{23}^2}{d_{12}^2}\right) \\[4pt] \left(\dfrac{d_{13}^4}{d_{12}^2} - \dfrac{d_{13}^2 d_{23}^2}{d_{12}^2} - d_{13}^2\right) \\[4pt] \dfrac{\left(d_{12}^2 - d_{23}^2 - d_{13}^2\right)d_{23}^2}{d_{12}^2} \\[4pt] \dfrac{d_{13}^2}{d_{12}^2} \\[4pt] \left(144\dfrac{v_t^2}{d_{12}^2} + d_{13}^2 d_{23}^2\right) \end{bmatrix} \qquad b = \begin{bmatrix} \left(d_{24}^2 - d_{34}^2\right)\left(d_{34}^2 - d_{14}^2\right) \\ \left(d_{25}^2 - d_{35}^2\right)\left(d_{35}^2 - d_{15}^2\right) \\ \vdots \\ \left(d_{29}^2 - d_{39}^2\right)\left(d_{29}^2 - d_{19}^2\right) \end{bmatrix}$$

From the above representation, after finding X_1, X_2, X_3, X_4, X_5 and X_6 we calculate d_{12}, d_{13} and d_{23} as following:

$$d_{12}^2 = \frac{X_3}{\left(1 - X_4 - X_5\right)}, \quad d_{13}^2 = \frac{X_3 X_5}{\left(1 - X_4 - X_5\right)}, \quad d_{23}^2 = \frac{X_3 X_4}{\left(1 - X_4 - X_5\right)}$$

If we let the coordinates of the submerged sensors S_1, S_2 and S_3 are $(0,0,0)$, $(0, y_2, 0)$ and $(x_3, y_3, 0)$ respectively, then the inter-sensors distances could be written with respect to coordinates of the sensors as following:

$$d_{12}^2 = y_2^2, \quad d_{13}^2 = x_3^2 + y_3^2, \quad d_{23}^2 = x_3^2 + \left(y_3 - y_2\right)^2$$

From the above values we get the unknown variables computed as following:

$$y_2 = d_{12}, \quad y_3 = \frac{d_{12}^2 + d_{13}^2 - d_{23}^2}{2 d_{12}}, \quad x_3 = \sqrt{\left(d_{13}^2 - \left(\frac{d_{12}^2 + d_{13}^2 - d_{23}^2}{2 d_{12}}\right)^2\right)}$$

where d_{12}, d_{13} and d_{23} are known computed distances. Table 1 summarizes the coordinates of the sensors for our problem domain.

Table 1. Coordinates of the sensors with known measurements

	S_1	S_2	S_3	
Coordinates	$(0,0,0)$	$(0, d_{12}, 0)$	$\left(\sqrt{\left(d_{13}^2 - \left(\dfrac{d_{12}^2 + d_{13}^2 - d_{23}^2}{2 d_{12}}\right)^2\right)}\right.$	$\left.\dfrac{d_{12}^2 + d_{13}^2 - d_{23}^2}{2 d_{12}}, 0\right)$

5   Analysis

The proposed method is for a specific configuration and scenario, where a single beacon node is necessary to determine the coordinates of the sensors. Usually numerous
beacons and sensors are deployed to localize and most methods rely on distance
measurements, thus the precision of distance measurement is one of the prime factors
for accurate coordinate determination.

In our approach the number of beacon is only one that floats on the surface of the
water and minimum of three sensors - a recognized number in monitoring or analyzing environment with sensors.

Our distance measurement technique is unique for underwater localization where
both acoustic and radio signal is used. Despite propagation limitation of radio signal,
we have tactically chosen to serve our purpose because radio signal can propagate up
to 1.8-323m with an approximate speed of 2.25×10^6m/s [5].   By using this tremendous speed we can measure the flight time of acoustic signal from beacon to sensors
precisely; moreover, with a single message exchange beacon node will come to know
the flight time as well as other variables associated at the sensors. Hence, average
sound speed determination procedure incorporates on-site values of variables that will
incorporate less error in inter nodes distance measurements.

In this approach of coordinate determination multipath fading will have minimal impact. The sensors are only detecting the presence of signals instead of accessing its
strength; as soon as the shortest Euclidean distance is travelled - signal can be detected
and timed. Hence, the obvious notion of multipath fading in the radio and acoustic signals propagation will not affect timing in our method. With this, the precise acoustic
velocity would give an accurate distance measurement resulting in coordinates estimates
with less error. Simulation results in Table 2 suggest that with true Euclidean distances
our method determines coordinates of the sensors with negligible error.

6   Simulation Results and Discussions

In order to validate the mathematical model, the proposed method has been simulated
using Matlab. A group of three sensors are placed at (0,0,0), (0,75,0), and (80,40,0)
and the mobile beacon moved randomly in a plane, which is parallel to the bottom
plane where the sensors are. Simulation for coordinate determination is done for
100m water column with 30°C surface and 15°C bottom temperature having salinity
variation of 0.5ppt. We have incorporated Gaussian noise ($\mu=0$, $\sigma=1$) to 15°C bottom
temperature and to the flight time $T_{ij(travel)}$ as both of them have more uncertainties.

Fig. 3 shows the variations in 'average sound speed' due to the aforesaid conditions
for 100 iterations with standard deviation 1.97.

We have also simulated the effect of temperature, water depth and salinity on the
average sound speed for different combinations of variables and shown in the following figures.

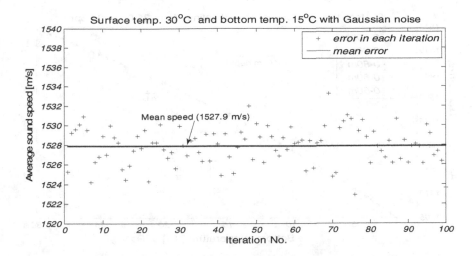

Fig. 3. Average speed variations due to additive Gaussian noise to bottom temperature

Mean of 'average sound speed' is found to be 1527.9m/s with Gaussian noise for the problem domain. Fig. 4 shows the variations in 'average sound speed' for various combinations of surface and bottom temperature without incorporating noise. The 'average sound speed' found to be 1526.56m/s for the same conditions as in Fig. 3.

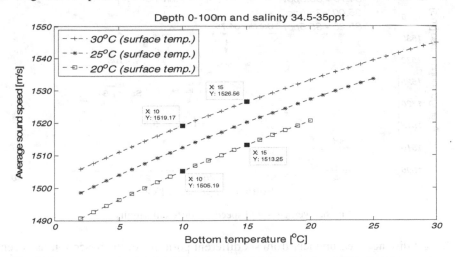

Fig. 4. Average sound speed for different combinations of surface and bottom temperature

It can be seen from Fig. 4 that average sound speed is 7.39 m/s less when the bottom temperature changes from $15°$ C to $10°$ C for $30°$ C surface temperature and it is 8.06 m/s less when surface temperature is $20°$ C. Result suggests that temperature is predominant for the water column. Fig. 5 shows the 'average sound speed' change due to depth variations which is not as significant as temperature variations.

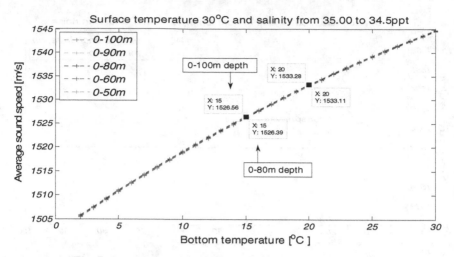

Fig. 5. Average sound speed for different depth of water column

Average sound speed changes due to salinity variation are even smaller than depth variations as shown in Fig.6.

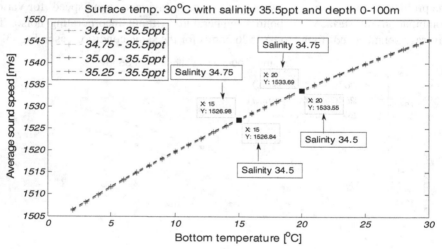

Fig. 6. Average sound speed for different salinity

To get distance measurement from six different positions of the beacon, it has been moved around in six different coordinates randomly within close proximity. However, mobility of the sensors is not considered in the proposed mathematical model. Errors in coordinates for S_2 and S_3 are shown in Fig. 7 and Fig. 8 respectively for 100 iterations. It should be noted that sensor S_1 is placed at the reference coordinate (0,0,0); hence producing no error in coordinates determination for S_1.

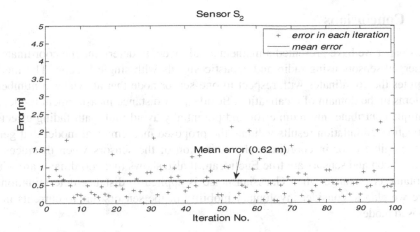

Fig. 7. Distance errors of sensor S_2 from original position

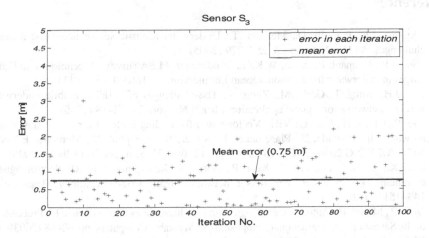

Fig. 8. Distance errors of S_3 from the original position

Table 2 compares the positional error of sensor S_2 and S_3 when distances between the beacon and sensors are with and without Gaussian noise. Positional error with true Euclidean distance is almost negligible which validates the proposed mathematical model. Besides, error with Gaussian noise is within acceptable range considering 100m water column and usual physical sizes of the sensors. It also shows that the mean distance error is 0.62m and 0.75m with standard deviation of 0.478 and 0.603 respectively.

Table 2. Distance errors in coordinate determination

Sensors	Distance Error (m) (Due to linearization, without noise)	Distance Error (with Gaussian noise)	
		μ (m)	σ
S_2	1.06×10^{-10}	0.62	0.478
S_3	2.75×10^{-10}	0.75	0.603

7   Conclusions

In this paper we have presented a mathematical model to determine the coordinates of submerged sensors using radio and acoustic signals with single beacon. The method computes the coordinates with respect to one sensor node that alleviates a number of problems in the domain of localization. Besides, our distance measurement model and technique contribute minimum error and potentially avoid multipath fading effects in localization. Simulation results validate the proposed mathematical model that generates negligible error in coordinate determination of the sensors when distances between beacon and sensors are true Euclidean. It also shows that coordinates are within acceptable error range when Gaussian noise is applied to distance determination. In future we plan to consider involuntary mobility of the sensors due to currents in the proposed model.

References

1. Akyildiz, I.F., Pompili, D., Melodia, T.: Underwater acoustic sensor networks: research challenges. Ad Hoc Networks 3, 257–279 (2005)
2. Tan, H.P., Diamant, R., Seah, W.K.G., Waldmeyer, M.: A survey of techniques and challenges in underwater localization. Ocean Engineering 38, 1663–1676 (2011)
3. Cui, J.H., Kong, J., Gerla, M., Zhou, S.: The challenges of building mobile underwater wireless networks for aquatic applications. IEEE Network 20, 12–18 (2006)
4. Xie, P., Cui, J.-H., Lao, L.: VBF: Vector-based forwarding protocol for underwater sensor networks. In: Boavida, F., Plagemann, T., Stiller, B., Westphal, C., Monteiro, E. (eds.) NETWORKING 2006. LNCS, vol. 3976, pp. 1216–1221. Springer, Heidelberg (2006)
5. Che, X., Wells, I., Dickers, G., Kear, P., Gong, X.: Re-evaluation of RF electromagnetic communication in underwater sensor networks. IEEE Communications Magazine 48, 143–151 (2010)
6. Duff, P., Muller, H.: Autocalibration algorithm for ultrasonic location systems. In: Proceedings of the Seventh IEEE International Symposium on Wearable Computers, pp. 62–68 (2003)
7. Olson, E., Leonard, J., Teller, S.: Robust range-only beacon localization. In: 2004 IEEE/OES, Autonomous Underwater Vehicles, pp. 66–75 (2004)
8. Guevara, J., Jimenez, A.R., Morse, A.S., Fang, J., Prieto, J.C., Seco, F.: Auto-localization in Local Positioning Systems: A closed-form range-only solution. In: 2010 IEEE International Symposium on Industrial Electronics (ISIE), pp. 2834–2840 (2010)
9. Guevara, J., Jiménez, A., Prieto, J., Seco, F.: Auto-localization algorithm for local positioning systems. Ad Hoc Networks 10, 1090–1100 (2012)
10. Stojanovic, M.: Underwater Acoustic Communications: Design Considerations on the Physical Layer. In: Fifth Annual Conference on Wireless on Demand Network Systems and Services, WONS 2008, pp. 1–10 (2008)
11. Mackenzie, K.V.: Nine-term equation for sound speed in the oceans. The Journal of the Acoustical Society of America 70, 807 (1981)
12. Fofonoff, N.P., Millard, R.C.: Algorithms for computation of fundamental properties of seawater. UNESCO Tech. Papers in Marine Sci. 44, 53 (1983)

A Gateway Discovery Approach Using Link Persistence Based Connected Dominating Sets for Vehicular Ad Hoc Networks[*]

Chi Trung Ngo and Hoon Oh[**]

School of Electrical Engineering
University of Ulsan
Ulsan Metropolitan City, South of Korea
chitrung218@gmail.com, hoonoh@ulsan.ac.kr

Abstract. In roadside gateway (RG) based vehicle ad hoc networks, the vehicles have difficulty in maintaining a stable route to the RG due to their high mobility. In addition, the redundant transmissions of an RG advertisement message transmitted to inform vehicles of the presence of RG can cause high control overhead. A new gateway discovery approach using tree topology is proposed to reduce control overhead as well as to cope with the high mobility of vehicles. The proposed approach employs the notion of connected dominating set (CDS) such that all nodes can receive the advertisement message if every vehicle in the CDS retransmits the message once. A prediction function of link persistence time is used in building the CDS, which replaces the node Id as primary key, to increase the stability of the route from each vehicle to an RG. The proposed protocol was evaluated comparatively with the traditional approaches, by resorting to simulation with various mobility scenarios generated by the VanetMobisim tool.

Keywords: Gateway discovery, Internet connectivity, Connected dominating set, Link persistence time, Route persistence time, VANETs.

1   Introduction

Recently, research on the provision of Internet connectivity service for the Intelligent Transportation System (ITS) is gaining much attention from the research community. In a network that consists of Roadside Gateways (RGs) and moving vehicles, the moving vehicles can access the Internet or communicate with wired hosts via the RGs for automotive safety or amusement. However, the frequent change of topology and the communication barriers around the roads make it challenging to perform the stable and efficient communication between two parties.

---

[*] This research was supported by 2013 Special Research Fund of Electrical Engineering at University of Ulsan.
[**] Corresponding author.

J. Cichoń, M. Gębala, and M. Klonowski (Eds.): ADHOC-NOW 2013, LNCS 7960, pp. 305–316, 2013.

RGs can be either installed at fixed positions along the roads or mobile as they are placed inside taxies or buses with WLAN or WWAN (e.g. WiMAX or 3G) [1]. Thus, moving vehicles can directly connect to the RGs and then reach the Internet or exchange data with wired hosts. However, the limited transmission range and the high mobility of nodes can often cause the frequent failures of connectivity to RGs. Thus, this approach may not be a feasible solution for urban low density roads due to the high cost that comes from the deployment of a large number of RGs. The use of multi-hop communication capability of vehicle ad hoc networks (VANETs) can be considered as a potential solution to improve connectivity between vehicles and RGs.

Connecting vehicles to the RGs in multiple hops requires a gateway discovery mechanism so that vehicles can discover RGs whenever they want Internet access. The gateway discovery mechanism needs to work without any prior configuration [2]. Particularly, many gateway discovery protocols have been proposed in the literature [1], [2], [3], [4], [5]. Basically, they are classified into three categories: (1) Proactive [3], where an RG floods the network with an advertisement message periodically, (2) Reactive [1], where a requester floods the network with a solicitation message, and (3) Hybrid [2], [4], [5], where an RG limits the flooding range of the advertisement message to a certain number of hops; requesters located within the advertisement zone use Proactive approach, while other requesters residing out of this range use Reactive approach. Proactive scheme achieves good connectivity but suffers from high overhead and established paths to an RG can be broken frequently owing to the high mobility of vehicles. Reactive scheme incurs low overhead for small networks but suffers from poor scalability and latency in discovering RGs. Hybrid scheme is supposed to be more efficient than Proactive and Reactive scheme; however, the selection of advertisement zone is a main consideration since it totally depends on the dimension of the network.

In this paper, we consider a new gateway discovery that is based on tree topology to reduce control overhead as well as to increase tree stability against the high mobility of vehicles. The new approach eliminates the use of a flooding by maintaining tree topology. It also employs the concept of *connected dominating set* (CDS) [6] to reduce the redundant retransmissions of the advertisement message which is issued periodically from an RG such that all vehicles can receive an advertise message if every node in CDS retransmits the message once.

Furthermore, when constructing a CDS, nodes compete for being included in the CDS by comparing their *Ids* such that a node with a higher *Id* takes the priority. In this paper, we use a *key* value that is determined by means of the predicted link persistence time instead of *Id*. Every vehicle takes a node in a CDS as its parents and maintains a robust and reliable upstream path to the RG, thus whenever it needs to communicate with another party outside the VANETs or to access the Internet, it only sends data or request along the established upstream path to the RG.

The proposed approach was implemented and evaluated in the network simulator QualNet and a realistic mobility model derived from the traffic simulation tool, VanetMobisim, respectively. According to simulation results, the proposed method can improve network performance significantly in terms of delivery ratio, end-to-end delay and control overhead compared with Proactive and Hybrid approaches.

In what follows, Section 2 gives the network model, followed by the motivation, and notation and message definitions. Section 3 reviews some related works. Section 4 details the gateway discovery using CDS. The performance of our approach is evaluated in Section 5. Finally, Section 6 concludes the paper.

2   Preliminary

2.1   The Network Model

We consider a VANET environment that consists of multiple stationary RGs and a number of mobile vehicles or nodes. The wireless transmission range of RG and vehicle are same. Two vehicles are said to have a link if they are located within their mutual transmission range. Two nodes that have a link are said to be each other's neighbor.

Each RG has its own unique ID and is equipped with two network interface cards, one to communicate with wired hosts (wired nodes or RG) and another to communicate with wireless nodes. The RGs communicate each other using either wired or wireless networks with the high speed and low error rate. An RG acts as a bridge between the Internet and VANET. Each vehicle has an onboard communication unit for wireless communication and a global positioning system (GPS) that can keep track of its geographical position, velocity and moving direction.

2.2   Motivation

Bechler et al. proposed a gateway discovery technique, DRIVE [3], which is based on the Proactive approach. The DRIVE specifies an advertisement interval to inform the presence of a gateway. However, it is not easy to set an appropriate value of the interval. A small interval can incur high overhead while a large interval causes vehicles to have the stale information about the presence of RG, thereby making it difficult to respond to the fast changing VANET topology.

In order to limit control overhead and cope with the high mobility of VANETs, Benslimate et al. proposed the use of the distributed contention process [7] in which when an RG or a node broadcasts an advertisement message, each receiver associates a timer value in inverse proportion to link stability between itself and the sender. Then the receiver with the shortest timer rebroadcasts the received message first, and suppresses other nodes from rebroadcasting the same message. This approach basically targets highway environment where two lanes on a road have the same direction, RGs are placed at regular distances, and the advertisement messages are forwarded in the opposite direction of node movement within the restricted broadcast zone of an RG. However, it can cause some problems in urban environment since the suppression of transmission can result in a partial reception failure such that some vehicles do not receive the advertisement message.

Let us take a look at Fig. 1 that shows the typical distribution of vehicles in urban roads. We assume that link (RG, A) has the highest link stability among three links (RG, A), (RG, B), and (RG, C). Then, node A will rebroadcast the advertisement message first, thus causing the retransmissions of nodes B and C to be suppressed. In consequence, nodes D and E fail to receive the advertisement message.

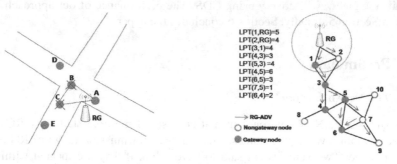

Fig. 1. The partial reception failure of an advertisement message by transmission suppression

Fig. 2. An example by the modified CDS algorithm

The problem of the partial reception failure can be avoided by building connected dominant set (CDS) [6] that is defined as a minimum number of nodes whose re-broadcasting can cover all nodes in the considered network. In Fig. 1, CDS = {B, C}. To compute CDS, a heuristic algorithm can be used since the computation problem of CDS is NP-Complete [8].

However, the problem is that the CDS does not take into account link stability which is of high importance in dealing with the gateway discovery in VANET. In this paper, we predict link persistence time and route persistence time between neighboring nodes and exploit those predicted values in building a robust CDS.

2.3   Notations and Messages

We define some notations used in this paper as follows

- *MsgId:* A message identification number that is uniquely assigned to a message by a node that generates the message
- *SeqNo:* A message sequence number
- *HopsToRG:* the distance in hops to the RG
- *sPos:* The current geographic position of a sender (a node or an RG)
- *sSpeed:* the current speed of a sender
- *sDir:* the moving direction of a sender
- *RPT:* A route persistence time
- *LPT:* A link persistence time
- *AdvZ:* The broadcast zone of an advertisement message issued by an RG.
- *P(i):* A parent of node i
- *F(i):* A node that has forwarded a message to node i
- *N(i):* A set of node i's neighbors;
- *N[i]:* $N(i) \cup \{i\}$;
- *d(u, v):* The distance between nodes u and v
- *r:* the transmission range of the RG and a node

We use some following messages for convenience of description.

— RG-ADV = (*MsgId, SeqNo, HopsToRG , sPos, AdvZ, RPT*): An RG broadcasts this message periodically to inform nodes of its presence. Initially, the RG sets the *RPT* to a large value, and then *RPT* is gradually updated by the receiving nodes as the message is propagated throughout the network.

— Hello = (*MsgId, SeqNo, sPos, sSpeed, sDir*): A vehicle broadcasts this message periodically to share its location information with its neighbors.

3   Related Works

3.1   Link and Route Persistence Time Estimation

In [9], Su et al. studied the method to estimate the *link persistence time* of two mobile nodes i and j, denoted as $LPT(i, j)$, based on their GPS information. Let (x_i, y_i) and (x_j, y_j) be the coordinates of mobile hosts i and j at time t, respectively, and let v_i and v_j be the speeds, and θ_i and θ_j be the moving directions of node i and j, respectively. Then, $LPT(i, j)$ is given as follows:

$$LPT(i,j) = \frac{-(ab + cd) + \sqrt{(a^2 + c^2)r^2 - (ad - bc)^2}}{a^2 + c^2} \tag{1}$$

where,

$$a = v_i cos\theta_i - v_j cos\theta_j$$
$$b = x_i - x_j$$
$$c = v_i sin\theta_i - v_j sin\theta_j$$
$$d = y_i - y_j$$

Note that $LPT(i, j)$ becomes ∞ if $v_i = v_j$ and $\theta_i = \theta_j$.

Then, consider a route $(n_1, n_2, n_3, ..., n_k)$ from node n_1 to node n_k. The *route persistence time* of the route, denoted as $RPT(n_1, n_2, ..., n_k)$, can be defined as the minimum value of all links on the route.

$$RPT(n_1, n_2, ..., n_k) = min\{LPT(n_i, n_{i+1})\}, i = 1, ..., k - 1 \tag{2}$$

3.2   Connected Dominating Sets

Let $G(V, E)$ be a graph where V is the set of nodes in the network and E is the set of links or edges between every pair of nodes in V. A subset D of V is said to be a *dominating set* if every vertex v, $v \in (V–D)$, is a neighbor of at least one member in D. D is a *connected dominating set* (CDS) if it satisfies two following properties.

(1) D is a dominating set; and

(2) All nodes in D form a connected component.

The algorithm to find the minimum CDS is known to be NP-Complete [8]. Wu et al. [6] devised an efficient heuristic algorithm for constructing a CDS such that a node can determine whether it is a member of the CDS by exploiting either the knowledge of its one-hop neighbors with their geographical positions or its two-hop topology information.

Each node x is given a function $id(x)$, a unique identification number of node x. Then, id(x) becomes a *competition resolution key* to determine which node will take priority to be included in CDS. A node is said to be an *intermediate node* if a node has at least one pair of neighbors that are not connected to each other. In addition, the following two rules, *Rule1* and *Rule2*, are used in order to build CDS.

Rule1: Given intermediate nodes u and v, if $N[v] \subseteq N[u]$ and $id(v) < id(u)$, CDS = CDS $\cup$ {u}.

After applying *Rule1*, the nodes in the CDS become *inter-gateways*.

Rule2: Suppose that after applying *Rule1*, nodes u and w are two inter-gateway neighbors of an inter-gateway node v. If $N[v] \subseteq N[u] \cup N[w]$ and $id(v) ==$ $min\{id(v), id(u), id(w)\}$, CDS = CDS – {$v$}.

Rule2 is used to reduce the number of inter-gateways and after applying *Rule 2*, the nodes in CDS become *gateways*.

The computational complexity of the algorithm for each node to determine whether it belongs to a CDS is $O(k^3)$, where k is the number of its neighbors. This method does not need any additional message.

4 Gateway Discovery Using CDS

4.1 A Modified CDS Algorithm Using LPT and RPT

In order to cope with the rapidly varying mobility (i.e., high acceleration and deceleration) of vehicles, we introduce a new competition resolution key, $key(x, y)$, that is generated when node x receives a message from node y as a function of $LPT(x, y)$ and RPT as follows:

$$key(x, y) = f(LPT(x, y), RPT) = 1 - e^{\frac{-LPT(x,y)}{RPT}} \qquad (3)$$

where, $LPT(x, y)$ and RPT are defined in Eq. (1) and Eq. (2), respectively, and $0 <$ key(x, y) ≤ 1. In case of ties, node Id is used to resolve.

Based on the modified key, we modify *Rule1* and *Rule2* to *MRule1* and *MRule2* as follows.

MRule1: When intermediate nodes u and v receive RG-ADV from node y, if $N[v] \subseteq N[u]$ and $key(v, y) < key(u, y)$, CDS = CDS $\cup$ {u}.

After applying *MRule1*, the nodes in CDS become *inter-gateways*.

MRule2: Suppose that after applying MRule1, u and w are two inter-gateway neighbors of an inter-gateway node v. If $N[v] \subseteq N[u] \cup N[w]$ and $key(v, F(v)) =$ $= min\{key(v, F(v)), key(u, F(u)), key(w, F(w))\}$, CDS = CDS – {$v$}.

After applying *MRule2*, the nodes in CDS become *gateways*.

Based on these rules, a modified CDS construction algorithm is described as follows. The RG-ADV message is issued by an RG or a node, referred to as node y in the description of this part. Upon receiving this message, if node x is an intermediate node, it initiates the process to determine whether or not it belongs to a CDS. Node x uses the motion parameters in the received RG-ADV and its own GPS information to calculate its $LPT(x, y)$ of link (x, y) using Eq. (1), and then obtain $RPT(x, y, ..., RG)$ according to Eq. (2). Then, node x sets a timer with a predefined short amount of time to check whether it receives RG-ADV from some nodes $k \neq y$ such that node y and node k have the same distance in hops to the RG and $RPT(x, k, ..., RG) > RPT(x, y, ...,$ $RG)$. When the timer expires, among senders of received RG-ADV messages node x chooses the sender from which it has the largest RPT to the RG and then uses motion parameters in RG-ADV broadcasted from that sender in order to calculate its key. Assuming that node y is chosen, node x obtains $key(x, y)$ according to Eq. (3). In addition, node x can obtain $LPT(i, y)$ and $key (i, y)$ for link (i, y), for all $i \in N(x)$ such that $i \neq y$, using the motion parameters obtained through the exchanged Hello messages. $key(y, y) = 1$ since $LPT(y, y) = \infty$. In this way, node x can get the $keys$ for all nodes in N[x]. After applying $MRule1$ and $MRule2$, if node x is a member of CDS, it is permitted to rebroadcast RG-ADV, after updating $RPT = min (RPT, LPT(x, y))$ and increasing $HopsToRG$ by 1. In this way, RPT is continuously maintained to have minimum LPT on the path from the receiving node of RG-ADV to RG.

Let us take a look at an example as illustrated in Fig. 2. We assume that each node generates the LPT values given in the figure upon receiving RG-ADV. Then, we get $key(1, RG) > key(2, RG)$, $key(5, 3) > key(4, 3)$, and $key(4, 5) > key(6, 5) > key(7, 5)$. Initially CDS = {RG}. RG broadcasts RG-ADV = (MsgID, HopsToRG = 0, SeqNo, sPos(RG), AdvZ, RPT = ∞) and nodes 1 and 2 receive RG-ADV. Since key(1, RG) > key(2, RG) and N[1] $\supseteq$ N[2], CDS = {RG, 1} (*by MRule1*). Only node 1 rebroadcasts RG-ADV with RPT = 5 since $LPT(1, RG) < \infty$. Since for any k = {3, 4, 5}, node k does not have any neighbor x with higher key such that N[x] $\supseteq$ N[k], CDS = {RG, 1, 3, 4, 5}. Nodes 3, 4, and 5 rebroadcast RG-ADV with $RPT = 4$, $RPT = 3$, and $RPT = 4$, respectively. When nodes 6 and 7 receive RG-ADV from node 5, since it does not hold that either N[5] $\supseteq$ N[6] or N[4] $\supseteq$ N[6], and either N[5] $\supseteq$ N[7] or N[6] $\supseteq$ N[7], CDS = {RG, 1 , 3, 4, 5, 6, 7} (*by Mrule1*). However, since N[5] $\cup$ N[6] $\supseteq$ N[7] and key(7, 5) < key (5, 5) and key(7, 5) < key(6, 5), CDS = CDS - {7} = {RG, 1, 3, 4, 5, 6} (*by Mrule2*). Therefore, node 6 is allowed to rebroadcast RG-ADV with $RPT = 2$.

4.2   Tree Path Establishment for Data Communication

The data communication can be performed in two ways - from vehicles to an RG and vice versus, then a tree has to be built with both upstream paths from vehicles to an RG and downstream paths from an RG to vehicles.

Upstream paths are constructed with the RG-ADV message issued periodically by an RG and rebroadcast by nodes in CDS. Upon receiving RG-ADV messages, a node takes the sender of RG-ADV as its parent from which it has the largest RPT to the RG.

Downstream paths in a tree are established by a data packet and a request message that a vehicle sends to an RG. Upon receiving data packet or request message, a vehicle is aware of both the originator and the forwarder of the message or the data. Then, it can create a pointer to the forwarder with respect to the originator. Eventually, every node can build a routing table such that each entry holds the pointer to next node (child) with respect to one of its descendants and the corresponding hop distance.

5   Performance Evaluation

5.1   Simulation Setup

Fig. 3. The street map

We used a commercial network simulator, QualNet 5.02 to evaluate the performance of the proposed approach. The traffic model were generated by traffic simulator VanetMobiSim, which produces the mobility traces for some network simulators such as Qualnet, ns-2 and GlomoSim.

We also used the real street map of the urban area of 1500 x 1500 (m) as shown in Fig. 3. An RG is placed at the middle of the top of the considered terrain.

Firstly, the Modified CDS construction algorithm (*ModCDS*) is compared with the original CDS one (*OrgCDS*) in terms of link stability and route stability, when they are applied to the proposed approach. Secondly, the performance of the Gateway Discovery using the Modified CDS (*GD-ModCDS*) is compared with the traditional approaches, Proactive and Hybrid, when vehicles in VANET are selected randomly to send data to a wired host.

GD-ModCDS was assessed with varying *maximum acceleration and deceleration (mAccelDecel)*, ranging from 0.5 to 2.5 m/s^2 instead of node speed because of some reasons explained in [10], with varying *number of sessions (nSessions)* from 10 to 25 sessions, and with the *number of nodes (nNodes)* fixed at 150 and 200 nodes. The other common parameters and their assigned values are summarized in Table 1.

The performance metrics such as *Control Overhead (CO)*, *Delivery Ratio (DR)*, and *End-to-end delay (E2ED)*, are used to examine the protocols' performance. All of the simulation results in the experiments were obtained by averaging the values of 10 independent runs with different seeds, along with the 95% confidence interval.

Table 1. Simulation Parameters

Parameter	Value
Dimension	1500 x 1500 m
Simulation time	600 seconds
Transmission range	250 m
Data type	CBR
Data Packet Size	128 bytes
Data rate	2 pkts/s
MAC protocol	802.11b
Wireless bandwidth	2 Mbps
Maximum speed	40 m/s
Desired speed	30 m/s
Hello interval	2 seconds
Gateway advertisement interval	2 seconds
Flooding range of Hybrid	$k = 2$

5.2 Effectiveness of the Modified CDS Algorithm

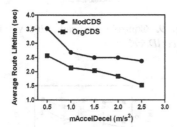

Fig. 4. Average route lifetimes with *ModCDS* and *OrgCDS*

Fig. 5. The number of broken links with *ModCDS* and *OrgCDS*

Since the modified CDS algorithm uses link persistence time prediction, its effectiveness can be evaluated by examining the number of broken links and the average route lifetime according to variation of *mAccelDecel*. In this simulation, *nNodes* was fixed at 150 and *mAccelDecel* varies from 0.5 to 2.5 m/s^2.

Fig. 4 and Fig. 5 show the average route lifetimes and the total number of broken links when the modified CDS algorithm and the original CDS algorithm are applied to the proposed approach for gateway discovery. It is natural that the average route lifetime decreases and the number of broken links increases according to the increase in mAccelDecel. It is shown that the use of the modified CDS algorithm increases the average route lifetime considerably compared to the use of the original CDS algorithm. This fact is also supported by the large improvement of the number of broken links when the modified CDS algorithm is applied as shown in Fig. 5.

5.3   Comparison of Different Gateway Discovery Approaches

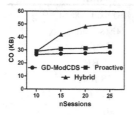

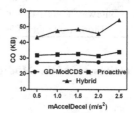

Fig. 6. Control overhead with varying *nSessisons*

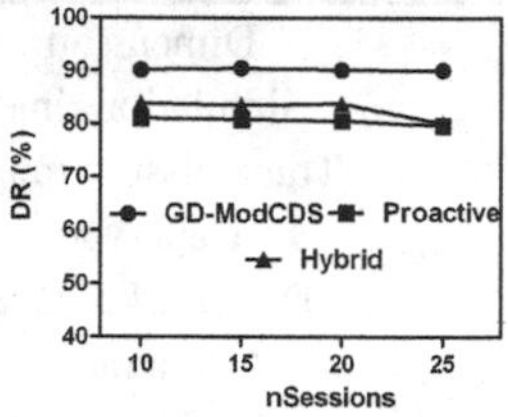

Fig. 7. Delivery ratio with varying *nSessions*

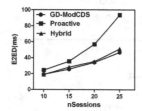

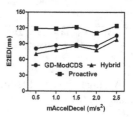

Fig. 8. End-to-end delay with varying *nSessions*

Fig. 9. Control overhead with varying *mAccelDecel*

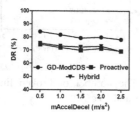

Fig. 10. Delivery ratio with varying *mAccelDecel*

Fig. 11. End-to-end delay with varying *mAccelDecel*

Sensitivity to Data Traffic: Fig. 6, Fig. 7 and Fig. 8 show control overhead, delivery ratio, and end-to-end delay of the different approaches according to variations of *nSessions* with *nNodes* and *mAccelDecel* fixed at 200 and 0.5 m/s², respectively.

According to Fig. 6, the control overhead of GD-ModCDS and the Proactive approach is not sensitive to the increase in nSessions while the Hybrid one shows a sharply increasing curve. This is because that with Hybrid, nodes undergo frequent link failures and thus issue a lot of solicitation messages to fix the failed routes to RG, thereby incurring a significant control overhead. The proposed approach, GD-ModCDS incurs the lowest and almost constant control overhead regardless of variation of data traffic. This supports the effectiveness of the modified CDS reinforced with link and route stability prediction functions.

Fig. 7 shows the graphs of delivery ratio according to the increase in nSessions. It is shown that the delivery ratios of three protocols are not sensitive to variation of nSessions. The reason is that the nature of Proactive is not sensitive to the delivery ratio with the increase in number of sessions except that in the Hybrid approach some nodes can use reactive manner for gateway discovery. Meanwhile, GD-ModCDS achieves the highest delivery ratio of nearly 90% overall due to the reduction of the redundant transmissions in advertisement messages and the increase in the lifetime of the routes to RG (see Fig. 4).

Finally, the result for average end-to-end delay per data packet is shown in Fig. 8. The end-to-end delay of Proactive approach shows the remarkably increasing curve with the increase in number of sessions, it is explained by the more congestion occurring among control messages and data packets. Whereas GD-ModCDS shows the linear increase in end-to-end delay and much less than the Proactive approach. This improvement comes from the utilization of the modified CDS algorithm to reduce the congestion among transmissions.

Sensitivity to Acceleration and Deceleration: Fig. 9, 10 and 11 show control overhead, delivery ratio, and end-to-end delay of the different approaches according to variation of *mAccelDecel* with *nNodes* and *nSessions* fixed at 150 and 20, respectively.

Fig. 9 shows the control overhead of the different approaches as we gradually increase the acceleration/deceleration of vehicles. Among three approaches the Hybrid scheme incurs the highest control overhead. The reason is that in the Hybrid scheme the frequent route failures make source nodes flood the network with more *solicitation messages* to repair the failed routes to the RG. As a result, the control overhead increases significantly. Meanwhile, since nodes can construct robust upstream tree paths to the RG, GD-ModCDS achieves the lowest control overhead, twice as low as the Hybrid scheme.

Referring to Fig. 10, the decrease in delivery ratios of three approaches makes sense because a route from a node to the RG is more likely to be broken easily with the increase in mobility and thus more data packets are dropped. However, the delivery ratio of GD-ModCDS remains higher than the other two's.

Fig. 11 shows average end-to-end delay per data packet. We can see that the Hybrid approach achieves the lowest end-to-end delay since when a broken link occurs on the route, a source node immediately finds an alternative route to the RG when it receives a route error message that was sent out by a detecting node, hence it can reduce the time that a data packet has to be in a buffer. On the contrary, in the Proactive approach source nodes have to wait for one interval to update the new route to the RG, thereby it incurs high end-to-end delay in a scenario with high mobility. We can see that GD-ModCDS deals with high mobility with an acceptable end-to-end delay.

6   Conclusion

In this paper, we proposed a new efficient gateway discovery approach for vehicles to access the Internet or communicate with wired hosts. The proposed approach uses the modified connected dominating set that is constructed based on link persistence time and route persistence time in order to reduce the number of relayed gateway advertisement messages. Since vehicles tend to change their speeds quickly and thus the acceleration and deceleration affect link stability largely, the modified CDS algorithm was able to deal with the problem of the routing stability in VANETs.

References

1. Ba, A.A., Hafid, A., Drissi, J.: Broadcast Control-Based Routing Protocol for Internet Access in VANETS. In: Wireless Communications and Mobile Computing Conference (IWCMC), pp. 1766–1771 (2011)
2. Boukerche, A., Abrougui, K., Pazzi, R.W.N.: Location-aided gateway advertisement and discovery protocol for VANets: Proof of correctness. In: The 34th Annual IEEE Conference on Local Computer Networks, pp. 778–785 (2009)
3. Bechler, M., Wolf, L., Storz, O., Franz, W.J.: Efficient discovery of Internet gateways in future vehicular communication systems. In: Vehicular Technology Conference, pp. 965–969 (2003)
4. Ruiz, P.M., Gomez-Skarmeta, A.F.: Adaptive gateway discovery mechanisms to enhance internet connectivity for mobile ad hoc networks. Ad Hoc and Sensor Wireless Networks 1, 159–177 (2005)
5. Ratanchandani, P., Kravets, R.: A hybrid approach to Internet connectivity for mobile ad hoc networks. In: Proceedings of the International Conference on Wireless Communications and Networking (WCNC), pp. 1522–1527 (2003)
6. Wu, J.: Dominating-set-based routing in ad hoc wireless networks. In: Handbook of Wireless Networks and Mobile Computing, pp. 425–450. John Wiley & Sons, Inc. (2002)
7. Benslimane, A., Barghi, S., Assi, C.: An efficient routing protocol for connecting vehicular networks to the Internet. Pervasive and Mobile Computing 7, 98–113 (2011)
8. Garey, M.R., Johnson, D.S.: Computers and Intractability; A Guide to the Theory of NP-Completeness. W. H. Freeman & Co. (1990)
9. Su, W., Lee, S.-J., Gerla, M.: Mobility prediction and routing in vehicular ad-hoc wireless networks. Int. J. Netw. Manag. 11, 3–30 (2001)
10. Jerome, H., Fethi, F., Bonnet, C.: On Meaningful Parameters for Routing in VANETs Urban Environments under Realistic Mobility Patterns. In: Proceedings of the 1st IEEE Workshop on Automotive Networking and Applications, AutoNet 2006 (2006)

Comparing Schedules in the SINR and Conflict-Graph Models with Different Power Schemes

Tigran Tonoyan

TCS Sensor Lab
Centre Universitaire d'Informatique
Route de Drize 7, 1227 Carouge, Geneva, Switzerland
`tigran.tonoyan@unige.ch`

Abstract. The scheduling problem in wireless networks asks to split a set of links (transmission requests) into the minimum number of "feasible" subsets. In this paper the theoretical gap between the schedule length in the SINR model and in the corresponding conflict-graph model is evaluated, considering three different power schemes. It is shown that this gap depends largely on the power scheme used and on the metric space where the network is located. While in metric spaces of small doubling dimension (such as Euclidean metrics) certain upper bounds can be proven for the difference between the two models, in general metrics the difference can be arbitrary.

1 Introduction

The topic of interference resolution in wireless networks has a rich history of research. A fundamental problem in this research is the scheduling problem: given a set of transmission requests, how can one organize them into subsets, as few as possible, so that the transmissions in each subset can be done simultaneously. This problem has been considered in several models such as simple conflict-graph models and models incorporating path-loss and fading. In this paper we work with two models: the path loss model with SINR (signal-to-interference-and-noise ratio) describing the interference and an approximation of this model, the Conflict-Graph model. The Conflict-Graph model can be described in terms of a conflict graph, where two links are adjacent or conflicting if they cannot transmit simultaneously with respect to the SINR model, i.e. at least one of them makes too much interference for the other one. The scheduling problem in this simplified model becomes equivalent to the well-known vertex coloring problem in the corresponding graph. A motivation for considering this variant of scheduling problem is the localized nature of graph-based models and simplicity of treatment of the problem, i.e. in order to resolve the interference for a given link one needs to consider only the adjacent links, which is convenient when considering e.g. topology-related problems [1]. However, the over-simplified structure resulting from graph-based models has the following two problems. First, the graphs

J. Cichoń, M. Gębala, and M. Klonowski (Eds.): ADHOC-NOW 2013, LNCS 7960, pp. 317–328, 2013.
© Springer-Verlag Berlin Heidelberg 2013

can miss the possibility of spatial reuse because of the rigid assumptions on the interference; this problem is investigated in [2], [3] by software simulations and examples. It has been found that the network throughput can suffer because of the lack of spatial reuse in case of graph-based models. This problem seems to not induce dramatic differences (at least asymptotically) between schedules in two models. The second problem, that can induce a larger gap between the two models, is that graph-based models do not take into account the accumulative nature of interference, i.e. the schedules based on graphs can be too optimistic. This problem is investigated experimentally in [4], where it is shown that a significant fraction of links in a schedule that is feasible in a graph-based model, has too much cumulative interference in SINR model. Another study of this problem for the uniform power scheme is done in [5] in a slightly different context, but their proofs seem problematic (in particular, the proof of Proposition 5.3).

In this paper we investigate the asymptotic gap between the two models from the scheduling aspect. The author is not aware of any publication that systematically treats this question. The main goal of this paper is to collect some known results together with new results that deal with the mentioned problem. Some of these results somehow appeared in publications, but the relation between conflict-graphs and SINR was considered merely as a tool towards solution of SINR-scheduling problems, while we believe that this relation is interesting on its own and is worth to be presented to the attention of the scientific community. The main question that we consider is how well does the Conflict-Graph model approximate the SINR model and how does the difference scale with the network size and topology. The answer to this question is different for different network settings. It has been shown in [6] that the difference is only a constant factor, when the lengths of all the links are close to each other. We extend this result by considering arbitrary sets of links and different power assignment schemes. In particular, we show that the extent of the gap depends on what power assignment is in use (considering the same network topology). The gap can also drastically depend on the properties of the metric space where the network nodes are located. For doubling metric spaces, we show that the difference between two models can be bounded by a factor of $\log \Delta$ for the *uniform* and *linear* power schemes (to be defined in Section 2.3) and by a factor of $\min\{\log n, \log \Delta\}$ for the *mean* power scheme, where Δ is the ratio of the lengths of longest and shortest links and n is the number of links. Hence, the mean power scheme is more scalable/flexible from this viewpoint. We also show that in general metric spaces the gap can be arbitrarily large. The main technical framework addressing the mean power scheme has been developed in [7] in the context of *power control* (not related to model comparison).

2   Preliminaries

2.1   The Path-Loss Model

Let $L = \{1, 2, \ldots, n\}$ denote the set of n links in the network that need to be scheduled. Each link represents a transmission request between two wireless nodes that are located on a metric space. Let the sender node of each link i be assigned a power level $P(i)$ according to some assignment policy or *power scheme.*

According to the path-loss propagation model [8], the receive signal of each link i is $P_i = P(i)/l_i^\alpha$, where l_i is the distance between the sender and the receiver of i and $\alpha > 0$ is the path-loss exponent. Similarly, the interference caused by link j at the receiver of link i is $I_{ji} = P(j)/d_{ji}^\alpha$, where d_{ji} denotes the metric distance from the sender node of link j to the receiver node of link i. We assume that the transmission of a link i is successful if and only if the SINR(signal-to-interference-and-noise ratio) is greater than a certain threshold $\beta \geq 1$:

$$\frac{P_i}{\sum_{j \in S \setminus \{i\}} I_{ji} + N} \geq \beta, \tag{1}$$

where the constant $N \geq 0$ denotes the noise and S is the set of links transmitting in the same time slot as i.

We call a subset S of L *feasible* if (1) holds for each link $i \in S$. Each partition of L into feasible subsets is called a *schedule*, and the number of subsets in such a partition is the *length* of the schedule. The *scheduling problem* asks to find a minimum length schedule for a given set of links.

In this paper we assume that $N = 0$. Note, however, that this assumption can be avoided if one is allowed to scale the power levels by a small multiplicative factor in the range $(1, 2]$. A formal proof of this statement is presented in [9].

For some of the results we also need the assumption that *doubling dimension* of the metric space be less than α (Theorem 2 and Lemma 3). The exact definition of doubling dimension can be found in [10]. Here we only need the fact that in a metric space with doubling dimension m, each ball of radius r contains at most $C \cdot (r/r')^m$ disjoint balls of a smaller radius r' where C is an absolute constant. It is known that the m-dimensional Euclidean space has doubling dimension m (see [10]), so for the euclidean plane we assume $\alpha > 2$, which is a common assumption in practice [8].

Now we can write the SINR condition as follows:

$$A(S, i) = \sum_{j \in S \setminus i} \min\{1, I_{ji}/P_i\} \leq 1/\beta. \tag{2}$$

Indeed, note that, since $\beta \geq 1$, $A(S, i) \leq 1/\beta$ if and only if (1) holds without the noise. This form of SINR condition has been considered in a number of papers (e.g. [11]) because it has the following additivity property: if there are two disjoint sets S_1 and S_2 then $A(S_1 \cup S_2, i) = A(S_1, i) + A(S_2, i)$.

Following [12] we say that a set S is a *p-signal set* if $A(S, i) \leq 1/p$. Similarly, a partition of the set of links is a p-signal partition if each subset is a p-signal set.

The following theorem shows that one can vary the threshold of the SINR condition without changing the schedule length much.

Theorem 1. *[12] There is a polynomial-time algorithm that takes a p-signal schedule and refines into a p'-signal schedule, for $p' > p$, increasing the number of slots by a factor of at most $\lceil 2p'/p \rceil^2$.*

2.2   The Conflict-Graph Model

We call two links i and j q-*adjacent* if

$$\text{either } A(\{i\}, j) \geq 1/q^\alpha \text{ or } A(\{j\}, i) \geq 1/q^\alpha,$$

otherwise we say that i and j are q-*independent*.

Using this definition we can define the q-adjacency graph $G_q(L)$ of the set of links L where the set of vertices of $G_q(L)$ is L and two links i and j are adjacent in $G_q(L)$ if and only if they are q-adjacent.

Note that if two links are q-adjacent then at least one of them interferes with the other one "too much", so $G_q(L)$ is a natural approximation of the stricter SINR model. The scheduling problem in the model associated with $G_q(L)$ is the famous vertex coloring problem in graphs, where the problem is to split the set of vertices of $G_q(L)$ into the smallest number of independent subsets. Following the standard notation, we denote that number $\chi(G_q(L))$.

The following lemma immediately follows from the definition of q-independence. It highlights an obvious relation between schedules in the two models.

Lemma 1. *A set of links that belong to the same q^α-signal slot in some schedule is q-independent.*

In the rest of this paper we study the relationship between the optimum schedule length (in the SINR model) and $\chi(G_q(L))$.

2.3   The Three Power Schemes

The motivation for considering power schemes is the fact that they are computable in a localized manner and do not depend on the whole network topology which makes them well-suited for decentralized wireless networks as opposed to complex power control algorithms.

We consider the following three power schemes. When the *uniform* power scheme is used, all the links are assigned the same power levels; hence, the sending power of each sender node is the same. When the *linear* power scheme is used, each link i is assigned a power level cl_i^α for a constant c. In this case the receive power of all links is constant (according to path-loss formula). When the *mean* power scheme is used, each link is assigned the power level $cl_i^{\alpha/2}$ for a constant c. Our analysis will be concentrated on these three power schemes.

3   The Uniform and Linear Power Schemes

Since the uniform power scheme and the linear power scheme are quite similar, we analyze them together. In this section we assume that the wireless nodes are placed on a doubling metric space of dimension $m < \alpha$ (this is used in Theorem 2).

Plugging the two power schemes in (2) we get that in the case of the uniform power scheme $A(S, i) = \sum_{j \in S \setminus i} (l_i/d_{ji})^\alpha$ and in the case of the linear power scheme $A(S, i) = \sum_{j \in S \setminus i} (l_j/d_{ji})^\alpha$.

It follows that two links i and j are q-independent with respect to the uniform power scheme if and only if

$$d_{ij} \geq ql_j \text{ and } d_{ji} \geq ql_i.$$

Similarly, for the linear power scheme the q-independence is equivalent to the following:

$$d_{ij} \geq ql_i \text{ and } d_{ji} \geq ql_j.$$

We say that a set of links is *nearly equilength* [6] if the lengths of any pair of links in the set differ by a factor of at most two. The following theorem together with Lemma 1 shows that q-independence is essentially equivalent to q^α-signal property for a given nearly equilength set of links.

Theorem 2. *[6] Suppose that $\alpha > m$. Let L be a q-independent set of nearly equilength links for a parameter $q > 2$. Then L is a $\Omega(q^\alpha)$-signal set when the powers are uniform.*

Remark. Note that this theorem is true not only for the uniform power scheme but also for the linear power scheme and mean power scheme because the power levels of the nodes in these cases differ just by factors of order 2^α from some uniform power.

Recall that if $q \in O(1)$ then, using Theorem 1, a $\Omega(q^\alpha)$-signal set can be transformed into a constant number of feasible subsets. So when the link-lengths are "almost equal" the optimal schedule length and the chromatic number of $G_q(L)$ differ by at most a constant factor; hence, in this case $G_q(L)$ is a good approximation for the SINR model. But what happens when the link-lengths are not close?

Theorem 3. *Suppose that $\alpha > m$. Let T denote the optimal schedule length of the set L in the SINR model, assuming that uniform, linear or mean power scheme is used. Then $T = O(\log \Delta \chi(G_q(L)))$ where $\Delta = max_{i,j \in L} l_i/l_j$ denotes the ratio of the lengths of the longest and the shortest links.*

Proof. We prove the claim only for the uniform power scheme, because the two other cases can be proven similarly (in virtue of the remark after Theorem 2). It is enough to show that each q-independent subset of L can be scheduled in $O(\log \Delta)$ subsets. Let $S \subseteq L$ be such a subset. S can be split into $\log \Delta$ subsets each of which is a nearly equilength set. Indeed, if l_1 is the length of the shortest link, then

$$S = \bigcup_{t=1}^{\log \Delta} \{i \in L : 2^{t-1} \leq l_i/l_1 < 2^t\}.$$

According to Theorem 2, each of these subsets can be scheduled into a constant number of feasible subsets, which gives $O(\log \Delta)$ subsets in total. $\qquad \square$

Note that in general the parameter Δ does not depend on the number of links, so it can theoretically be arbitrarily large with respect to n. On the other hand, the length of any schedule does not exceed n, the number of links. Next we show that the upper bound from Theorem 3 is tight for the uniform and linear power schemes. We show that there are examples of networks for which $\log \Delta = n$ and $T = \Theta(\log \Delta \chi(G_q(L)))$ when these power schemes are in use. These examples are the q-independent variants of exponential networks from [13].

Theorem 4. *Suppose that either the uniform power scheme or the linear power scheme is used, and let $q \in O(1)$. Then for each $n > 0$ there is a set of n links L on the line, such that $OPT(L) = \Theta(\log \Delta \chi(G_q(L)))$ and $\log \Delta = \Theta(n)$, where $OPT(L)$ is the minimal schedule length for L.*

Proof. Let us consider the following simple linear network. There are n links $\{1, 2, \ldots, n\}$ sequentially aligned on a straight line in the increasing order of numbers going from left to right. We set $l_i = 2^i$ for $i = 1, 2, \ldots, n$, i.e. $l_1 = 2$ is the length of the shortest link and $l_n = 2^n$ is the length of the longest one, hence $\log \Delta = \log l_n / l_1 = n - 1$. Each link has its sender on the left side and the receiver on the right side. For each $i > 1$ we define $d_{i,i-1} = q l_i$. Now we can calculate all the other distances. For each $i > 1$ we have:

$$d_{i1} = \sum_{t=2}^{i-1} (d_{t,t-1} + l_t) + d_{i,i-1} = \sum_{t=2}^{i-1} l_t + q \sum_{t=2}^{i} l_t =$$

$$= (q+1) \sum_{t=2}^{i-1} l_t + q l_i = (2q+1)l_i - 4(q+1),$$

where we used the fact that $l_i = 2^i$. Now if $i > j$ then $d_{ij} = d_{i1} - d_{j1} - l_j = (2q+1)(l_i - l_j) - l_j$ and $d_{ji} = d_{i1} - d_{j1} + l_i = (2q+1)(l_i - l_j) + l_i$. It is easy to check that this set of links is q-independent with respect to both uniform and linear power schemes.

Suppose that the uniform power scheme is used. Consider any *feasible* subset of links S of size k with the links $i_1 < i_2 < \cdots < i_k$. For the longest link i_k we have:

$$A(S, i_k) = \sum_{t=1}^{k-1} \frac{l_{i_k}^{\alpha}}{d_{i_t i_k}^{\alpha}} = \sum_{t=1}^{k-1} \frac{l_{i_k}^{\alpha}}{((2q+1)(l_{i_k} - l_{i_t}) + l_{i_k})^{\alpha}} \geq \frac{k-1}{(2q+2)^{\alpha}}.$$

We should have also that $A(S, i_k) \leq 1/\beta$ which allows us bound the number of links in S:

$$k \leq \frac{(2q+2)^{\alpha}}{\beta} + 1.$$

Hence, one cannot schedule the given set of links into less than $\frac{\beta n}{(2q+2)^{\alpha} + \beta} \in \Omega(n)$ feasible subsets.

For the case of the linear power scheme a similar argument works if we consider $A(S, i_1)$. $\qquad \square$

These results show that for the uniform and the linear power schemes the difference between the schedules in the SINR model and the Conflict-Graph model depends on the structure of the network with the factor $O(\log \Delta)$. In the next section we show that for the mean power scheme a better bound can be achieved.

4 The Mean Power Scheme

The following results have been proven in [7]. The proofs are presented here for reader's convenience, as they reflect the main ideas connecting Conflict-Graph and SINR models. As we saw in the previous section, the results for approximating SINR with conflict graphs can be unsatisfactory when using the uniform or the linear power scheme. It turns out that when one uses the mean power scheme, a better approximation is achieved. In this section we assume that the nodes are placed on a doubling metric space of dimension $m < \alpha$.

Theorem 5. *[7] Suppose that $\alpha > m$ and that the mean power scheme is used and let S be a 3-independent set of links. Then S can be scheduled into $O(\log n)$ subsets.*

It follows from Theorem 5 that the gap between schedule lengths in the Conflict-Graph model and the SINR model is more scalable for the mean power scheme than for the other two power schemes.

Corollary 1. *Suppose that $\alpha > m$. Let T denote the optimal schedule length of the set of links L in the SINR model assuming that the mean power scheme is used. Then $T = O(\min\{\log \Delta, \log n\}\chi(G_2(L)))$, where $n = |L|$.*

The proof of Theorem 5 is based on the following crucial lemma.

For each set of links S and each link $i \in S$, let $\gamma(S, i)$ denote the number of links $j \in S$ such that $l_j \geq n^2 l_i$, and either $A(i, j) \geq 1/(2n)$ or $A(j, i) \geq 1/(2n)$, where $n = |S|$.

Lemma 2. *[7] Suppose that the mean power scheme is used for a 2-independent set of links S. There is a constant $N = N(C, m)$ such that for each link $i \in S$, $\gamma(S, i) \leq N$.*

Let us postpone the proof of this lemma and turn to the proof of Theorem 5.

Proof (Proof of Theorem 5). It suffices to show that each 3-independent subset of L can be scheduled in $O(\log n)$ subsets. Let S be a 3-independent subset. Let us assume $\beta = 1$ for simplicity of expressions. S is scheduled in three stages.
Stage 1. First we split S into $O(\log n)$ subsets that have certain desired properties. Let l_1 be the length of the shortest link. Then $S = \cup_{t=1}^{\log \Delta} Q_t$, where

$$Q_t = \{i \in S : 2^{t-1} \leq l_i/l_1 < 2^t\}.$$

By rearranging the terms we can write

$$S = \bigcup_{t=1}^{2\log(2n)} B_t \text{ with } B_t = \bigcup_{k=0}^{\infty} Q_{t+2k\cdot\log(2n)},$$

i.e. the set B_t (for $t = 1, 2, \ldots 2\log(2n)$) is formed by taking each $2\log(2n)$-th set, starting from Q_t. It is not hard to check that for any two links $i, j \in B_t$ with $l_i \geq l_j$, either $l_i \leq 2l_j$ or $l_i > 2n^2 l_j$. Each set B_t is scheduled separately and the union of those schedules is taken.

Stage 2. Take B_1 w.l.o.g. Recall that B_1 is a union of non-intersecting sets Q_t, $t \in I$, where I is some set of indices. Moreover, each of these sets Q_t is nearly equilength and 3-independent by construction. By Theorem 2 (and the remark after it), Q_t is an $\Omega(2^\alpha)$-signal set and, by Theorem 1, it can be transformed into a 2-signal schedule $\{Q_t^1, Q_t^2, \ldots\}$ consisting of $O(1)$ subsets. Since $B_1 = \bigcup_{t \in I} Q_t$, by rearranging the terms we also have that

$$B_1 = \bigcup_k S_k \text{ with } S_k = \bigcup_{t \in I} Q_t^k,$$

i.e. S_k is the union of k-th subsets (in an arbitrary ordering) of schedules for all Q_t. Hence the number of different S_k is in $O(1)$. In the following stage each S_k is scheduled separately.

Stage 3. Take S_1 w.l.o.g. Recall that S_1 is 3-independent; hence Lemma 2 holds. Let N be the constant from Lemma 2. In order to schedule S_1, take $N+1$ subsets $R_1, R_2 \ldots, R_{N+1}$ (initially empty) in a fixed order. Consider the elements of S_1 in an increasing order of link-lengths and add each next link i to the first subset R_t such that for each link $j \in R_t$ with $l_j \geq n^2 l_i$,

$$A(i,j) < 1/(2n) \text{ and } A(j,i) < 1/(2n).$$

Note that such R_t exists for each link i, in virtue of Lemma 2. It remains to prove that each set R_t is feasible. Take R_1 w.l.o.g. and consider some link $i \in R_1$. By construction in Stage 1, R_1 can be split into two subgroups as follows: $R_1^1 = \{j \in R_1 : l_j/l_i \in [1/2, 2]\}$, $R_1^2 = \{j \in R_1 : l_j/l_i \geq 2n^2 \text{ or } l_i/l_j \geq 2n^2\}$. The construction in Stage 2 guarantees that $A(R_1^1, i) \leq 1/2$. The choice of R_1 in Stage 3 makes sure that for each link $j \in R_1^2$, $A(j,i) < 1/(2n)$. Using additivity of $A(.,i)$ yields that $A(R_1^2, i) < 1/2$. Using additivity of $A(.,i)$ once again yields $A(R_1, i) < 1$. It is easy to check that the union of all schedules computed consists of $O(\log n)$ subsets. This completes the proof.   □

In order to prove Lemma 2, we need another technical lemma which encapsulates the properties of doubling metric spaces that will be used in the proof.

Lemma 3. *Let $\{p_0, p_1, p_2, \ldots, p_k\}$ be a set of points in a metric space of doubling dimension m and let c_1, c_2, c_3 and $\{b_0, b_1, b_2, \ldots, b_k\}$ be positive real numbers, such that*

   a) $b_s \geq c_1 b_0$ for $s = 1, 2, \ldots, k$,

b) $d(p_0, p_s) \leq c_2 b_0 b_s$ for $s = 1, 2, \ldots, k$ and
c) $d(p_s, p_t) \geq c_3 b_s b_t$ for $s, t = 1, 2, \ldots, k, t \neq s$.

Then $k \leq C \left(\left(\dfrac{2c_2}{c_1 c_3} \right)^2 + 1 \right)^m + 1.$

Proof. For each pair of indices $s, t \in \{1, 2, \ldots, k\}$, the triangle inequality implies that

$$d(p_s, p_t) \leq d(p_s, p_0) + d(p_t, p_0).$$

Combining this with inequalities b) and c) results in the following expression

$$c_2 b_0 b_s + c_2 b_0 b_t \geq c_3 b_s b_t. \tag{3}$$

Assume w.l.o.g. that $b_s \leq b_t$. Then it follows from (3) that $b_0 \geq \dfrac{c_3}{2c_2} b_s$. If we fix $t = \arg\max_t b_t$ and apply the previous argument to all pairs s, t with $s \in \{1, 2, \ldots, k\} \backslash \{t\}$, we find that $b_0 \geq \dfrac{c_3}{2c_2} b_s$ for all indices $s \in \{1, 2, \ldots, k\} \backslash \{t\}$. Suppose w.l.o.g. that those indices are $1, 2, \ldots, k-1$. Plugging these inequalities in b) and using the assumption a) we find that the following inequalities hold for all indices $s, t \in \{1, 2, \ldots, k-1\}$ with $i \neq j$:

$$d(p_s, p_t) \geq c_1^2 c_3 b_0^2 \quad \text{and} \quad d(p_0, p_s) \leq \dfrac{2c_2^2}{c_3} b_0^2.$$

The first inequality asserts that the balls of radius $c_1^2 c_3 b_0^2 / 2$ with centers at points p_s for different $s \in \{1, 2, \ldots, k-1\}$ don't intersect. The second inequality asserts that those balls are contained in the bigger ball of radius $(2c_2^2/c_3 + c_1^2 c_3/2) b_0^2$ with the center at p_0. Then the property of the metric space mentioned before the lemma implies the following upper bound on the number of points: $k - 1 \leq$

$C \left(\left(\dfrac{2c_2}{c_1 c_3} \right)^2 + 1 \right)^m$, which completes the proof. $\square$

The proof of Lemma 2 follows.

Proof (Proof of Lemma 2). Suppose that the mean power scheme is used for a q-independent set of links S with $q \geq 2$. Let us fix some link and assign it the number 0. Assume w.l.o.g. that $R = \{1, 2, \ldots, k\}$ is the subset of links $j \in S$ such that

$$l_j \geq n^2 l_0, \tag{4}$$

and either $A(0, j) \geq 1/(2n)$ or $A(j, 0) \geq 1/(2n)$, the last statement being equivalent to the following:

$$\min \{d_{0j}, d_{j0}\} \leq (2n)^{1/\alpha} \sqrt{l_0 l_j}. \tag{5}$$

We need to show that $\gamma(S, 0) = |R| \leq N$ for a constant N. q-independence implies that $A(i, j) \leq q^{-\alpha}$ and $A(j, i) \leq q^{-\alpha}$ for all $i, j \in R$, hence

$$d_{ij} \geq q\sqrt{l_i l_j} \quad \text{and} \quad d_{ji} \geq q\sqrt{l_i l_j}. \tag{6}$$

Let us assume w.l.o.g. that $l_i \leq l_j$. Applying the triangle inequality gives $d(s_j, s_i) \geq d_{ji} - l_i \geq q\sqrt{l_i l_j} - l_i$, and since $l_i \leq l_j$,

$$d(s_j, s_i) \geq (q-1)\sqrt{l_i l_j}. \tag{7}$$

A similar argument with the inequality $d(r_j, r_i) \geq d_{ij} - l_i$ gives

$$d(r_j, r_i) \geq (q-1)\sqrt{l_i l_j}. \tag{8}$$

Let us choose a node p_0 (the sender or the receiver of the link 0) and a set of nodes P differently, depending on the following two cases:

Case 1. There is a subset $R_1 \subseteq R$ with $|R_1| \geq |R|/2$, such that $d_{0j} \leq (2n)^{1/\alpha}\sqrt{l_0 l_j}$ for all $j \in R_1$. In this case we take p_0 to be the sender node of the link 0, i.e. s_0, and P to be the set of receiver nodes of the links in R_1, i.e. $P = \{r_j | j \in R_1\}$.

Case 2. If the first case does not hold, then according to the pigeonhole principle (applied to (5)) there is a subset $R_2 \subseteq R$ with $|R_2| \geq |R|/2$, such that $d_{j0} \leq (2n)^{1/\alpha}\sqrt{l_0 l_j}$ for all $j \in R_2$. In this case we take p_0 to be the receiver node of the link 0, i.e. r_0, and P to be the set of sender nodes of the links in R_2, i.e. $P = \{s_j | j \in R_2\}$.

In both cases $|P| \geq |R|/2$, so upper-bounding $|P|$ yields an upper-bound for $|R|$.

Consider the first case. Let $|P| = k$, and w.l.o.g. $P = \{r_1, r_2, \ldots, r_k\}$. By definition, $d(p_0, r_j) \leq (2n)^{1/\alpha}\sqrt{l_0 l_j}$ for $j = 1, 2, \ldots, k$. On the other hand, from (8) we have $d(r_i, r_j) \geq (q-1)\sqrt{l_i l_j}$ for $i, j = 1, 2, \ldots, k$ and $i \neq j$. Hence, by denoting $b_0 = \sqrt{l_0}$, $p_t = r_t$ and $b_t = \sqrt{l_t}$ for $t = 1, 2, \ldots, k$, we get

$$d(p_0, p_t) \leq (2n)^{1/\alpha} b_0 b_t$$

$$d(p_s, p_t) \geq (q-1) b_s b_t, \text{ for } s, t = 1, 2, \ldots, k, s \neq t,$$

so Lemma 3 applies to the set of points $\{p_0, p_1, \ldots, p_k\}$, with positive real numbers $b_0, b_1, \ldots, b_k$ as defined above and $c_1 = \sqrt{n^2} = n$ (because of (4)) $c_2 = (2n)^{1/\alpha}$, $c_3 = (q-1)$. This application of Lemma 3 gives the needed upper bound:

$$|P| = k \leq C\left(\left(\frac{2(2n)^{1/\alpha}}{(q-1)n}\right)^2 + 1\right)^m + 1 \in O(1),$$

where last relation is due to the assumption that $\alpha > 1$ and $q \geq 2$. Thus, if the first case holds, the lemma is proven. With almost the same steps the lemma can be proven for the second case, this time using (7). That will complete the proof. □

5 General Metric Spaces

It is known that some approximation results for scheduling in the SINR model hold true in general metric spaces [11]. When a non-Euclidean path-loss appears in practice, there is no apparent reason for it to obey metric space constraints. However, this research is valuable at least from the theoretical viewpoint. Hence, there is a natural question: could one transfer the approximation results to arbitrary metric spaces, without assuming "nice" metric properties such as the doubling property? The answer is negative and is very easy to prove. To see this, let us consider an abstract network of n q-*independent equilength* links using the mean power scheme (the latter is not important because the links have the same length), where $q \in O(1)$. We will define the distances between the nodes in such a way that the metric space constraints hold true, but the difference between the schedule lengths in the SINR model and the conflict-graph model is $\Theta(n)$.

Let us number the links $\{1, 2, \ldots, n\}$. For each link i we define $l_i = 1$. Let s_i and r_i denote the sender and the receiver node of the link i, respectively. The distances between the nodes are defined as follows:

1. sender to sender distances: $d(s_i, s_j) = q(l_i + l_j) = 2q$,
2. sender to receiver distances: $d(s_i, r_j) = d(s_i, s_j) + l_j = 2q + 1$,
3. receiver to receiver distances: $d(r_i, r_j) = d(s_i, s_j) + l_i + l_j = 2q + 2$.

It is straightforward to check that such distances define a metric. Moreover, the whole set of links in this metric is q-independent with respect to all three power schemes considered in this paper. Let us consider any subset of k links $\{i_1, i_2, \ldots, i_k\}$, where $k > 0$ and $i_1 < i_2 < \cdots < i_k$. Then we have:

$$A(S, i_1) = \sum_{t=2}^{k} \left(\frac{\sqrt{l_{i_1} l_{i_t}}}{d_{i_t i_1}} \right)^{\alpha} = \sum_{t=2}^{k} \frac{1}{(2q+1)^{\alpha}} = \frac{k-1}{(2q+1)^{\alpha}}.$$

It follows that any feasible subset of links must contain $O(1)$ links; hence, the optimal schedule length in the SINR model is $\Theta(n)$. Thus we proved the following.

Theorem 6. *For any $n > 0$ and $q \in O(1)$, there is a q-independent set of n equilength links on a metric space for which the optimal schedule length in the SINR model is $\Theta(n)$.*

This result implies that Theorem 2, Theorem 3 and Theorem 5 are very far from being true in general metric spaces. Thus, the conflict-graph model is appropriate to use only in metrics that can transform the independence of the links into SINR feasibility, to which an example is doubling metrics.

A consequence that we can draw from [11] (Theorem 4.4) is the following theorem.

Theorem 7. *In any metric space, for any set of a nearly-equilength links, the schedule length using the mean power scheme is at most $O(\log n)$ times more than the schedule length using the best possible power assignment.*

Combining this theorem with Theorem 6 we have the following corollary.

Corollary 2. *For any $n > 0$ and $q \in O(1)$, there is a q-independent set of n equilength links on a metric space for which the optimal schedule length in the SINR model is $\Theta(n/\log n)$, even when using the best possible power assignment.*

6   Conclusion

This paper presented a set of results trying to evaluate the asymptotic difference between the SINR schedules and Conflict-Graph based schedules in wireless networks. These results indicate that this gap is bounded in doubling metric spaces of small dimension such as the Euclidean plane. For the case of the uniform and linear power schemes the upper bound is $O(\log \Delta)$ and is sharp. For the mean power scheme the gap is in $O(\min\{\log n, \log \Delta)$, so the upper bound for the mean power scheme scales better with the number of links and the topology of the network. In the case of the mean power scheme no example of network meeting the upper bound is known to the author, so this could be a subject of a future work. At last, it was shown that in general metric spaces the difference between the schedules in the two models can be arbitrary.

References

1. Cardieri, P.: Modeling interference in wireless ad hoc networks. IEEE Commun. Surveys Tuts. 12(4), 551–572 (2010)
2. Grönkvist, J., Hansson, A.: Comparison between graph-based and interference-based STDMA scheduling. In: 2nd ACM International Symposium on Mobile Ad Hoc Networking & Computing, pp. 255–258. ACM (2001)
3. Moscibroda, T., Wattenhofer, R., Yves, W.: Protocol design beyond graph-based models. In: 5th ACM SIGCOMM Workshop on Hot Topics in Networks, Irvine, California, USA. ACM (2006)
4. Behzad, A., Rubin, I.: On the performance of graph-based scheduling algorithms for packet radio networks. In: IEEE Global Telecommunications Conference, pp. 3432–3436. IEEE (2003)
5. Iyer, A., Rosenberg, C., Karnik, A.: What is the right model for wireless channel interference? IEEE Trans. Wireless Commun. 8(5), 2662–2671 (2009)
6. Halldórsson, M.M.: Wireless scheduling with power control. In: Fiat, A., Sanders, P. (eds.) ESA 2009. LNCS, vol. 5757, pp. 361–372. Springer, Heidelberg (2009)
7. Tonoyan, T.: Algorithms for scheduling with power control in wireless networks. In: Marchetti-Spaccamela, A., Segal, M. (eds.) TAPAS 2011. LNCS, vol. 6595, pp. 252–263. Springer, Heidelberg (2011)
8. Rappaport, T.S.: Wireless Communications: Principles and Practice, 2nd edn. Prentice Hall (2002)
9. Tonoyan, T.: On the capacity of oblivious powers. In: Erlebach, T., Nikoletseas, S., Orponen, P. (eds.) ALGOSENSORS 2011. LNCS, vol. 7111, pp. 225–237. Springer, Heidelberg (2012)
10. Heinonen, J.: Lectures on Analysis on Metric Spaces, 1st edn. Springer (2000)
11. Halldórsson, M.M., Mitra, P.: Wireless capacity with oblivious power in general metrics. In: 22nd Annual ACM-SIAM Symposium on Discrete Algorithms, pp. 1538–1548. SIAM (2011)
12. Halldórsson, M.M., Wattenhofer, R.: Wireless communication is in APX. In: Albers, S., Marchetti-Spaccamela, A., Matias, Y., Nikoletseas, S., Thomas, W. (eds.) ICALP 2009, Part I. LNCS, vol. 5555, pp. 525–536. Springer, Heidelberg (2009)
13. Moscibroda, T., Wattenhofer, R.: The complexity of connectivity in wireless networks. In: 25th IEEE International Conference on Computer Communications, Barcelona, Spain, pp. 1–13. IEEE (2006)

Author Index